MODERN ASPECTS OF ELECTROCHEMISTRY

No. 42

Modern Aspects of Electrochemistry

Topics in Number 41 include:

- Solid State Electrochemistry, including the major electrochemical parameters needed for the treatment of electrochemical cells as well as the discussion of electrochemical energy storage and conversion devices such as fuel cells
- Nanoporous carbon and its electrochemical application to electrode materials for super capacitors in relationship to the key role nanoporous carbons have played in the purification of liquids and the storage of energy
- The analysis of variance and covariance in electrochemical science and engineering
- The use of graphs in electrochemical reaction networks, specifically: (1) reaction species graphs, (2) reaction mechanism graphs, and (3) reaction route graphs

Topics in Number 40 include:

- Polymer Electrolyte Membrane (PEM) fuel cell bipolar plates, discussion of the difficulties associated with confronting bipolar plate development
- The use of graphs in electrochemical reaction networks with focus on analysis of variance (ANOVA) observation methods
- Nano-materials in lithium ion battery electrode design, presentation of a plasma-assisted method to create a carbon replica of an alumina template membrane
- Direct methanol fuel cells, extensive discussion and review of various types of fuel cells and advances made in the performance of DMFC's since their inception
- Direct simulation of polymer electrolyte fuel cell catalyst layers, presentation of a systematic development of the direct numerical simulation

MODERN ASPECTS OF ELECTROCHEMISTRY

No. 42

Edited by

CONSTANTINOS G. VAYENAS
University of Patras
Patras, Greece

RALPH E. WHITE
University of South Carolina
Columbia, South Carolina, USA

and

MARIA E. GAMBOA-ALDECO
Managing Editor
Superior, Colorado, USA

Springer

Constantinos G. Vayenas
Department of Chemical Engineering
University of Patras
Patras 265 00
Greece
cat@chemeng.upatras.gr

Maria E. Gamboa-Aldeco
1107 Raymer Lane
Superior, CO 80027
USA
mariagamboa06@msn.com

Ralph E. White
Department of Chemical Engineering
University of South Carolina
Columbia, SC 29208
USA
white@engr.sc.edu

ISBN: 978-1-4419-2383-7 e-ISBN: 978-0-387-49489-0
DOI: 10.1007/978-0-387-49489-0

Softcover reprint of the hardcover 1st edition 2008

Printed on acid-free paper.

9 8 7 6 5 4 3 2 1

springer.com

Preface

This volume analyzes and summarizes recent developments in several key interfacial electrochemical systems in the areas of fuel cell electrocatatalysis, electrosynthesis and electrodeposition. The six Chapters are written by internationally recognized experts in these areas and address both fundamental and practical aspects of several existing or emerging key electrochemical technologies.

The Chapter by R. Adzic, N. Marinkovic and M. Vukmirovic provides a lucid and authoritative treatment of the electrochemistry and electrocatalysis of Ruthenium, a key element for the development of efficient electrodes for polymer electrolyte (PEM) fuel cells. Starting from fundamental surface science studies and interfacial considerations, this up-to-date review by some of the pioneers in this field, provides a deep insight in the complex catalytic-electrocatalytic phenomena occurring at the interfaces of PEM fuel cell electrodes and a comprehensive treatment of recent developments in this extremely important field.

Several recent breakthroughs in the design of solid oxide fuel cell (SOFC) anodes and cathodes are described in the Chapter of H. Uchida and M. Watanabe. The authors, who have pioneered several of these developments, provide a lucid presentation describing how careful fundamental investigations of interfacial electrocatalytic anode and cathode phenomena lead to novel electrode compositions and microstructures and to significant practical advances of SOFC anode and cathode stability and enhanced electrocatalysis.

The electrocatalysis of the electrochemical reduction of CO_2, a reaction of great potential importance for the future, is treated in an authoritative chapter by Y. Hori, who for many years has been a leading expert in this area. The Chapter reviews critically the plethora of experimental investigations of CO_2 reduction on different metals and provides deep and useful insight about the fundamental mechanisms leading to dramatically different product selectivity on different metals.

The Chapter by E. Gileadi and N. Eliaz provides a lucid and thorough treatment of the interfacial phenomena of electrodeposition and codeposition. The authors, who are internationally renowned experts in this important field, provide a deep and concise survey of experimental and theoretical findings and point out several important codeposition phenomena which defy existing theoretical treatments and show the necessity for novel theoretical analyses of the electrode-electrolyte interface, accounting for ion rather than electron transfer at the electrochemical interface.

The great usefulness of scanning tunneling microscopy (STM) for a better understanding of catalysis, electrocatalysis and electrodeposition at the fundamental level is presented by M. Szklarczyk, M. Strawski and K. Bieńkowski in a concise historical review which summarizes key landmarks in this important area and presents some of the almost limitless opportunities for the future.

The key role of electrochemistry in several important emerging technologies, such as electrodeposition and electroforming at the micro and nano level, semiconductor and information storage, including magnetic storage devices, and modern medicine, is described lucidly by M. Schlesinger in an authoritative Chapter. These new "high tech" electrochemical applications, presented by an author with great relevant experience open numerous challenges and opportunities for the electrochemist of the 21st century.

C.G. Vayenas
University of Patras
Patras, Greece

R.E. White
University of South Carolina
Columbia, South Carolina, USA

Contents

Chapter 1

SOME RECENT STUDIES IN RUTHENIUM ELECTROCHEMISTRY AND ELECTROCATALYSIS

N. S. Marinkovic, M. B. Vukmirovic, and R. R. Adzic

Chapter 4

INDUCED CODEPOSITION OF ALLOYS OF TUNGSTEN, MOLYBDENUM AND RHENIUM WITH TRANSITION METALS

Noam Eliaz and Eliezer Gileadi

Chapter 5

25 YEARS OF THE SCANNING TUNNELING MICROSCOPY
20 Years of Application of STM in Electrochemistry

Marek Szklarczyk, Marcin Strawski, and Krzysztof Bieńkowski

Chapter 6

MODERN APPLICATIONS OF ELECTROCHEMICAL TECHNOLOGY

M. Schlesinger

List of Contributors, MAE 42

R. R. Adzic
Department of Chemistry, Brookhaven National Laboratory, Upton, NY 11973

Krzysztof Bieńkowski
Laboratory of Electrochemistry, Department of Chemistry, Warsaw University, ul. Pasteura 1, 02-093 Warsaw, Poland

Noam Eliaz
Biomaterials and Corrosion Laboratory, School of Mechanical Engineering, Tel-Aviv University, Ramat Aviv, Tel-Aviv 69978, Israel, Ph: +972(3)640-7384; *Fax:* +972(3)640-7617; *neliaz@eng.tau.ac.il*

Eliezer Gileadi
School of Chemistry,Faculty of Exact Sciences,Tel-Aviv University, Ramat Aviv, Tel-Aviv 69978, Israel
Ph: +972(3)640-8694; Fax: +972(3)642-1982; gileadi@post.tau.ac.il

Y. Hori
Shirakawa 4-9-13-1201, Koto-ku, Tokyo, 135-0021 Japan
Phone/Fax +81-3-3630-4086, y-hori@white.plala.or.jp

N. S. Marinkovic
University of Delaware, Department of Chemical Engineering, Newark, DE 19716

M. Schlesinger
Department of Physics, University of Windsor, Windsor, Ontario Canada, N9B 3P4

Marek Szklarczyk
Laboratory of Electrochemistry, Department of Chemistry, Warsaw University, ul. Pasteura 1, 02-093 Warsaw, Poland

Marcin Strawski
Laboratory of Electrochemistry, Department of Chemistry, Warsaw University, ul. Pasteura 1, 02-093 Warsaw, Poland

Hiroyuki Uchida
Interdisciplinary Graduate School of Medicine and Engineering, University of Yamanashi, Kofu 400-8511, Japan

M. B. Vukmirovic
Department of Chemistry, Brookhaven National Laboratory, Upton, NY 11973

Masahiro Watanabe
Clean Energy Research Center, University of Yamanashi, Kofu 400-8511, Japan

1

Some Recent Studies in Ruthenium Electrochemistry and Electrocatalysis

N. S. Marinkovic,* M. B. Vukmirovic, and R. R. Adzic

Department of Chemistry, Brookhaven National Laboratory, Upton, NY 11973
**Present address: University of Delaware, Department of Chemical Engineering, Newark, DE 19716*

I. INTRODUCTION

Ruthenium is a metal of a considerable importance in electrochemical science and technology. It is a catalyst or co-catalyst material in Pt-Ru alloys for methanol- and reformate hydrogen-oxidation in fuel cells, while ruthenium oxide, a component in chlorine-evolution catalysts, represents an attractive material for electrochemical supercapacitors. Its facile surface oxidation generates an oxygen-containing species that provides active oxygen in some reactions. Ru sites in Pt-Ru catalysts increase the "CO tolerance" of Pt in the catalytic oxidation-reaction in direct methanol fuel cells (DMFC) and in reformate hydrogen-oxidation in proton exchange membrane fuel cells (PEMFC). The mechanism of Ru action is not completely understood, although the current consensus revolves around the so-called "bifunctional mechanism" wherein Ru provides oxygenated species to oxidize CO that blocks Pt sites, and has an electronic effect on Pt-CO interaction.

Modern Aspects of Electrochemistry, Number 42, edited by C. Vayenas et al., Springer, New York, 2008.

a) Ru(0001) b) Ru(10$\bar{1}$0)

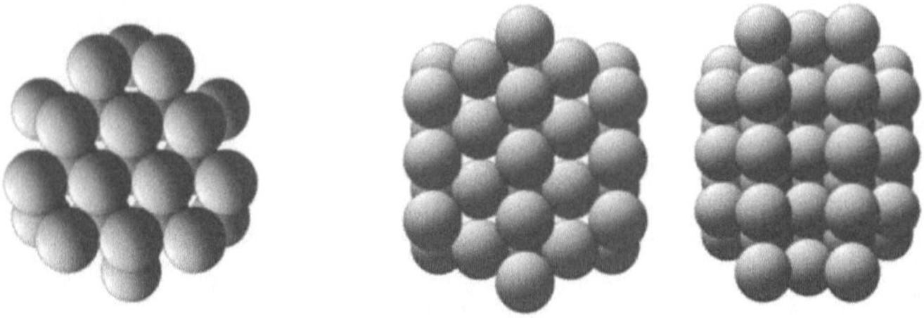

Figure 1. Structural model of (a) Ru(0001); (b) two terminations of the Ru(10$\bar{1}$0) surface.

While various studies of polycrystalline Ru go back several decades,[1,2,3] those involving single crystal surfaces and the structural sensitivity of reactions on Ru surfaces emerged only recently. Using well-ordered single crystalline surfaces brings useful information as the processes on realistic catalysts are far too complex to allow identification of the microscopic reaction steps. In this article, we focus on progress in model systems and conditions, such as electrochemistry and electrocatalysis on bare and Pt-modified well-ordered Ru(0001) and Ru(10$\bar{1}$0) single-crystal surfaces. We also review the current understanding of the mechanistic principles of Pt-Ru systems and a new development of a Pt submonolayer on Ru support electrocatalyst.

Ruthenium crystallizes in a hexagonal close-packed structure, (*hcp*). Figure 1 shows the two single crystal surfaces of Ru. The Ru(0001) surface possesses the densest, i.e., hexagonal arrangement of atoms, Fig. 1a. The other plane, Ru(10$\bar{1}$0), can have one of the two terminations of the surface atoms, Fig. 1b. One termination can be described as a stepped surface with a trigonal arrangement of atoms in two-atom-long terraces with a step of the same orientation; the other termination is a square-symmetrical arrangement of atoms in two-atom-long terraces with the same orientation of atoms in steps. In the faced-centered cubic (*fcc*) system, these three structures are uniquely defined and labeled as (111), (110), and (210), respectively.

II. PREPARATION OF WELL-ORDERED Ru SINGLE-CRYSTAL SURFACES

For over three decades, a procedure has been known for preparing well-ordered Ru single-crystal surfaces in ultra-high vacuum (UHV).[4] It involves Ar^+ sputtering at room temperature, followed by several cycles of oxygen adsorption (at 800 K) and desorption (at 1700 K), both in an oxygen atmosphere of 10^{-7} Torr; finally, the crystal is flash heated to 1700 K at UHV to remove traces of oxygen from the Ru surface.[4] One of the first reports on voltammetry and *in situ* infrared (IR) spectroscopy of CO on the well-ordered Ru(0001) surface, appearing at the turn of the millennium, used this method for obtaining Ru(0001) single-crystal surfaces.[5] A simpler method developed recently involved heating the Ru crystal in an H_2 stream;[6] it produced a well-ordered Ru(0001) surface as verified by Scanning Tunneling Microscopy (STM). This method shortened preparation time from days (UHV) to a few hours.[7-9] A modified method, subsequently reported further reduced the preparation time[10] in which crystals were inductively heated to 1700 K for 30–60 seconds in a stream of Ar-15% H_2. STM pictures and voltammetry profiles on single-crystals thus obtained were essentially the same as those from the other two methods.

III. ELECTROCHEMISTRY OF SINGLE-CRYSTAL Ru SURFACES

1. Voltammetry Characterization

Figure 2a displays the voltammetry curves of polycrystalline and Ru(0001) surfaces in 1 M H_2SO_4. The broad, featureless oxidation-process at the polycrystalline electrode's surface involves currents about one order-of-magnitude larger than that of the single-crystal surface, and was attributed to the continuous oxidation of Ru in a process encompassing more than one electron per atom.[11-13] The potential regions of hydrogen adsorption and surface oxidation are generally acknowledged to almost overlap, since the Ru oxidation

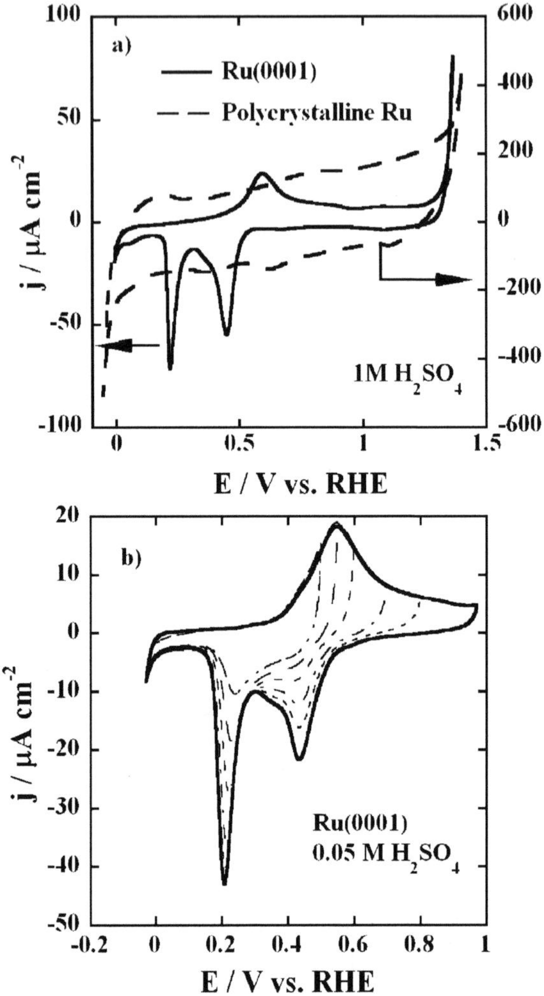

Figure 2. Voltammetry curves for (a) polycrystalline Ru and Ru(0001) in 1 M H_2SO_4, and, (b) oxidation with different positive potential limits in 0.05 M H_2SO_4. Sweep rates: (a) 10 mV s^{-1}, and (b) 20 mV s^{-1}.

starts very early in the potential scale (0.2 V vs. RHE).[†] The electrochemical processes are more easily identified for electrodeposited Ru films[11,14,15] than for bulk metal.

Figure 2b shows the voltammetry curves for the surface oxidation of Ru(0001) in 0.05 M H_2SO_4. Similar curves are presented in several publications.[5,9,16-18] Before starting a sweep in the positive direction, the potential was held sufficiently long at the negative limit to ensure a negligible reduction current originating from the previous potential cycle. The voltammetry curves show a single anodic peak with a long tail extending to the onset of bulk oxidation, and two major cathodic peaks correlated to the reduction processes, which begin at 0.6 V. Surface oxidation occurs above 0.4 V, and the integrated anodic charge reaches levels required for an one-electron oxidation of the Ru(0001) surface (260 $\mu C\ cm^{-2}$) in the sweep up to 1 V. Increasing the sweep rate up to 500 $mV\ s^{-1}$ caused a linear increase in the current density, but repeated potential cycling between 0 and 1.2 V did not significantly change the voltammetry curves. These observations suggest that the oxidation of the Ru(0001) surface is limited to the top layer with one electron per atom exchange at the potential below the onset of bulk oxidation. This can be represented by the following reaction,

$$\mathrm{Ru{-}A^- + H_2O \rightarrow RuOH + H^+ + e^- + A^-} \quad \text{at} \quad E > 0.40\text{V} \qquad (1)$$

Several groups have published voltammetry studies of the Ru(0001) surface in solutions containing non-specifically adsorbed ions.[5,8,9,16-28] The voltammetric profile of the Ru(0001) surface in 0.1 M $HClO_4$ has an integrated charge between 0.1 and 1 V of 230 $\mu C\ cm^{-2}$, somewhat smaller than that required for an one-electron process, (dashed line in Fig. 3a).[17] A comparison of the processes in the two electrolytes indicates an effect of anion adsorption on the oxidation of the Ru surface. Strongly adsorbed bisulfate anions prevent OH adsorption due to water-oxidation at low potentials, and promote the complete removal of the oxygen-containing species in the cathodic sweep. Comparing the electrochemical processes with the opening of the anodic limit in the two acid

---

[†]Unless otherwise stated, all potentials are expressed vs. reversible hydrogen electrode, RHE.

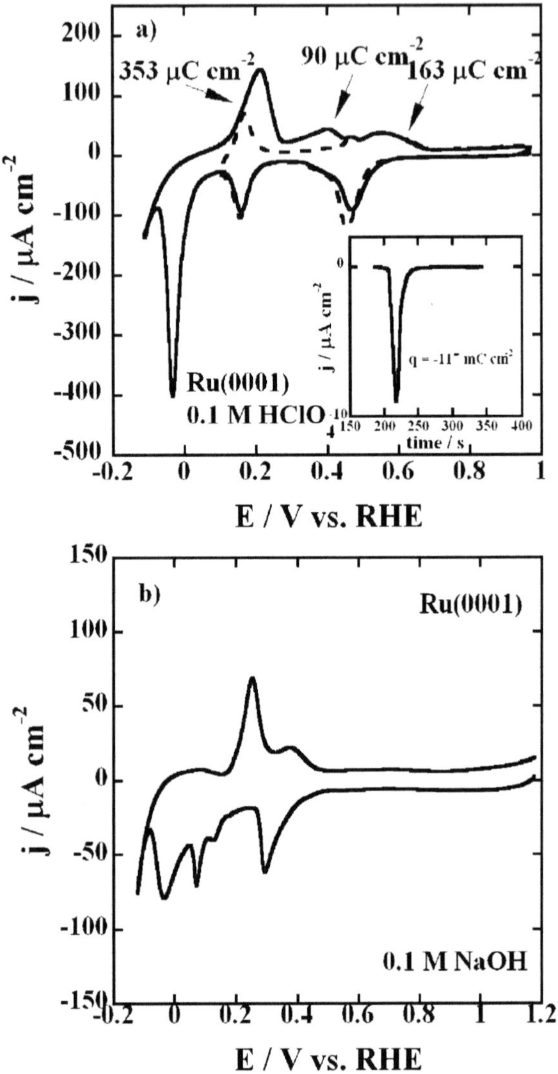

Figure 3. Voltammetry curves for the Ru(0001) surface oxidation in (a) 0.1 M $HClO_4$ and (b) 0.1 M NaOH. Sweep rates: (a) 50 mV s^{-1} and (b) 20 mV s^{-1}. The inset in (a) shows the charge associated with the displacement of adsorbed species at 0.12 V by the adsorption of CO.

solutions suggests that the reduction process in perchloric acid solution cannot be completed without extending the sweep into the hydrogen adsorption/evolution region.[17] Indeed, extending the cathodic limit into the H_2 evolution region generates a large peak around –0.05 V (full line in Fig. 3a). The peak's associated charge is over 300 μC cm^{-2}, suggesting that it reflects a combination of at least two processes, one being hydrogen evolution. The other process, however, is puzzling. Since the same peak occurs in hydroxide solutions (see below), it is not due to perchlorate anion reduction. Also, it cannot be attributed to impurities in solution (like Cl^-)[28] because it is not supported with adequate voltammetry response. The probable explanation is the reduction of adsorbed OH, as also suggested in a recent publication.[29] The subsequent positive sweep shows a much greater peak around 0.2 V, and another, smaller one around 0.4 V. A total charge of over 600 μC cm^{-2} is associated with the scan between –0.1 and 0.7 V.

To identify the nature of the species adsorbed at potentials from 0.1 to 0.25 V, a charge displacement technique was used, which was proven useful in identifying species on the Pt(111) electrode.[30] The electrode potential was held at 0.12 V and CO was introduced into the cell; there was no significant CO oxidation on Ru(0001). The charge associated with the resulting displacement process was negative, adding up to –117 μC cm^{-2} (see inset in Fig. 3a). If H_{ad} was the adsorbed species in that potential region, the charge would be positive. Therefore, contrary the conclusions from other work,[23] the reaction must be associated with some oxygen-containing species. This finding was confirmed recently using the same approach.[28] A plausible reaction to account for this negative charge is

$$\mathrm{RuOH} + \mathrm{CO_{soln}} + e^- \rightarrow \mathrm{RuCO_{ad}} + \mathrm{OH^-} \tag{2}$$

The measurement at 0.25 V revealed again a negative charge, albeit a smaller one, of –31 μC cm^{-2} implying that some transformation took place in the species existing at 0.12 V. The difference between the charges at 0.12 and 0.25 V is 79 μC cm^{-2}, in good agreement with 86 μC cm^{-2}, the voltammetric charge in that potential interval. Interestingly, measurements in H_2SO_4 solution do not show this process;[17] it is probably precluded by strong bisulfate adsorption on the hexagonal structure of Ru(0001). As discussed later in this chapter, infrared spectroscopy and X-ray diffraction studies indicate that bisulfate adsorption on Ru(0001) is essentially at saturation coverage between 0.0 and 0.5 V, and that water chemisorbs on Ru(0001) in the absence of chemisorbed anions.[16,17]

Anion effects on the electrooxidation of Ru(0001) were explored by introducing Cl^- and Br^- into the 0.1 M $HClO_4$ solution.[17] A sharp rise of anodic current occurs near 0.2 V, which is at a more negative potential than the onset of surface oxidation in sulfuric acid. Hence, even though halide ions are strongly adsorbed, they do not better protect the Ru from surface oxidation than do bisulfate ions. It is likely that a different redox process occurs with halide ions because they form compounds with Ru in several different oxidation states.

Voltammetry of the oxidation of Ru(0001) in 0.1 M NaOH, Fig. 3b, shows a curve that has some similarities to that recorded in perchloric acid. The curve is dominated by a strong anodic peak at 0.25 V and a strong cathodic peak occurring at –0.05 V, the same potential as in perchloric acid. Its origin is not understood, as discussed. The multitude of peaks indicates the complexity of the oxidation/reduction processes. The symmetric peaks in this potential range on polycrystalline Ru were attributed to processes involving hydrogen adsorption and/or Ru oxidation.[13,31,32]

The voltammetry curve for the Ru($10\bar{1}0$) surface in 0.05 M H_2SO_4 solution (Fig. 4a) reveals a remarkable difference between the oxidation processes for Ru(0001) and Ru($10\bar{1}0$). The oxidation of this face is more facile than that of Ru(0001), as indicated by the onset of the reaction at lower potentials and by increase of the charge with each potential cycle. This difference most likely is the consequence of the more open structure of the Ru($10\bar{1}0$). A pair of peaks at 0.12 and 0.3 V is reminiscent of hydrogen adsorption on Pt metals. However, CO displacement showed a negative charge of $-354\ \mu C\ cm^{-2}$. Thus, the peaks probably represent partial Ru oxidation to RuOH, wherein OH is the predominant adsorbed species, perhaps with some co-adsorption of bisulfate.

Unlike the behavior of the Ru(0001) surface, the gradual increase in the positive potential limit has deleterious effects on the ordering of Ru($10\bar{1}0$). The increase of the charge associated with the voltammetry curve in each subsequent cycle indicates the oxi-

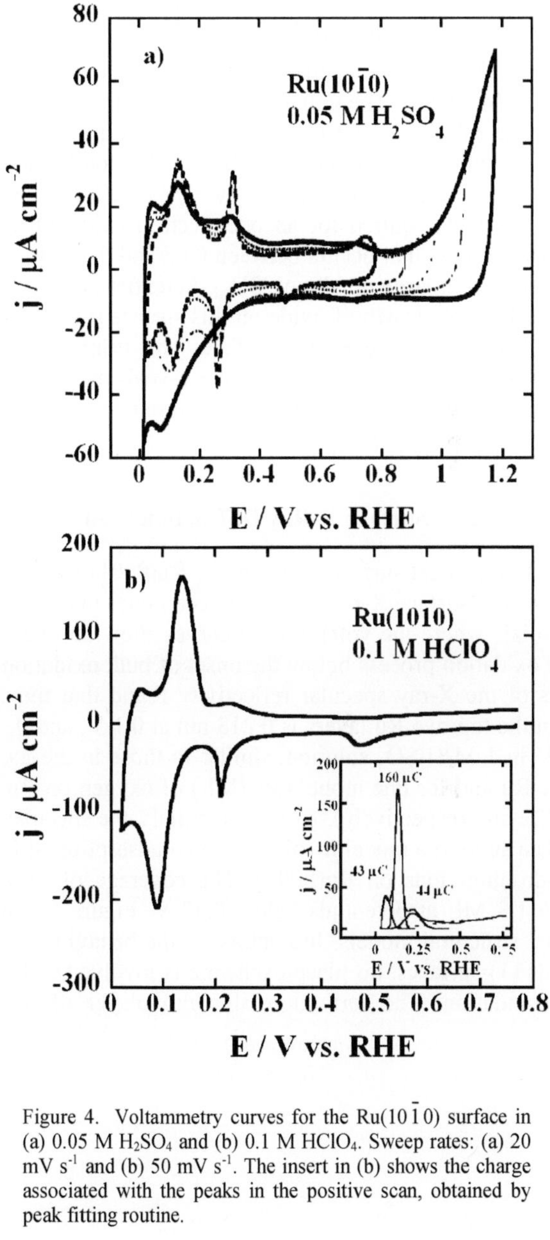

Figure 4. Voltammetry curves for the Ru(10$\bar{1}$0) surface in (a) 0.05 M H_2SO_4 and (b) 0.1 M $HClO_4$. Sweep rates: (a) 20 mV s^{-1} and (b) 50 mV s^{-1}. The insert in (b) shows the charge associated with the peaks in the positive scan, obtained by peak fitting routine.

dation of several of the crystal's top atomic layers, similar to the behavior of polycrystalline Ru.[11-13]

Figure 4b shows the cyclic voltammetry of the Ru($10\bar{1}0$) surface in 0.1 M $HClO_4$. The voltammetry profile differs from that in sulfuric acid solution suggesting that sulfate absorption /desorption contributes to the peaks observed in Fig. 4a. The total charge in the positive sweep between 0 and 0.25 V is 247 μC cm^{-2}, which is greater than that required for an one-electron oxidation (159 μC cm^{-2}). The process that occurs between 0.05 and 0.3 V is probably associated with a large uptake of OH_{ad}. Potential cycling to 0.8 V causes additional growth of oxide on this surface (cf. Fig. 4a), unlike the behavior of the Ru(0001) face. The striking differences between the two Ru single-crystal planes reveals the large stability of the hexagonal Ru(0001) surface and a pronounced reactivity of the rectangular Ru($10\bar{1}0$) face.

2. Surface X-Ray Diffraction Study

The electrochemical surface oxidation of Ru(0001) was characterized by *in situ* surface X-ray scattering techniques in acid solutions at potentials where the voltammetry curves show an one-electron surface oxidation process below the onset of bulk oxidation.[17] The analysis of the X-ray specular reflectivity found that the spacing between the top two Ru layers is 0.213 nm at 0.1 V, and 0.220 nm at 1.0 V in 1 M H_2SO_4 solution, similar to those in the gas phase for bare Ru and for one monolayer (ML) of oxygen on Ru (0.210 and 0.222 nm, respectively). At low potentials, the specular reflectivity data support a model involving the co-adsorption of bisulfate and hydronium ions on Ru(0001). The coverage of bisulfate is close to 1/3 ML at potentials below 0.57 V. Figure 5 shows the proposed structural models. In contrast to the behavior of Pt(111) and Au(111) surfaces, no place exchange is involved in Ru(0001) surface oxidation. The formation of a monolayer of ruthenium

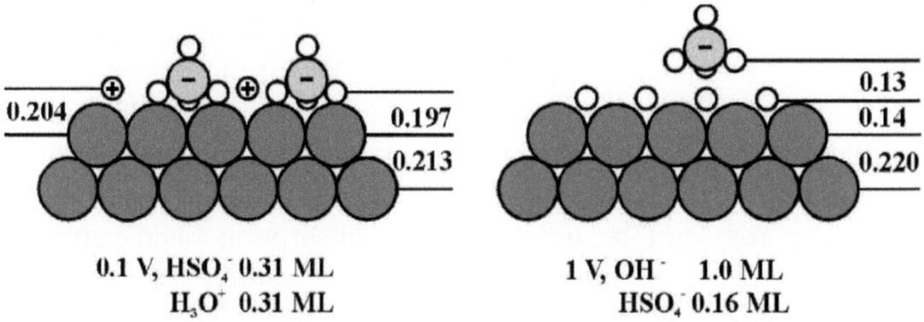

Figure 5. Proposed structural models for Ru(0001) oxidation, where the O, S, and Ru atoms are represented by the open, lightly-shaded, and heavily- shaded circles, respectively. The layer spacings are given in nm, and coverage is given in monolayers. Reprinted with permission from [17]. Copyright (2001) American Chemical Society.

oxide induces partial desorption of bisulfate, in agreement with the Fourier Transform infrared (FTIR) results (see Section III.3). Strikingly, oxygen penetration is completely blocked on Ru(0001) at potentials below the bulk oxidation potential, in contrast to the high degree of surface oxidation of polycrystalline Ru that occurs between the onset of hydrogen evolution to the onset of bulk oxidation. Since gas-phase studies demonstrated that subsurface oxygen plays a major role in the activity of Ru for CO oxidation, the lack of subsurface oxygen on the Ru(0001) electrode might explain its inactivity for CO electrooxidation; this interpretation is discussed below.

3. Infrared Spectroscopy and Anion Adsorption

The adsorption of anions on solid surfaces is of considerable interest, mainly because of its effect on the kinetics of electrochemical reactions. Several *in-situ* techniques have been applied toward this purpose.[33] Infrared measurements were used to identify adsorbed species, estimate anion adsorption isotherms, and to gain information on anion interaction with electrode surfaces.[34] Sulfuric acid anions are possibly the commonest anion adsorbates because of their specific adsorption on metal surfaces. Depending on the metal, its surface orientation, and the concentration of anion, either sulfate or bisulfate can be specifically adsorbed on the surface. Identifying the predominant adsorbate on platinum-group metals has engendered some controversy. While STM studies show that sulfate and/or bisulfate on Pt metals of (111) orientation form a (√ 3 x √ 7) overlayer,[35] questions remain about the nature of the species. Even though the symmetry differences of sulfate and bisulfate ions should, in principle, reveal the preferentially adsorbed species, the interpretation of the mostly equivalent *in situ* IR spectra may not be unique.[36] Two absorption bands around 1200 and 1100 cm^{-1} observed on a polycrystalline Pt electrode were attributed to adsorbed bisulfate and sulfate, respectively.[37] In addition, a third band at 950 cm^{-1} on polycrystalline Pt, led to the conclusion that both bisulfate and sulfate on the surface give rise to three bands.[38] For Pt (111), Nart et al. concluded that the adsorbate is sulfate coordinated via three oxygens presenting a C_{3v} symmetry,[34] while Faguy et al. argued that the adsorbed species is not described adequately as either sulfate or bisulfate but rather as an $H_3O^+ – SO_4^{2-}$ ion pair.[39]

Recent data confirmed that the preferred species on Pt (111),[40] Pd[41] and Ir[42] is bisulfate, while sulfate adsorbs on Ag(111)[43] and on Au(111).[44-46] Apparently the *sd* metals (Rh, Pt, Pd) with (111) orientation adsorb bisulfate, whereas the *sp* metals (Cu, Ag, Au) adsorb sulfate. The IR study on Ru(0001) seemed to support this conclusion (see below), as the latter has the same orientation of surface atoms as the *fcc* metals of (111) orientation, and the adsorbed species is bisulfate.

(*i*) ***Polycrystalline Ru Electrode***

As discussed in the previous Sections, electrochemical oxidation of polycrystalline Ru involves about one order-of-magnitude larger currents than that of Ru(0001), starting as early as 0.2 V.[16] Polycrystalline Ru is covered with hydroxyl ions from water very early in the potential scale even in acidic solutions, thus blocking the surface from anion adsorption by the supporting electrolyte. This conclusion is supported by the *in situ* IR spectra presented below.

Specific adsorption of sulfate-bisulfate generally displays spectral features that are blue shifted (i.e., to higher frequencies) with higher electrode potentials, as observed for both polycrystalline Pt[37,38,47] and the Pt(111) surface.[36] The shift is explained in terms of one or more of the following mechanisms: electron donation between the adsorbate and the metal surface (vibronic coupling); coupling of the electric field that exists in the double layer with the dipole moment of the adsorbates (Stark effect); and, the interaction from dipole–dipole coupling due to an increase in coverage.[48,49] The absence of these mechanisms in the IR spectra shown in Fig. 6 for polycrystalline Ru–H_2SO_4 system suggests that there is no specific adsorption of sulfuric acid anions.

Five distinctive potential-independent peaks are observed at 1100, 980, 1205, 1051, and 885 cm^{-1} from the polycrystalline Ru surface in 0.05 M H_2SO_4 (see Fig. 6a). The first two are associated with sulfate and the rest with the bisulfate species.[50] Positive-going bands in the $-\Delta R/R$ representation show an increase of the species at the electrode's surface and/or in the solution layer sampled by the IR light at the sample's potential relative to that at the reference potential. However, adsorbed species at the electrode surface are sensitive only to the *p*-polarized light due to the surface-selection rule, while both polarizations interact equally with randomly oriented species in solution.[51] The fact that the intensity of both the sulfate and bisulfate bands rises with increasing electrode potential, and that their features are observed with both *s*- and *p*-polarized light, suggests strongly that the spectra represent solution species, i.e., the migration of ions into the diffuse part of the double layer to compensate for the charge buildup at the electrode surface. Therefore, a marked interaction of OH with polycrystalline Ru electrode at low potentials prevents sulfate-bisulfate adsorption throughout the whole potential region from hydrogen evolution to oxygen evolution.

(*ii*) *Ru(0001) and Ru(10$\bar{1}$0) Single-Crystal Electrode Surfaces*

Figure 6b shows a set of spectra taken at the Ru(0001) electrode in 0.05 M H_2SO_4 solution. There is a well-defined bipolar peak with the positive lobe centered at 1280 cm^{-1} that shifts with increasing potential. A similar band has been observed for several single crystal surfaces with hexagonal surface orientation.[52] On the basis of *ab initio* calculations for anion adsorption at the Pt(111) surface, Sawatari et al. concluded that the totally symmetric stretch of bisulfate at 1051 cm^{-1} should undergo a large frequency shift.[53] In the IR spectrum given in the $-\Delta R/R$ representation, this phenomenon should be visible as a positive-going, potential-

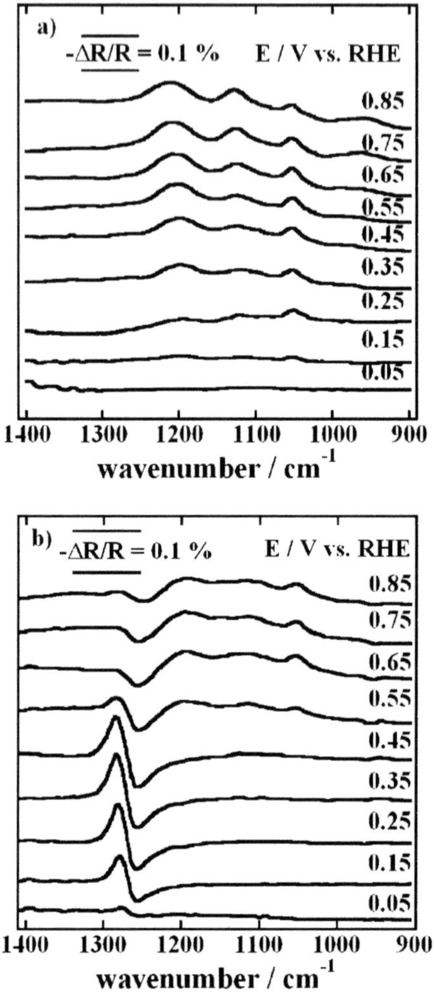

Figure 6. *In-situ* infrared spectra obtained from (a) polycrystalline Ru, and (b) Ru(0001) electrode in 0.05 M H_2SO_4. The reference spectrum is obtained at 0.03 V, and sample spectra are taken every 0.1 V, from 0.05 to 0.85 V. 4096 scans were co-added in 16 cycles, 256 scans each; the resolution was 8 cm^{-1}. Spectra are offset for clarity. Reprinted from [16], Copyright (2001) with permission from Elsevier.

dependent peak around 1250 cm^{-1}, together with a negative-going 1051 cm^{-1} band representing the loss of HSO_4^- species in solution.[39] The two bands were observed in the Pt(111)/H_2SO_4 system, and the strongly blue-shifted band was assigned to the absorption of the bisulfate or sulfate-hydronium ion pair.[39] Consequently, the band at 1280 cm^{-1} in the present case arises from bisulfate adsorption at Ru(0001). However, the bipolarity of the peak is unusual, as it has not been noted in any other sulfate-bisulfate adsorption studies.[39,43,52] Furthermore, the peak for bisulfate ions in solution is visible only above 0.45 V. Both phenomena appear to be related to the bisulfate adsorption at the reference potential. The absence of the solution band at 1051 cm^{-1} at potentials below 0.45 V indicates the lack of a measurable change in the bisulfate coverage at the sample and reference potentials, i.e., that a sizeable coverage of bisulfate is attained already at the reference potential of 0.03 V. This conclusion is supported by the bipolar shape of the band for adsorbed bisulfate at 1280 cm^{-1} because such a band in subtractive normalized interfacial FTIR (SNIFTIR) spectra is apparent when a species is absorbed at both the reference and sample potential, and the frequency of its band center depends upon potential.[54]

Figure 6b also reveals that the positive lobe of the bipolar band around 1280 cm^{-1} decreases at potentials above 0.45 V; this decline coincides with the onset of the surface oxidation in the voltammetry of Ru(0001) (c.f., Fig. 2). Adsorption of the OH species is followed by the desorption of bisulfate and a concurrent increase in the bisulfate species in the double layer. This effect becomes visible in the IR spectrum by the appearance of the positive-going solution-phase bands for the bisulfate anion at 1051 and 1200 cm^{-1} at potentials equal to, or higher than 0.55 V. The most pronounced feature in the IR spectra above 0.55 V is the negative lobe of the bipolar band centered at 1248 cm^{-1}, which represents adsorbed bisulfate at the reference potential.

The open structure of the Ru($10\bar{1}0$) surface (c.f., Fig. 1b) is expected to react with H_2O at very low potentials and become covered with OH species. UHV data indicate that oxygen adsorption on this surface produces c(2x4)-2O and (2x1)p2mg-2O overlayers with O atoms occupying the three-fold *hcp* sites formed by two atoms in the first atomic layer and one atom in the second layer.[55] These three-fold sites are the only likely candidates for sulfate or bisulfate adsorption on either of the two terminations of the Ru(10$\bar{1}$0) surface. Since the latter are occupied, it is not surprising that this plane shows no sulfate or bisulfate adsorption in IR spectra (not shown). The surface remains inactive for anion adsorption even after extensive hydrogen evolution, which was shown to free some oxygen-occupied sites in the first atomic layer.[7]

4. Surface-Oxide Formation

The specific catalytic properties of polycrystalline and single crystal surfaces have prompted extensive research on their oxidation in electrochemical- and gas- phase environments.[17,55,56] Recent developments in fuel cell technology have renewed efforts to improve Pt-Ru electrocatalysis for both reformate hydrogen- and methanol-oxidation.[57] In the following Section, we discuss the oxidation of single crystal surfaces in both UHV- and electrochemical-environments.

(*i*) *Gas-Phase Oxidation*

In general, the catalytic activity of transition metal surfaces for certain oxidation reactions (e.g., CO) is determined by the propensity of the metal's surface to dissociate oxygen molecules and is counterbalanced by the bond strength of the active oxygen species on the metal's surface.[58] The transition metals with half-filled *d*-bands reveal the highest activity, where the dissociation probability is not too low and the adsorption energy is not too high. Accordingly, Pt, Pd, and Rh are efficient metal catalysts, while Ru is a poor one due to its very high oxygen-binding energy.

On the other hand, UHV measurements showed that the Ru surface can be used as a kind of storage, able to accommodate large amounts of atomic oxygen.[59,60] Other transition metals also exhibit this ability, but the exceptional property of Ru surfaces is due to the fact that oxygen can be completely removed by simply heating the sample up to about 1700 K without irreversibly incorporating oxide in the bulk. The oxidation of the Ru(0001) surface in UHV at low O_2 pressure facilitates the formation of (2 x 2)-O and (2 x 1)-O superstructures at coverages of 0.25 and 0.5 ML, respectively.[4] Both superstructures have minimal catalytic activity for CO oxidation. Even the (1 x 1)-O overlayer on Ru(0001), prepared at elevated temperatures by the enrichment of the (2 x 1)-O phase due to the decomposition of NO_2[61] is inactive a catalyst. The high reactivity of the Ru(0001) surface for CO oxidation is only reached when the total O_2 concentration exceeds the equivalent of about 3 ML, i.e., when at least 2 ML of oxygen are located in the subsurface region and the reaction probability increases by two orders-of-magnitude.[62,63] The active part of this "O-rich" ruthenium phase, prepared by directly exposing the Ru(0001) surface to high doses of O_2 at elevated temperatures, was demonstrated to be RuO_2, growing epitaxially with its (110) plane parallel to the Ru(0001) surface.[64] Its extraordinary activity towards CO oxidation was reported in several papers.[65]

Similar growth of RuO_2(110) on Ru(0001) can be achieved by annealing the substrate in Ar containing ca. 100 ppm of O_2 at around 700 K.[10] The epitaxial growth of RuO_2 at elevated temperatures on Ru(10$\bar{1}$0) was demonstrated with its (100) face oriented parallel to the substrate.[23] An electrochemical STM image (ECSTM) of RuO_2 prepared by chemical oxidation of Ru(0001) in Ar/O_2 (115 ppm), obtained in 0.05 M H_2SO_4 under an open-circuit potential shows stripes with step sizes of about 0.3 nm (i.e., corresponding to one monolayer of oxide) growing unidirectionally at 60° with respect to the steps' direction, Fig. 7a.[10] An atomi-

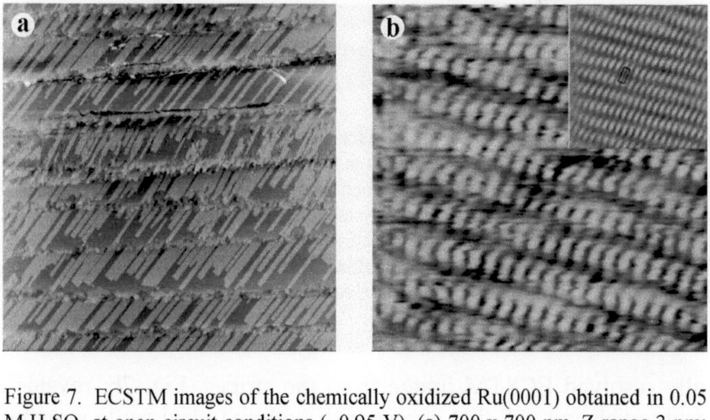

Figure 7. ECSTM images of the chemically oxidized Ru(0001) obtained in 0.05 M H_2SO_4 at open circuit conditions (~0.95 V). (a) 700 x 700 nm, Z range 3 nm; and, (b) 5.8 x 5.8 nm. Reprinted from [10], Copyright (2004) with permission from Elsevier.

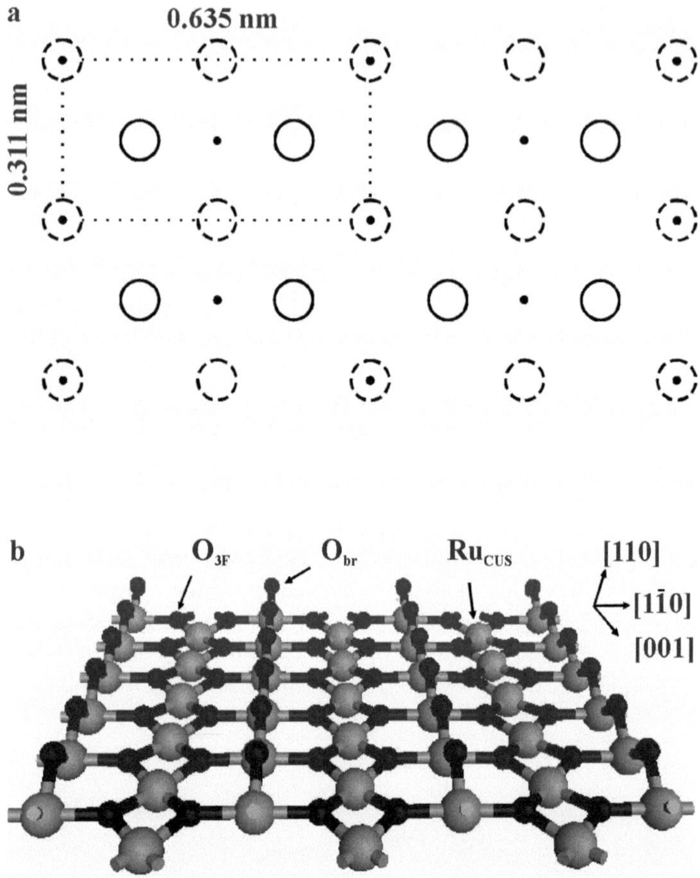

Figure 8. a) Schematic view of the surface structure of an ideal $RuO_2(110)$ single-crystal, where solid circles represent Ru atoms in the surface plane, open circles O atoms in the surface plane, and dotted open circles O atoms below surface plane; b) model of an ideal $RuO_2(110)$ single-crystal. The $RuO_2(110)$ surface contains two kinds of coordinatively unsaturated (CUS) atoms: two-fold bridging O (O_{br}) and five-fold Ru (Ru_{CUS}). The O_{3F} is the O atoms that lay in the plane in the Ru atoms and posses its bulk-like three-fold coordination.

cally resolved ECSTM image from the stripes, Fig. 7b, reveals a rectangular unit cell that agrees well with the ideal rutile structure of RuO_2 with an (110) orientation.[66] The model of ideal the

$RuO_2(110)$ single-crystal surface is depicted in Fig. 8. LEED and STM data for UHV oxidation of Ru(0001) suggest that RuO_2 overlayer is not pseudomorphic with the Ru substrate.[64,67] Between the RuO_2 domains, the Ru surface is covered with a monolayer of RuOH, which is the precursor to Ru oxidation.

(ii) Electrochemical Oxidation

The initial phase of the electrochemical oxidation of Ru(0001) in 0.05 M H_2SO_4 is a hump at *ca.* 0.57 V (cf., Fig 2). At lower potentials the surface is protected by bisulfate adlayer.[16,17] Figure 9 shows ECSTM images obtained by progressively increasing the electrode's potential.

At 1.0 V, Fig. 9a, the Fourier transform of an atomically resolved ECSTM image obtained from a terrace reveals a hexagonal array of dark spots at distances of 0.270 nm, corresponding to the inter-atomic distance of the Ru atoms. The dark contrast of the spots suggest that they are associated with RuOH, as shown in the UHV studies of chemisorbed O atoms in the (1 x 1) overlayer on the Ru(0001) surface.[23,64] Identical large-frame images are obtained at any potential between 0.05 and about 1.10 V. The first observable changes, which appear as a difference in contrast at the step edges, occur at a potential of 1.17 V, indicating their coverage by oxide (Fig. 9b). The steps' edges are covered with irregularly spaced oxide islands separated by non-oxidized segments. At the onset of bulk oxidation, ECSTM image shown in Fig. 9c reveals smooth RuOH domains, oxide islands along meandering step edges, and large oxide islands concentrated mainly near the steps' edges. The latter suggests that the Ru atoms necessary for oxide growth come from the dissolution of the step edges, rather than terraces. The same type of topographical features is observed at 1.35 V (Fig. 9d) but with more large oxide- islands, twisting step-edges, and smaller area of smooth RuOH domains. The dimensions of the large oxide islands are the same as in Fig. 9c, indicating that they do not grow with increasing potential; rather, only their number increases. These randomly distributed islands represent RuO_2 domains, in accordance to the Pourbaix diagrams[68] and an *ex situ* study of Ru(0001) oxidation in a similar potential region;[23] their nucleation process is instantaneous, as verified by potential step experiments.[10]

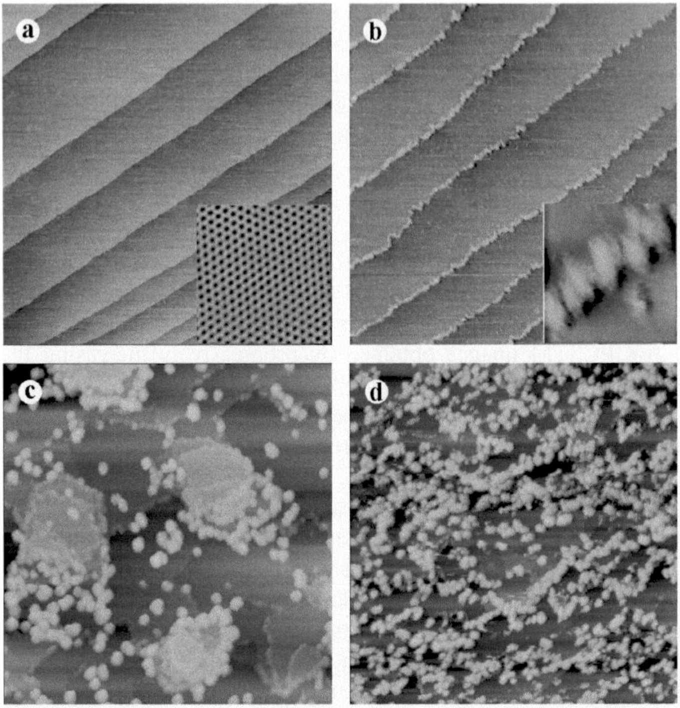

Figure 9. ECSTM images of the oxidation of Ru(0001) in 0.05 M H_2SO_4 at (a) 1.0 V, (b) 1.17 V, (c) 1.27 V, and (d) 1.35 V. Image (a) 500 x 500 nm, Z range 2 nm, inset of (a) 4 x 4 nm; (b) 230 x 230 nm, Z range 2 nm, inset of (b) 20 x 20 nm, Z range 2 nm; (c) 165 x 165 nm, Z range 10 nm; and, (d) 250 x 250 nm, Z range 10 nm. Reprinted from [10], Copyright (2004) with permission from Elsevier.

IV. ELECTROCATALYSIS ON Ru SINGLE-CRYSTALS AND NANOPARTICLE SURFACES

The ruthenium surface shows certain activity in hydrogen evolution, oxygen reduction, and CO oxidation; it is not active for methanol oxidation because methanol is not adsorbed on oxygen-covered Ru. The unique activity of Pt-Ru catalysts towards methanol is briefly discussed in Section V.

1. Hydrogen Oxidation and Evolution Reactions

While it is expected that electrocatalytic reactions on Ru surfaces should be strongly structure-sensitive, the first report on structural effects on hydrogen oxidation and evolution reactions appeared only recently.[7] The structural effects in the hydrogen oxidation reaction (HOR) and the hydrogen evolution reaction (HER) may be factors affecting the performance of hydrogen fuel cell anodes.

Hydrogen oxidation on Ru is likely to proceed through a mechanism involving two or three of the following reactions, as on Pt:[69,70]

$$\begin{aligned} H_2 &\rightarrow H_{ad} + H_{ad} \\ H_2 &\rightarrow H^+ + H_{ad} + e^- \\ H_{ad} &\rightarrow H^+ + e^- \end{aligned} \tag{3}$$

For the hydrogen-evolution reaction, these processes are referred to as the recombination reaction, or Tafel reaction, the ion-plus-atom reaction, or Heyrovski reaction, and the charge transfer reaction, or Volmer reaction, respectively. As mentioned before in Section III.1, there are disagreements about the existence of underpotential deposition of hydrogen at Ru surfaces.[18,23] If the peaks preceding the hydrogen evolution at Ru single crystal surfaces were due to the adsorption/desorption of OH, as the above discussion indicates, then a rate expression for the hydrogen-oxidation reaction would have to include the coverage of both adsorbed species, i.e., θ_{OH} and θ_H, and it would be difficult to obtain an indication of the rate-determining step of these three reaction processes from the Tafel slopes.

The kinetics of the HOR on polycrystalline Ru and carbon-supported nanoparticles is about two orders-of-magnitude smaller than that on Pt or Pt-Ru alloys, and it usually is assumed that the Ru contribution to the H_2 oxidation current of fuel cell anodes is negligible.[71,72] However, at temperatures at which PEMFCs operate (60-80 °C) the kinetics of HOR on Ru is considerably faster than at room temperature,[71] so that the effect of Ru surfaces may be of importance in PEMFC catalysis.

The hydrogen oxidation reaction was found to be under kinetic control at both Ru(0001) and Ru($10\bar{1}0$) surfaces.[7] Dependence on the rotation rate is negligible on the (0001) surface, while that on $(10\bar{1}0)$ is significant, but the limiting current density is not linear with the square root of the rotation rate. On the more open $Ru(10\bar{1}0)$ surface, the peak current densities are about twice as large than those on hexagonal Ru(0001) surfaces at room temperature, as well as at higher temperatures in 0.05 M H_2SO_4, Fig. 10. On both surfaces the reaction rate rapidly falls with increasing overpotentials, causing almost complete inhibition at potentials above 0.6 V, due to the growing oxide layer.[73] Purportedly, hydrogen can still be adsorbed on the ridges of $(10\bar{1}0)$ surface despite the early coverage by oxygen-containing particles, while the latter causes complete inhibition on the smooth Ru(0001) surface. In perchlorate solutions, the more open $(10\bar{1}0)$ surface again exhibits higher currents for H_2 oxidation, but considerably lower than those in sulfuric acid solution. As perchlorate ions are not considered to be specifically adsorbed on metal surfaces, the lower kinetics in $HClO_4$ is probably due to the larger oxide build-up at low potentials.[17]

The apparent electrochemical-activation energy for the HOR, $\Delta H^{0\#}$, on the Ru(0001) is about tenfold higher than on the Pt surface with identical atomic arrangement, Pt(111). Similarly, $\Delta H^{0\#}$,on the $Ru(10\bar{1}0)$ surface is about one order-of-magnitude higher than that on Pt(110), in agreement with the large difference in the observed reaction rates on Ru and Pt.[72,73]

The structural effects on hydrogen evolution kinetics on Ru are small, as inferred from the HER curves for (0001)- and $(10\bar{1}0)$-oriented surfaces obtained in perchloric and sulfuric acid solutions.[7] The similarities in the reaction kinetics in the two indicate that hydrogen evolution proceeds on bare Ru surface, i.e., a surface not covered with either OH or, in the case of sulfuric-acid solutions, with bisulfate ions.

2. CO Oxidation

Carbon monoxide is one of the best characterized adsorbates in catalysis because of its specific role in many catalytic reactions. Since it is a strongly adsorbed species, it usually blocks the catalyst's surface for desired reactions, for instance in the oxidation of methanol or hydrogen reformate.

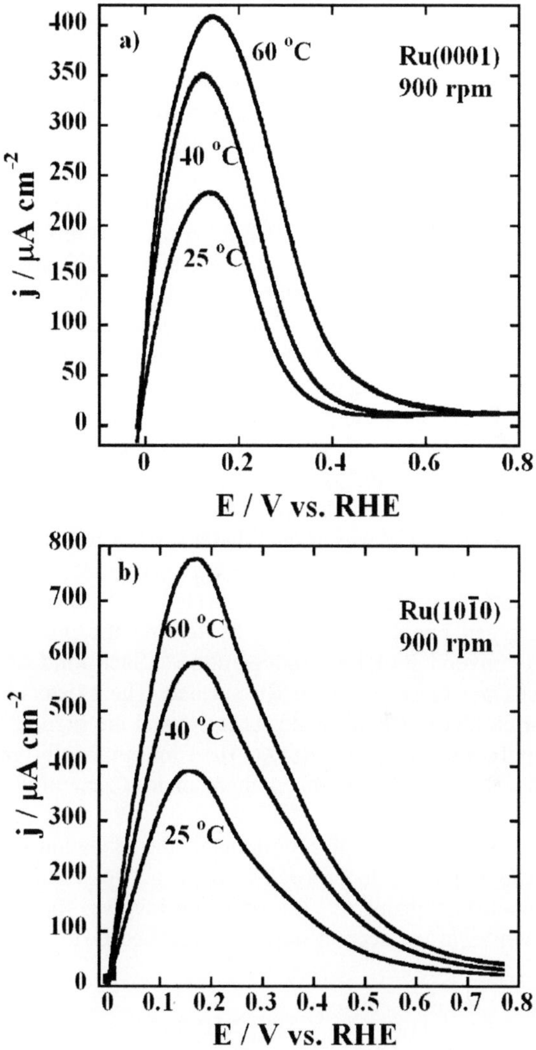

Figure 10. Hydrogen oxidation on rotating Ru(0001) and Ru(10$\bar{1}$0) electrodes in an H_2-saturated 0.05 M H_2SO_4 solution at three different temperatures. Sweep rate 20 mV s^{-1}; rotation rate 900 rpm. Reprinted from [7], Copyright (2003) with permission from Elsevier.

Gas-phase oxidation of CO on the Ru surface has been the topic of numerous investigations.[6,61,74-77] While bare Ru(0001) surfaces and those with up to a monolayer coverage with oxygen have low reactivity, oxygen coverages above 3 ML play a role in the high reactivity of Ru(0001). The efficiency was proposed to arise from the subsurface oxygen layer, with an integral reaction yield about two orders-of-magnitude higher than for samples with an adsorbed oxygen layer only.[62,78] Two explanations are offered for the catalytic enhancements that include modification of the structure and the electronic properties by the oxygen dissolved or incorporated in the Ru surface.[78] In the first model, the redistribution of electron density in the topmost Ru layer induced by subsurface oxygen lowers the activation barrier for the reaction with CO. The second concept postulated that the subsurface-oxygen phase consists of mobile oxygen atoms that can participate in CO oxidation only via thermal diffusion after reaching the topmost surface layer, the latter process being the rate-limiting step in CO oxidation.[62] In either case, it was argued that a surface containing a large amount of oxygen in the subsurface region provides new adsorption sites for oxygen at the topmost surface layer, thereby creating a new oxygen phase, and raising by tenfold the probability for CO/CO_2 conversion at room temperature.[79] Such kind of oxygen bond does not exist at a bare Ru surface. The existence of this oxygen-rich phase was disputed recently, and the high CO-oxidation activity attributed to a $RuO_2(110)$ film epitaxially grown on Ru(0001).[64] This interpretation is more or less generally accepted today.

In the electrochemical environment, the oxidation of pre-adsorbed CO is known to proceed through a two-step (Langmuir-Hinshelwood) mechanism.[80] The first step involves the adsorption of OH, while, in the second step, the adsorbed OH causes oxidation of CO,

$$\begin{aligned} H_2O + * &\rightarrow OH_{ad} + H^+ + e^- \\ OH_{ad} + CO_{ad} &\rightarrow 2* + CO_2 + H^+ + e^- \end{aligned} \tag{4}$$

where * denotes an empty site on the metal. On Pt, the rate-determining step is the first reaction, as electro-oxidation starts as soon as OH adsorption sets in. However, the second step appears to be the rate-determining one on Ru, as the metal's surface is covered with OH_{ad} very early in the potential scale, before the onset of oxidation (*vide supra*).

The first demonstration of electrochemical oxidation of CO on Ru(0001) by cyclic voltammetry and SNIFTIRS was made only recently.[5] Simultaneously, a first observation was reported of CO adsorption using FTIR and STM.[6] Unlike the polycrystalline Ru surface where only on-top linearly bonded CO (CO_L) is visible,[5, 81] CO adsorption on Ru(0001) shows contributions from both CO_L and multiple-bonded CO (CO_H).[5,6] While a certain amount of activity towards CO oxidation to CO_2 was seen on the surface of polycrystalline Ru, Ru(0001) exhibited almost none, judged by the absence of the FTIR peak[5] around 2350 cm^{-1}. At low CO doses, the STM image showed a ($\sqrt{3} \times \sqrt{3}$) R 30° CO overlayer with a coverage of 0.33 ML, similar to the structures found with UHV; a further increase in CO doses produced a new c(2 x 2)-2CO structure as the saturation phase, where CO occupied both the on-top and the three-fold hollow sites, with coverage of 0.5 ML.[6] A combined electrochemical, STM and FTIR study of CO on bare and Pt-modified Ru(0001) and Ru($10\bar{1}0$) surfaces followed.[82]

This study revealed that the electrochemical oxidation of CO at the Ru(0001) surface occurs at negligible rates, probably only at surface imperfections. Figure 11 shows STM image of the Ru(0001) surface after slight mechanical polishing with 0.05 μm alumina slurry, starting from a well-ordered one, shown in Fig. 9a. A comparison of Figs. 11 and 9a reveals that the mechanical polishing results in significant reduction of terrace widths and introduction of kinks in step edges. Interestingly, this surface shows voltammetry curve almost identical to the one obtained with well-prepared crystal.[83] Apparently, the cyclic voltammetry is insensitive to the surface imperfections in Ru(0001). This can be explained if Ru atoms in edges and kinks are more extensively oxidized than those in terraces. Since such Ru atoms cannot be easily reduced, thus they do not contribute to the charge in voltammetry and behave as passivated surface. This assumption seems quite plausible because the surface depicted in Fig. 9b, having highly oxidized steps, shows a similar voltammetry curve as the surface presented in Fig. 9a. We speculate that this oxide reacts with CO in the same manner as RuO_2 does it in gas phase.[74]

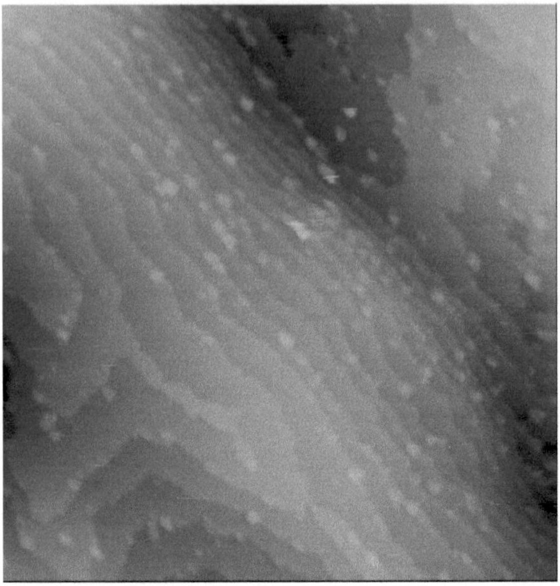

Figure 11. STM image of the Ru(0001) surface after slight mechanical polishing with 0.05 μm alumina slurry, starting from a well-ordered single-crystal surface. Image size 100 x 100 nm, Z range 8 nm.

Thus, it appears that strongly oxidized steps are the cause for the observed activity of mechanically polished Ru(0001) for CO oxidation (see below).

The cyclic voltammetry of CO oxidation on two Ru single crystal surfaces is presented Fig. 12. The more open structure of Ru($10\bar{1}0$) has a sizeable activity, several times larger than that of Ru(0001). Interestingly, it is also larger than that of the polycrystalline Ru in the positive scan (not shown). Fig. 12 shows that a well ordered Ru(0001) is very inactive for CO oxidation.[5] Introduction of steps enhances oxidation, as observed by Jin et al.[84] and confirmed by Lee et al.[27] No oxidation is apparent on Ru($10\bar{1}0$) in the CO-saturated solution in the negative scan if the upper limit is above 0.8 V, as confirmed by progressively opening the upper

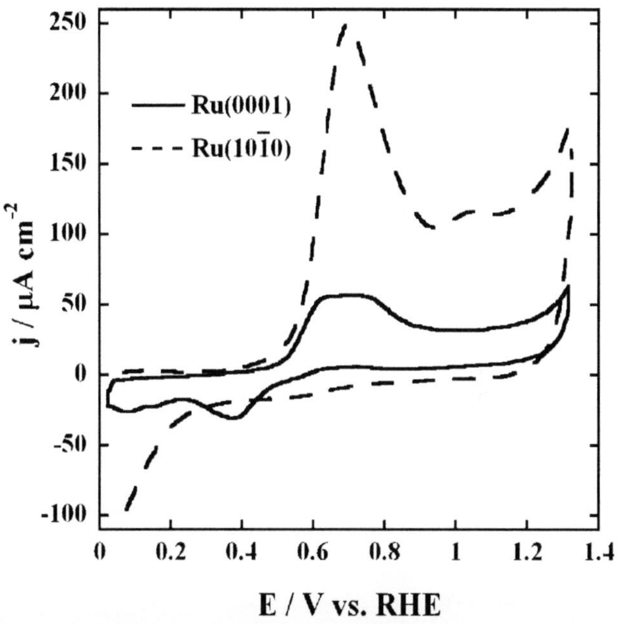

Figure 12. Voltammetric curves for CO oxidation on the Ru(10$\bar{1}$0) and Ru(0001) surfaces in CO-saturated 0.1 M H_2SO_4 solution. Sweep rate 20 mV s^{-1}.Reprinted from [82], Copyright (2002) with permission from Elsevier.

limit.[82] This phenomenon, also observed at polycrystalline Ru electrodes, could be related to either too high a surface oxide-coverage at the electrode's surface, or the different oxidation state of Ru at these potentials. Furthermore, the adsorption rate of CO on an oxide-covered surface produced at high potentials probably differs from that at the RuOH-covered surface at low potentials.

The surface imperfections in Ru(0001) from mechanical polishing appear responsible for a sizable activity of CO oxidation that resulted in the spectra with peaks around 2346, 2010, and 1780 cm^{-1}. They correspond to CO_2 in solution, CO_L and CO_H, respectively Fig. 13.[84] Interestingly, in the FTIR spectra for the

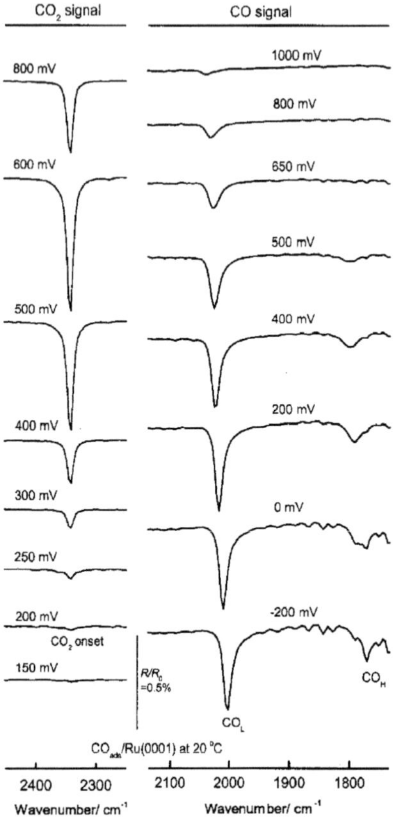

Figure 13. *In situ* FTIR spectra collected from the Ru(0001) electrode in 0.1 M $HClO_4$ solution at 20 °C during a potential step experiment after the adsorption of CO. The CO was pre-adsorbed at –200 mV , after which the solution was sparged with N_2, the potential was then stepped up to +1100 mV in 25 mV increments, with further spectra collected at each step. The spectra showing the CO_2 absorption were normalized to the first spectrum, collected at–200 mV. The spectra showing the CO absorption were normalized to a spectrum taken after holding the potential at +1100 mV for 2 min at the end of the experiment, to ensure the electrode surface was free of adsorbed CO. Some of the spectra collected are omitted for the sake of clarity. Potentials are referenced against Ag/AgCl,Cl^- electrode. Reprinted from [84], Copyright (2004) with permission from Elsevier.

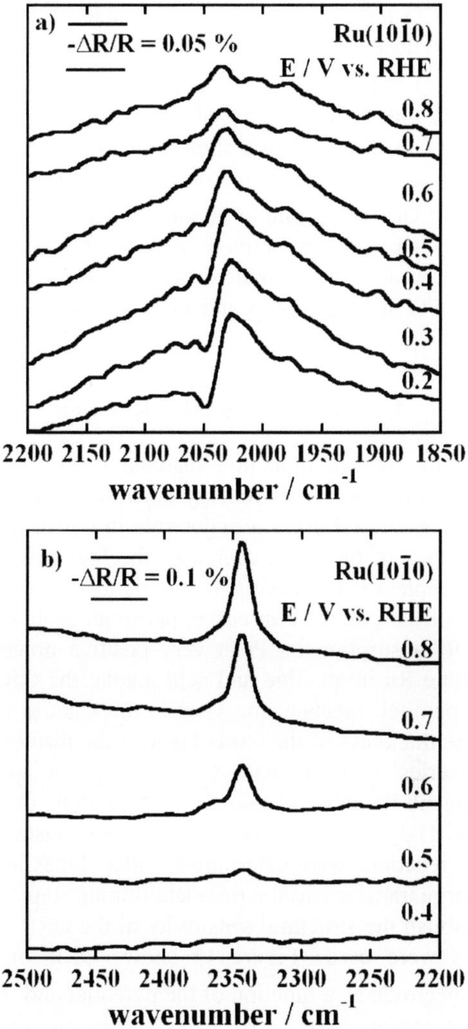

Figure 14. *In situ* IR spectra of CO oxidation at the Ru(10$\bar{1}$0) surface at different potentials in CO-saturated 0.1 M H_2SO_4 solution. 128 scans with 4 cm^{-1} resolution are collected in a single step. Background scan was obtained at 1.1 V. Reprinted from [82], Copyright (2002) with permission from Elsevier.

oxidation of CO on a Ru($10\bar{1}0$) single crystal electrode, Fig. 14, the CO_H is not observed.[82] While the three-fold hollow sites can be found on a well-ordered Ru($10\bar{1}0$) surface (c.f., Fig. 1b), the steric effects of the energetically favored CO_L probably prevent their adsorption of CO. The CO adsorption tuning rates ($\Delta\nu(CO_L) / \Delta E$) on both single crystal surfaces is lower than that on any platinum-group metal.[52] The tuning rate of 20 and 26 cm^{-1} V^{-1} for Ru($10\bar{1}0$) and Ru(0001) surfaces, respectively, may be correlated in the former to the presence of the oxide at all potentials, or to the high coverage of CO at both on-top and three-fold hollow sites in the latter case.

3. Oxygen Reduction Reaction

Electrocatalytic oxygen reduction reaction (ORR) is of a great importance to electrochemical energy conversion in fuel cells and metal-air batteries, and plays a major role in corrosion. The slow reaction kinetics of the ORR decreases the fuel cell's efficiency. The major problem is the large potential deficiency during the initial portion of the polarization curve, partly attributed to the inhibition of OH adsorption on Pt at very positive potentials. On a polycrystalline Ru in alkaline and acid media, the ORR proceeds through a 'parallel' mechanism, wherein the kinetics strongly depends on the thickness of the oxide layer in the former, and a predominant 'series' pathway with the exchange of approximately four electrons in the latter solution.[13,31] No catalytic decomposition of H_2O_2 and HO_2^- species occurs, and the rate constants for these two species reactions were determined. Tafel slopes indicated that the first charge transfer was the rate-determining step.[13,31]

To establish the structural sensitivity of the ORR, single crystal surfaces were used.[8] Figure 15 shows ORR on a rotating Ru($10\bar{1}0$) electrode as a function of the potential and rotation rate. The reaction appears to be under mixed kinetic and diffusion control over a wide potential range. At the most negative potentials, the ring current decreases to a negligible value, suggesting a complete four-electron reduction of O_2 in that region. The Tafel slope obtained was –128 mV per decade, surprisingly close to the expected value of –120 mV per decade for the slow, first charge-transfer step, even though the Ru surface is covered with OH. The

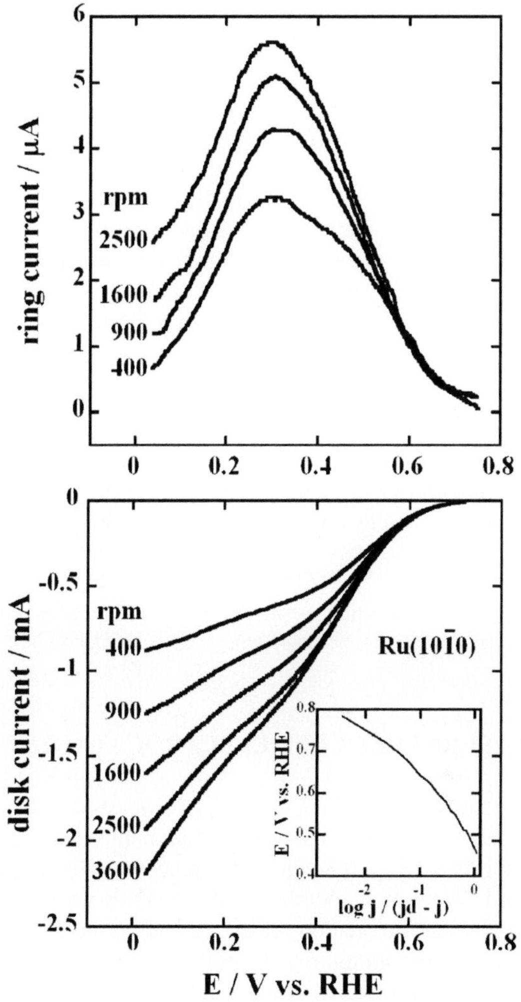

Figure 15. Rotating disk-ring electrode measurements for O_2 reduction on Ru(10$\bar{1}$0) in 0.1 M $HClO_4$. Sweep rate 20 mV s^{-1}; ring potential E = 1.2 V; disc area 0.282 cm^2; rotation rates are indicated in the graph. The insert shows the log j /(j_d- j) vs. E plot obtained for 1600 rpm. Reprinted from [8], Copyright (2002) with permission from Elsevier.

number of electrons, *n,* exchanged per reduced O_2 molecule was calculated assuming first-order kinetics for the dissolved O_2. The experimental value of the slope obtained from a Koutecky-Levich plot agrees with the calculated value for $n = 4$. A small dependence of n on potential is related to the generation of small amount of H_2O_2.

Oxygen reduction on Ru(0001) (not shown) appears to be under considerable kinetic control.[8] A diffusion-limiting current is not reached until the potential of hydrogen evolution, as seen for polycrystalline Ru in acid solutions.[31] Unlike the behavior of Ru($10\bar{1}0$), sweeps in anodic direction almost retrace the curves in cathodic sweeps, suggesting that the behavior of the surface processes are reversible under these conditions. A Tafel slope of –135 mV per decade was obtained for 1600 rpm. The limiting current for Ru($10\bar{1}0$) is not reached before the most negative potentials, and is taken for j_d. A slope larger than –120 mV points to an additional drop in voltage at the oxide layer, which is a plausible explanation for the reaction on the Ru(0001) surface covered by a monolayer of RuOH. Therefore, the observed slope is an indication of the slow, first charge-transfer rate determining step, as in the case of Ru($10\bar{1}0$).

V. Pt-Ru FUEL CELL ELECTROCATALYSTS

Reviewing the work on the Pt-Ru electrocatalysts is beyond the scope of this article. We will briefly comment on some key advances in this area. Although early discovery by Petrii,[85] and Bockris and Wroblowa[86] established the catalytic activity of Pt-Ru alloys for methanol oxidation, despite of active investigation that followed, even the optimum composition of Pt–Ru is yet to be firmly settled.[71] An early explanation for the mechanism by which bimetallic catalysts improve upon the performance of pure Pt, that is, the bifunctional mechanism proposed by Watanabe and Motoo,[87] was recently challenged.

The bifunctional mechanism is explained in terms of the independent function of atoms of different metals: methanol adsorption and decomposition takes place on Pt, while the alternative metal atoms provide preferred sites to bind OH. Several metals, such as

Ru, Sn and Mo,[88-90] were assessed in combination with Pt. Recent works in developing fuel cell technology have renewed the efforts to improve Pt-Ru electrocatalysts for reformate hydrogen- and methanol-oxidation, in which Ru provides active oxygen for oxidizing strongly bound CO on Pt. It was noted that the atoms substituted for Pt alter the surface electronic structure. Accumulated evidence suggests that substitution changes the binding energy of adsorbates, and so the bifunctional mechanism should be altered to reflect changes in the adsorption bond. Krausa and Vielstich examined the oxidation of small molecules on Pt and a Pt–Ru alloy by differential electrochemical mass spectrometry (DEMS), and noted a cooperative effect on the alloy's electrode, which they interpreted as evidence of a modification in the Pt electronic structure.[91] McBreen and Mukerjee recorded a decrease in Pt–Pt bond length in X-ray absorption studies of Pt–Ru electrode surfaces that they correlated with an increase in *d*-band vacancies.[92] They concluded that the bifunctional mechanism needs to be modified to account for a cooperative electronic effect in the Pt–Ru catalyst. On the other hand, Gasteiger et al. argued that the binding energy of CO to pure Ru under UHV is not different enough from the binding energy on pure Pt to mandate such alterations.[71]

1. Pt Submonolayers on Ru Single-Crystal Surfaces

Existing fuel cell technology suffers from at least two problems. One is the low CO tolerance of anodes for the oxidation of reformated hydrogen, impure hydrogen, or methanol. Small concentrations of CO are inevitable in H_2 produced by reforming methanol or other hydrocarbons, and the performance of Pt-based catalysts is strongly impaired by the presence of small amounts of CO. The other problem is the high Pt loading, which is the major constituent of Pt-Ru catalysts.

Adzic and coworkers proposed a radically new approach in electrocatalysis and catalysis[93] that can alleviate both problems. It is based on a catalyst consisting of only a submonolayer Pt deposited on carbon-supported Ru nanoparticles. The Pt submonolayer on Ru ($PtRu_{20}$) electrocatalyst demonstrated higher CO tolerance than commercial catalysts in rotating disk experiments. Tests of the long-term stability of the fuel cells detected no loss in performance over 870 h, even though the Pt loading was approximately 1/10 of the standard loading. *In situ* X-ray adsorption spectra (XAS) indicated an increase in the *d*-band vacancy of deposited Pt, which may facilitate partly the reduced susceptibility to CO poisoning for this catalyst. Below, we describe the single crystal experiments leading to the new catalyst, the synthesis of high-surface-area catalyst, and its characterization and performance.

The synthesis of the $PtRu_{20}$ was facilitated by the discovery of electroless (spontaneous) deposition of Pt on Ru, which was not observed for noble-metal substrates. In the electroless deposition of Pt on an Ru(0001) single crystal surface[94] the surface morphology, and the amount and the uniformity of the Pt deposited can be varied by changing the concentration of $[PtCl_6]^{2-}$ or $[PtCl_4]^{2-}$ ions in the solution and also the time of immersion. Columnar-shaped Pt clusters of relatively uniform size can be obtained, with coverage from submonolayer to multilayer. Figure 16 is a representative ECSTM image of a Pt deposit obtained by immersing a freshly prepared Ru single crystal in an $[PtCl_6]^{2-}$ solution.[94] Clusters of

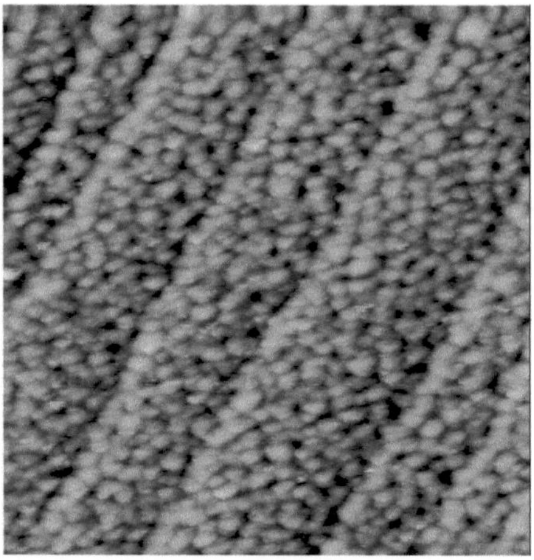

Figure 16. STM image of an electroless (spontaneous) Pt adlayer deposited on Ru(0001) in 0.01 M H_2PtCl_6 + 0.1 M H_2SO_4 solution. Image recorded at open circuit potential in 0.1 M H_2SO_4. Image size 100 x 100 nm, *Z* range 2 nm. Reprinted from [94], Copyright (2001) with permission from Elsevier.

2–6 nm were obtained, with a slight preferential deposition of Pt on the Ru edges. The average height on Ru(0001) terraces is about 2 ML, while that on the edges is about 3 ML, yielding the total coverage of about 92%. Electroless deposition occurs only on freshly prepared Ru surfaces, a feature ascribed to the strongly bound OH groups that prevent or reduce electroless deposition when the Ru surface contacts an aqueous solution. The electroless process is tentatively ascribed to the local cell mechanism involving RuOH formation.[95] Interestingly, the driving force of the electroless deposition reaction is the difference in the equilibrium potentials of $[PtCl_6]^{2-}$, or $[PtCl_4]^{2-}$ reduction and Ru oxidation. It is interesting that in the electroless deposition of Pd on Ru(0001) an atomic resolution of Pd(111) was obtained.[96]

Wieckowski et al. reported the spontaneous deposition Ru adlayers from RuO^{2+} solutions on three low-index Pt surfaces.[57] The maximum coverage of Ru on these adlayer is about 20%,[97] and potential must be applied to reduce the Ru adlayer to metallic Ru. The Ru-decorated Pt nanoparticles showed considerable catalytic activity in the methanol- oxidation reaction.[57,98] We will discuss the catalytic properties of the Ru-decorated Pt nanoparticles in Section V.3.

(*i*) *Adsorption Properties of Pt Submonolayers on Ru(0001)*

IR spectroscopy was used to obtain insights on the carbon monoxide absorption and oxidation mechanism on Pt-Ru electrocatalysts.[93] Figure 17 shows the SNIFTIR spectra of CO on submonolayer Pt deposits on Ru(0001). Two bipolar bands are clearly visible at potentials from 0.1 to 0.8 V. Analyses of IR spectra (*vide supra*) attributed the bipolar band at lower frequencies to blue-shifted CO (i.e., moved to higher frequency) on polycrystalline Ru,[5,81] whereas the higher-frequency bipolar band represents red-shifted CO on Pt(111).[39]

Blyholder gave the first particle model of charge migration in the binding of CO.[99] The charge flows from the highest occupied CO orbital, 5σ, to empty *d*-orbitals on the metal atom. A negative formal charge on the metal atom is avoided by a postulated back-donation from the metal d_π orbitals into the lowest unoccupied CO

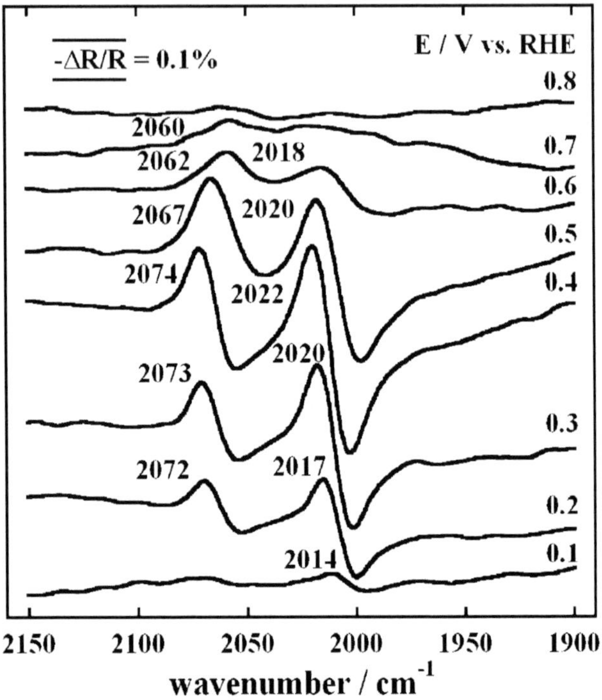

Figure 17. SNIFTIRS spectra for a Ru(0001) electrode with a submonolayer of Pt in a CO-saturated 0.1 M H_2SO_4 solution. The reference spectrum is obtained at 0.075 V and the sample spectra are taken from 0.10 V incremented by 0.1 V up to 0.80 V. 8192 scans were co-added in 16 cycles, 512 scans each; the resolution was 8 cm^{-1}. Spectra are offset for clarity. Reproduced with permission from [93]. Copyright (2001), The Electrochemical Society.

orbitals, $2\pi^*$. Theoretical methods (see below) later showed this simple model to be essentially valid, and it has proved useful in interpreting the adsorption of CO on Pt.

The modulation in the vibrational frequency of CO on Pt–Ru is attributed to the charge transfer from Ru to Pt, due to difference in work function that weakens the Ru–CO bond and strengthens the Pt–CO bond.[81] The weaker bonding of Ru–CO on Pt–Ru makes CO more reactive here than on pure Ru, and accordingly, electro-oxidation sets in earlier in the potential scale. On the other hand, no significant charge transfer was found in a theoretical study of the electronic mechanism underlying the adsorption of CO on pure Pt, Ru, and on mixed Pt-M metal surfaces (M=Ru, Sn, Ge) with the relativistic density-functional self-consistent field method on clusters of ten atoms.[100] The presence of the metal weakened the Pt-C bond, therefore increasing CO tolerance, and also slightly lowered the CO stretching frequency of adsorbed carbon monoxide on Pt, although it was acknowledged that weakening the Pt-C bond would increase vibrational frequency by donation back-bonding. The promoting mechanism for bifunctional catalysis in alloying Pt with Ru, Sn, or Ge was attributed to lowering the dissociation energy of water, as well as modifying Pt–CO binding energy.[100]

Density functional theory (DFT) demonstrated that the lowest CO bonding occurs in a Pt monolayer on an Ru(0001) surface.[101] Hammer and Nørskov's model revealed a weaker Pt–CO bond,[102] which includes a large transfer of d electrons from Pt to Ru, with the consequent shift of *d*-bands that lessens CO adsorption due to decreased back donation from Pt to antibonding CO orbitals. A parallelism was found in the change of adsorption energy with *d*-band center shifts for CO and H_2, was suggesting some weakening of the H_2 chemisorption bond for a Pt monolayer on Ru surface that could reduce the kinetics of HOR.[103] The temperature-programmed desorption data for CO on Pt on Ru(0001) also indicate a decrease in the bonding strength of CO to Pt.[104] Furthermore, reactivity scales well with shifts in the center of *d*-band for strained crystals and overlayers,[102] which could be operative for Pt islands given the ~ 4% lattice mismatch between Pt and Ru.

2. Pt Deposition on Ru Nanoparticles

The concept of a Pt monolayer catalyst was first verified with a Pt submonolayer on Ru substrate.[93] This approach radically changed the design of the Pt-Ru catalysts and it is likely to similarly affect a broad range of catalysts. It facilitates an ultimate reduction of Pt loadings in Pt-Ru catalysts by depositing Pt only at the surface of Ru nanoparticles, so that the most of the Pt atoms become available for the catalytic reaction. Ru (10%) nanoparticles on Vulcan XC-72 carbon were heated in an H_2 atmosphere at ~ 300°C for 2 h. This temperature is much lower than that required for bulk Ru preparation. After cooling down to room temperature, they were immersed in a solution of $[PtCl_4]^{2-}$ ions for 30–60 minutes to obtain a bulk Pt:Ru ratio ranging from 1:20 to 1:5. The entire procedure was carried out in atmosphere of Ar or H_2, and the amount of Pt available for electroless deposition was controlled by the concentration and volume of the immersing solution. The area ratio of surface atoms to total atoms was calculated to be roughly 0.45 for 2.5 nm Ru particles,[105] so that the coverage ratio of Pt to Ru ranges from 1:9 to 1:2.

(*i*) *EXAFS and TEM Characterization*

Structural and electronic information can be inferred from *in situ* extended X-ray absorption fine structure (EXAFS) measurements.[106] Figure 18a presents Fourier transform of the Pt L_3 edge of the $PtRu_{20}$ catalyst at 0.1 V compared with that of carbon-supported Pt nanoparticles at 0.48 V in 1 M $HClO_4$. The potential of the Pt-Ru catalyst was chosen so that the interaction of Pt with the adsorbed H_2O and anions is the smallest, and hence, the perturbation of its electronic properties was expected to come predominantly from interaction with Ru atoms. The general appearance of the spectra is very different, with the Pt–Ru catalyst exhibiting the first large peak shifted to lower r-values due to Pt-Ru bonding. Fit of the data indicates that Pt is deposited in two-dimensional (2D) islands with Pt atoms coordinated with 3-4 Ru atoms, and the Pt-Ru length (0.269 nm) is the same as in Pt-Ru alloys. The latter finding further implies that Pt is bonded to Ru rather to RuOH.

The two-dimensional deposition of Pt was confirmed by comparing the Pt L_3 edge X-ray absorption near edge spectroscopy (XANES) spectra of the catalyst, held in the double layer region (at 0.40 V) with those of a reference Pt foil, Fig. 18b.[107] The absorption peaks (white lines) of the Pt submonolayer on Ru nanoparticles at the L_3 and L_2 edges are larger than the corresponding peaks for the Pt foil, indicating an increased *d*-band vacancy caused by the interaction between the Pt atoms and the Ru nanoparticle surface. Pt 5*d*-band vacancies of 0.345 and 0.3 for $PtRu_{20}$ and Pt foil, respectively, were calculated.[108] The large increase in *d*-band vacancy confirms 2D growth because a smaller increase would be expected for three-dimensional (3D) clusters

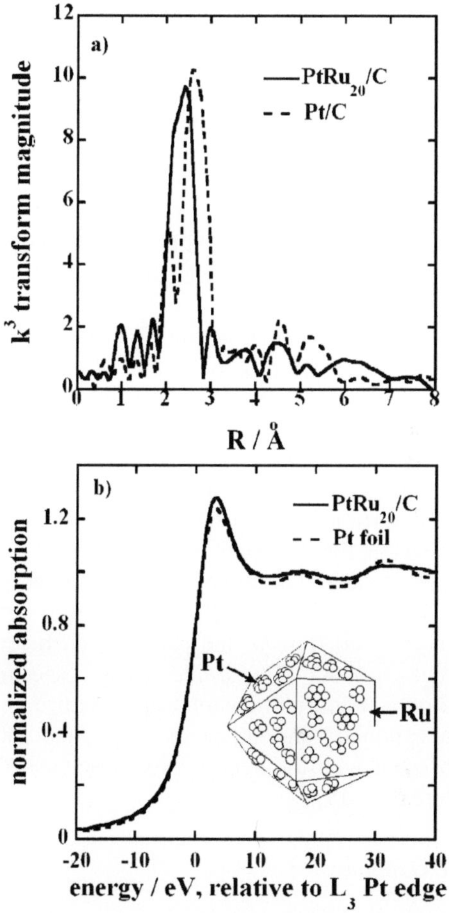

Figure 18. (a) EXAFS FT (k^3-weighted) of Pt L_3-edge of the catalyst sample compared with *fcc* Pt. Reprinted from [106], Copyright (2003) with permission from Elsevier. (b) *In situ* XANES spectra at the Pt L_3 of the $PtRu_{20}$ electrocatalyst held at 0.40 V in 1 M $HClO_4$ electrolyte solution. The spectra were obtained in a fluorescence mode. The spectra of the Pt foil used as a reference and in the calculation of the *d*-band vacancies; Insert in (b) shows a cubo-octahedral particle model for the electrocatalyst consisting of the Ru particle with two-dimensional Pt islands on its surface. Reprinted from [107] Copyright (2004) with permission from Elsevier.

due to the effect of the atoms inside the clusters that are not in contact with Ru.

A prevailing view of carbon-supported metal nanoparticles is that they are in a form of cubo-octahedral or icosohedral structures.[105,109] One such model for the active electrocatalyst with submonolayer Pt coverage, a cubo-octahedral Ru particle with Pt islands on its surface, is shown as an inset in Fig. 18b. Pt atoms are in 2D islands as expected from EXAFS measurements and the Ru nanoparticles are supported on a high surface area Vulcan XC-72 carbon.

Transmission electron microscope (TEM) measurements determined the size of the metal particles after Pt electroless deposition.[18] Because of the small amount of Pt relative to Ru, no attempt was made to locate the Pt atoms by using TEM chemical- and diffraction-analyses. Measurements of over a hundred metal particles from TEM images yielded an average diameter about 2.5 nm, only slightly larger than the 2 nm value as listed in E-TEK's specification for the Ru/C samples. Clearly, there is no significant agglomeration due to heating and Pt deposition. In the high-resolution images, ordered atomic structures are seen. To identify the crystal structure of the particles, diffraction analysis was carried out on about 170 particles using a special computer routine based on fast Fourier transform of the lattice images of the particles. The symmetry and lattice spacing of the dominant features are consistent with the hexagonally close-packed Ru single crystal structure (Fig. 19).

(ii) H_2/CO Oxidation

In addition to having a good CO tolerance, Pt-Ru electrocatalysts must also have a high activity for H_2 oxidation. Comparison of the mass-specific activity of a $PtRu_{20}$ electrocatalyst with a commercial Pt-Ru 1:1 alloy electrocatalyst for the oxidation of pure H_2 showed that its activity is three times that of the commercial alloy. This indicates that even for a low Pt coverage on Ru, its activity for H_2 oxidation is preserved, a prerequisite for an active CO tolerant catalyst.[93] Comparing the CO tolerance of the $PtRu_{20}$ electrocatalyst with that of two commercial Pt-Ru alloy electrocatalysts for the oxidation of 1095 ppm CO in H_2 confirmed the exceptional stability of the former (Fig. 20); the measurements

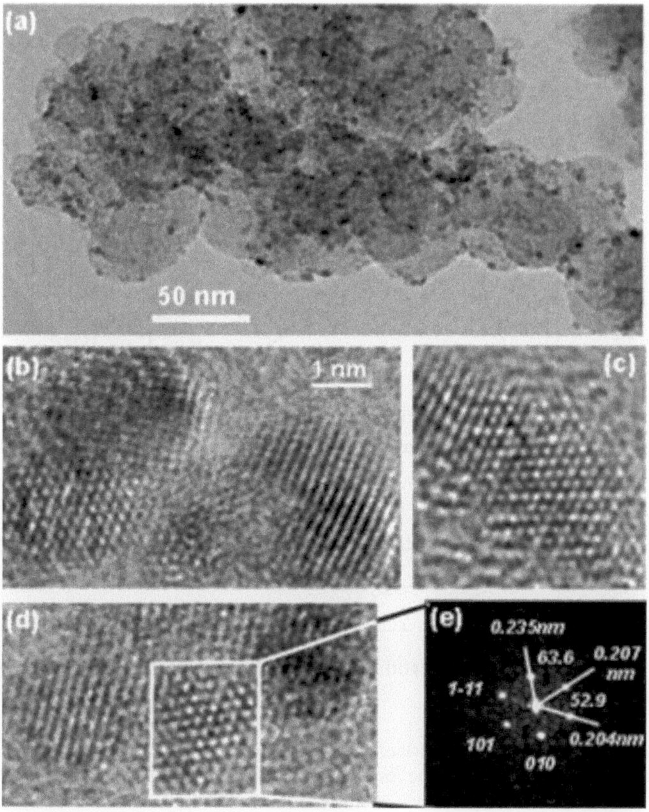

Figure 19. Electron micrographs of the $PtRu_{20}$ catalyst made by spontaneous deposition of Pt on Ru nanoparticles. (a) Low magnification morphology of the metal particles (black dots, average size 2.5 nm) on carbon spheres (average size 50 nm). (b - d) High-resolution images showing atomic resolved lattice structures. (e) Diffractogram obtained from the high-resolution image shown in (d) with measured the angles and lattice spacings that are consistent with *hcp* Ru single crystal structure. Reproduced with permission from [72]. Copyright (2003), The Electrochemical Society.

were obtained using a thin film rotating disk electrode at 2500 rpm at 60°C in 0.5 M H_2SO_4. The current drops to half the initial value after about 4 hours for $PtRu_{20}$, while for the two commercial catalysts it happens in less than an hour. This enhanced CO tolerance of $PtRu_{20}$ is apparently due to a weakened CO adsorption on Pt/Ru

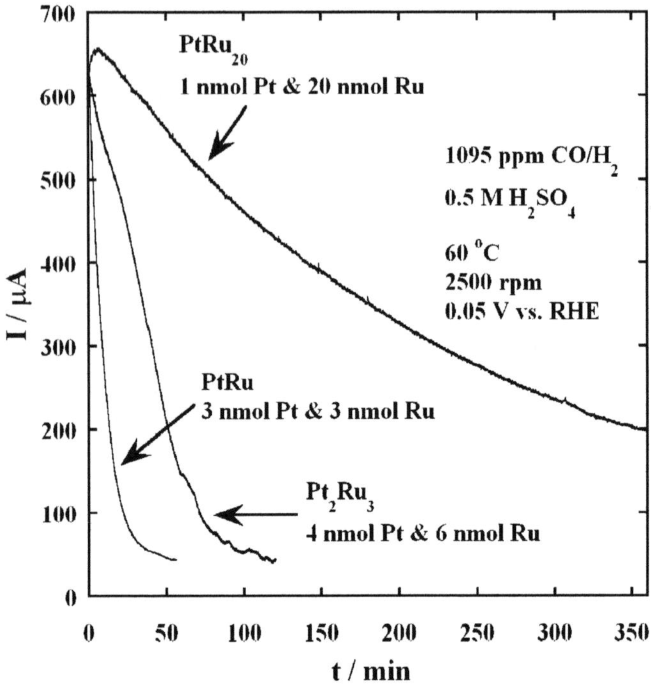

Figure 20. Comparison of the CO tolerance of three catalysts based on the current as a function of time for the oxidation of H_2 with 1095 ppm of CO at 60 °C for the $PtRu_{20}$, and two commercial electrocatalysts at 0.05 V with the loadings indicated in the graph. A considerably larger CO tolerance is seen for the $PtRu_{20}$ (1% Pt, 10% Ru on C) electrocatalyst. Reprinted from [106], Copyright (2003) with permission from Elsevier.

and an efficient spill-over of CO from Pt sites to the surrounding Ru sites.

(*iii*) *Fuel Cell Tests*

Fuel-cell tests offer the ultimate verification of the usefulness of an electrocatalyst by determining its long-term stability under real operating conditions. They were performed on single cells using electrodes of 50 cm^2 and an anode catalyst loading of 0.2 mg cm^{-2} total metal, corresponding to 0.18 mg Ru cm^{-2} and 0.018 mg Pt cm^{-2}; hence, the amount of Pt is approximately 1/10 of what is considered a standard anode loading for these tests. The cathode was a standard O_2 cathode with the Pt/C electrocatalyst. The fuel was clean H_2 or H_2 with 50 ppm of CO and 3% air (adding air is a routine procedure that helps CO oxidation at the electrode's surface). Excellent performance stability was seen. After 900 h of the constant current (0.4 A cm^{-2}) test, no losses were observed with the

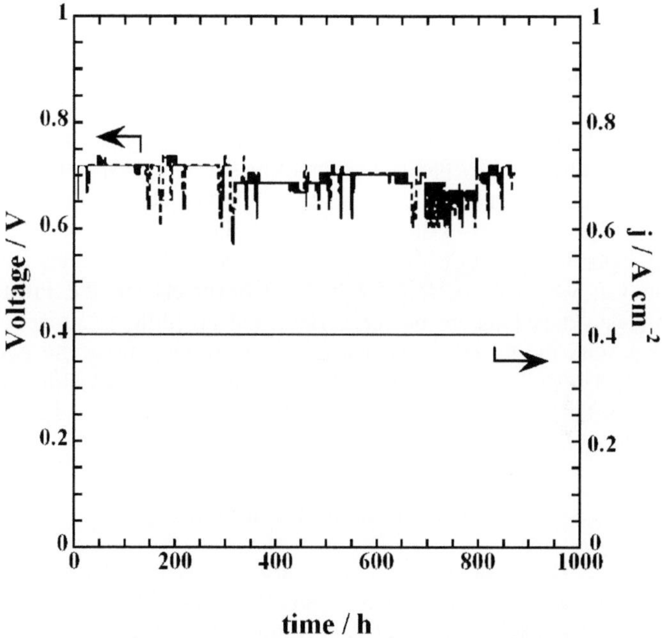

Figure 21. Long-term test of the performance stability of the $PtRu_{20}$ electrocatalyst in an operating fuel cell. The fuel cell voltage at constant current of 0.4 A cm^{-2} is given as a function of time for the electrode of 50 cm^2 with an anode containing to 0.18 mg Ru cm^{-2} and 0.018 mg Pt cm^{-2}; (approximately 1/10 of the standard Pt loading) and a standard air cathode with a Pt/C electrocatalyst. The fuel was clean H_2 or H_2 with 50 ppm of CO and 3% air; temperature 80 °C.

cell operating on clean H_2, or H_2 + 50 ppm CO + 3% air mixture. The cell was run with the latter fuel for a third of the total time. Switching between H_2 and H_2/CO fuels entails small sharp changes in the cell's voltage, which produced spikes in the plot (Fig. 21). The effect of air bleed on CO tolerance was such that the catalyst worked well, even at contamination level of 100 ppm.

XANES data show that Pt in the $PtRu_{20}$ has 0.345 5d vacancy/atom. The back-bonding from the Pt to CO may decrease in this case, the likely reason for this catalyst's good CO tolerance. The weaker CO bonding on $PtRu_{20}$ than on Pt or the Pt-Ru alloy was confirmed by CO striping voltammetry. These data show that the electronic effect in Pt-Ru electrocatalysts may be more important than previously assumed.

These results clearly demonstrate that Pt submonolayers on Ru nanoparticles are stable during the fuel cell's operation and that this system represents a real catalyst. In addition to the high activity for H_2 oxidation and weak CO bonding, the strong segregation of Pt and Ru is a key factor that determines this catalyst's stability. Pt-Ru alloys represent a very strongly segregated system[110] in which Pt segregates to the surface, and is the origin of their inherent instability because the segregation essentially generates a Pt catalyst, which is poisoned by CO. Conversely, in the $PtRu_{20}$ catalyst, Pt islands are on the surface and no further segregation takes with usage; this explains the full retention of activity over 900 h. The data provide a powerful illustration of the possibility of nanoparticle surface modification by Pt monolayers as a method of obtaining low noble-metal loading of catalysts.

3. Methanol Oxidation

As we pointed out for the Pt-Ru electrocatalysts, reviewing the work on methanol oxidation is beyond the scope of this article. We will briefly discuss some selected results with the electrocatalysts comprising a Pt submonolayer on Pt nanoparticles[111] and Ru-decorated Pt nanoparticles.[97] The direct methanol fuel cell is a potential candidate as a noteworthy power source for mobile devices because of its high energy density per unit volume and immediate reuse by refueling. Methanol has a high energy density, with full oxidation to CO_2 occurring through a six-electron reaction, and is a promising choice for energy applications, for instance, in portable electronics. Its attractive properties include availability, ease of handling and storage. The obstacles against the implementing DMFCs are the insufficient activity of the catalysts employed so far as well as the high cost of Pt in platinum-based bimetallic catalysts.

Bewick et al.[112] identified CO as the species that acts as a catalytic poison and inhibits further oxidation of methanol on Pt electrodes. The reactive intermediate is a formate species, HCOO that generates asymmetric COO vibration around 1300 cm^{-1}, leading to an increase in the methanol oxidation current after CO oxidation.[113] Recently, water molecules were detected adsorbed on the Ru sites on Ru and Pt-Ru (but not on Pt) catalysts, and were assigned as the oxygen donor to the methanol adsorbates that promote methanol oxidation.[114] This was considered as directly supporting the "bi-functional mechanism" of Pt-Ru catalysts for the methanol-oxidation reaction.[87]

Methanol oxidation proceeds with the progressive dehydrogenation of the organic molecule on the anode surface, viz,

$$\begin{aligned} CH_3OH_{ad} &\rightarrow CO_{ad} + 4H^+ + 4e^- \\ H_2O_{ad} &\rightarrow OH_{ad} + H^+ + e^- \\ CO_{ad} + OH_{ad} &\rightarrow CO_2 + H^+ + e^- \end{aligned} \qquad (5)$$

Pt has the highest adsorption of methanol on its surface, but its catalytic properties are low due to the formation of 'poison' species (most notably CO) that can be oxidized only after the Pt is covered with OH. Platinum-based bimetallic electrocatalysts, such as Pt-Ru alloys and Ru-decorated Pt materials, are the most active ones. The bi-functional mechanism is to a large extent operative in these catalysts. Most commercial Pt-Ru catalysts are based on 1:1 Pt-Ru alloy. While the alloys typically show enhanced activity in comparison with pure Pt, there is significant Pt loading in the bulk of the alloy in which catalysis does not proceed because the sites are inaccessible for methanol adsorption; hence, the need for reducing the Pt content.

Figure 22 shows the oxidation of methanol on a submonolayer of Pt on Ru, $Pt_{3.9}Ru_{10}/C$ (3.9 $\mu g\ cm^{-2}$ Pt), and commercial PtRu/C (10 $\mu g\ cm^{-2}$ Pt) electrocatalysts. The Pt-mass specific activity (current) of the monolayer-level electrocatalyst is several times higher than that of a commercial sample.

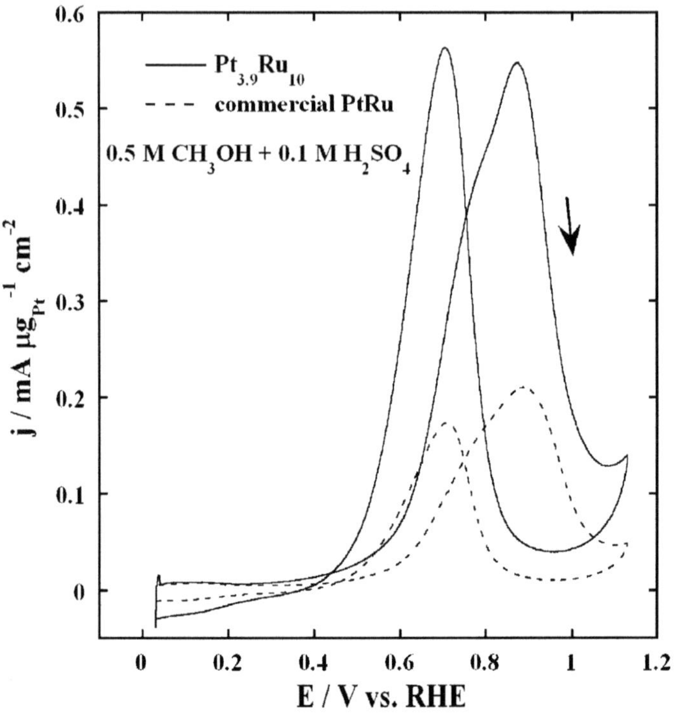

Figure 22. Oxidation of methanol on a submonolayer of Pt on Ru, i.e., $Pt_{3.9}Ru_{10}/C$ (3.9 µg cm^{-2} Pt) and comercial PtRu/C (10 µg cm^{-2} Pt) electrocatalysts in 0.5 M CH_3OH + 0.1 M H_2SO_4; sweep rate of 50 mV s^{-1}; room temperature. The currents presented are normalized by Pt mass.

Wieckowski et al.[97,98] investigated in detail properties of the Ru-decorated Pt surfaces obtained by spontaneous deposition. These studies were reviewed recently;[115] here we give a brief summary. Several methods were used to obtain Ru-modified Pt surfaces, including electrodeposition of Ru ad-atoms,[87] electrochemical,[116,117] spontaneous deposition,[57,118] UHV deposition[119] and organometallic chemistry.[120] However, these various methodologies produce different amounts of metallic vs. oxidized Ru atoms, engendering different electrochemical activity of the Pt-Ru system. The amount of Ru deposited was monitored by AES, XPS or *in-situ* STM, or by following the difference of the voltammetric peak currents in the double-layer range.[57] The increase of electrochemical activity of Ru-decorated Pt surfaces relies on the bifunctional mechanism and the decrease of CO-binding energy on Pt atoms that are in close proximity to the Ru islands. In contrast to FTIR measurements on optimized Pt-Ru alloys where only one linearly bonded CO peak is visible, the Ru submonolayers on Pt(hkl) show two linearly bonded CO peaks (one for CO-Pt and one for CO-Ru).[121] It was later confirmed that Ru is deposited in form of islands with a minimal size of about eight atoms; this minimal cluster size was shown to be needed for the appearance of the Ru-CO IR peak.[122] The existence of the two peaks is attributed to slow CO oxidation kinetics, but it is unclear whether the reaction limiting step reflects the slow diffusion of CO towards the reactive Pt-Ru sites, or the slow change in the oxidation state of Ru that, in turn, reduces the speed of delivery of the OH species required for the reaction.[115]

The growth of Ru islands on Pt(hkl) was found to be substrate-dependent, so that the Ru layer is almost completely in the form of a monolayer on Pt(110), whereas the two- and three-dimensional growth is facilitated on the other two low-index Pt surfaces, especially on Pt(111).[123] The Ru-Pt(111) is more effective catalyst for methanol oxidation[124] than the other two surfaces decorated with Ru, because the edge of a Ru island is the active site in methanol oxidation; therefore, controlling the extent of the multidimensional islands is of a particular importance for fuel cell catalysis.[125]

VI. CONCLUSIONS

Knowledge of the electrochemistry and electrocatalysis of Ru has significantly progressed since the turn of the millenium when a quick method for preparing well-ordered single crystal Ru electrodes was introduced. Considerable structural effects were observed for several reactions on single crystal and polycrystalline Ru surfaces. The densest packing of Ru atoms, i.e., the (0001) surface that is protected from oxidation by chemisorbed bisulfate species at low potentials, stays impervious to oxygen penetration, and has low catalytic activity towards CO oxidation, hydrogen oxidation, and oxygen reduction. In sharp contrast, the more open structure of $(10\bar{1}0)$ shows no measurable coverage of sulfate/bisulfate species, the shape of the voltammetry profiles dramatically change upon cycling, and its catalytic activities are several times higher than on the Ru(0001) surface.

Initial stages of Ru oxidation, involving RuOH formation by the oxidation of H_2O and the distinction of that reaction from H adsorption seems now well understood. Random formation of RuO_2 islands on the smooth OH-covered Ru(0001) surface under electrochemical conditions was observed without place exchange by *in situ* STM and X-ray scattering measurements.

Fuel cell electrocatalysis also has advanced significantly with innovations in the preparation of active Pt-Ru catalysts. A new type of electrocatalyst was developed, consisting of a Pt submonolayer on Ru nanoparticles. It has high CO tolerance and a very low Pt content. Its synthesis was facilitated by the discovery of electroless deposition of Pt on Ru nanoparticles that can be controlled so that most (> 90%) Pt atoms become available for the catalytic reaction. The catalytic activity of $PtRu_{20}$ prepared by this method affords considerable advantages in the oxidation of H_2, CO, and CH_3OH compared with commercial Pt-Ru alloys.

While the experimental data has mounted, theoretical elucidations of the observed phenomena also made significant progress. There are, however, still open basic questions. The difference in behavior of single-crystal surfaces is not well understood. Similarly, there is a need to alter the simple bifunctional mechanism to reflect changes in the adsorption bond of metal-CO, as well as in the electronic structure of the Pt-Ru catalyst.

The rich chemical properties of Ru are replicated in its fascinating surface electrochemical and catalytic properties. For this reason, Ru and its alloys are likely to remain in focus of the research in catalysis and electrocatalysis for the foreseeable future.

REFERENCES

[1]S. Trasatti, W. E. O'Grady in: *Advances in Electrochemistry and Electrochemical Engineering,* Vol. 12, Ed. by H. Gersher, C. Tobias, Wiley, New York, 1982.

[2]S. Gottesfeld, T. Zawodzinski in: *Advances in Electrochemical Science and Engineering*, Vol. 5, Ed. by D.M. Kolb, R. Alkire, VCH, New York, 1982.
[3]P. N. Ross in: *Electrocatalysis*, Ed. by J. Lipkowski, P. N.Ross, Wiley-VCH, New York, 1998.
[4]T. E. Madley, H. A. Engelhardt and D. Manzel, *Surf. Sci.* **48** (1975) 304.
[5]N. S. Marinkovic, J. X Wang, H. Zajonz and R. R Adzic, *Electrochemical and Solid State Letters* **3** (2000) 508.
[6]N. Ikemiya T. Senna and M. Ito, *Surf.Sci.* **464** (2000) L681.
[7]H. Inoue, J. X. Wang, K. Sasaki and R. R. Adzic, *J. Electroanal. Chem.* **554** (2003) 77.
[8]H. Inoue, S.R. Brankovic, J.X. Wang and R.R.Adzic, *Electrochim.Acta* **47** (2002) 3777.
[9]P. C. Liu, C. H.Yang, S. L. Yau and M. S. Zei, *Langmuir* **18** (2002) 754.
[10]M. B.Vukmirovic, R. L.Sabatini and R. R.Adzic, *Surf. Sci* **572** (2004) 269.
[11]S. Hadzi-Jordanov, H. Angerstein-Kozlowska, M. Vukovic and B. E. Conway, *J. Electrochem. Soc.* **125** (1978) 1569.
[12]D. Michell, D. A. J. Rand and R. Woods, *J. Electroanal. Chem.* **89** (1978) 11.
[13]N. A. Anastasijevic, Z. M. Dimitrijevic and R. R. Adzic, *J. Electroanal. Chem.* **199** (1986) 351.
[14]V. Horvat-Radosević, K. Kvastek, M. Vuković, B. Čukman, *J. Electroanal. Chem.* **482** (2000) 188.
[15]R. C. Walker, M. Bailes and L. M. Peter, *Electrochim. Acta* **44** (1998) 1289.
[16]N. S. Marinkovic, J. X. Wang, H. Zajonz and R. R. Adzic, *J. Electroanal. Chem.* **500** (2001) 388.
[17]J. X. Wang, N. S. Marinkovic, H. Zajonz, B. M. Ocko and R. R. Adzic, *J. Phys. Chem. B* **105** (2001) 2809.
[18]S. R. Brankovic, J. X. Wang, Y. Zhu, R. Sabatini, J. McBreen and R. R. Adzic, *J. Electroanal. Chem.* **524-525** (2002) 231.
[19]W. F. Lin, P. A. Christensen, A. Hamnett, M. S. Zei and G. Ertl, *J. Phys. Chem. B* **104** (2000) 6642.
[20]W. F. Lin, P. A. Christensen and A. Hamnett, *Phys. Chem. Chem. Phys.* **3** (2001) 3312.
[21]W. F. Lin, P. A. Christensen and A. Hamnett, *J. Phys. Chem. B* **104** (2000) 12002.
[22]E. Y. Cao, D. A. Stern, J. Y. Gui and A. T. Hubbard, *J. Electroanal. Chem.* **354** (1993) 71.
[23]W. F. Lin, M. S. Zei, Y. D. Kim, H. Over and G. Ertl, *J. Phys. Chem. B* **104** (2000) 6040.
[24]M. S. Zei and G. Ertl, *Phys. Chem. Chem. Phys.* **2** (2000) 3855.
[25]W. B. Wang, M. S. Zei and G. Ertl, *Phys. Chem. Chem. Phys.* **3** (2001) 3307.
[26]W.B. Wang, M.S. Zei and G. Ertl, *Chem.Phys.Lett.* **355** (2002) 301.
[27]J. Lee, W. B. Wang, M. S. Zei and G Ertl, *Phys. Chem. Chem. Phys.* **4** (2002) 1393.
[28]A. M. El-Aziz and L. A. Kibler, *Electrochem.Commun.* **4** (2002) 866.
[29]H. Hoster, B. Richter and R. J. Behm, *J.Phys.Chem B* **108** (2004) 14780.
[30]R. R. Adzic, F. Feddrix and E. Yeager, *J. Electroanal. Chem.* **341** (1992) 287.
[31]N. A. Anastasijevic, Z. M. Dimitrijevic and R. R. Adzic, *Electrochim. Acta* **31** (1986) 1125.
[32]J. Prakash and H. Joachin, *Electrochim. Acta* **45** (2000) 2289.
[33]H. D. Abruna (Ed.), *Electrochemical Interfaces*, VCH New York 1991.
[34]F. C. Nart, T. Iwasita and M. Weber, *Electrochim.Acta* **30** (1994) 961.

[35]P. C. Lu, C. H. Yang S. L. Lau and M. S. Zei, *Langmuir* **18** (2002) 754 and references therein.
[36]P. W. Faguy, N. S. Marinkovic and R. R. Adzic, *Langmuir* **12** (1996) 243.
[37]K. Kunimatsu, M.G. Samant and H. Seki, *J. Electroanal. Chem.* **258** (1989) 163.
[38]F. C. Nart and T. Iwasita, *J. Electroanal. Chem.* **322** (1992) 289.
[39]P.W. Faguy, N. S. Marinkovic and R. R. Adzic, *J. Electroanal .Chem.* **407** (1996) 209.
[40]W. Savich, S.G. Sun, J. Lipkowski and A. Wieckowski, *J. Electroanal. Chem.* **388** (1995) 233.
[41]M. Nakamura, Y. Sakurai and M. Ito, *J. Electroanal. Chem.* **563** (2004) 63.
[42]T. Senna, N. Ikemiya and M. Ito, *J. Electroanal. Chem.* **511** (2001) 115.
[43]N. S. Marinkovic, J. S. Marinkovic and R. R. Adzic, *J Electroanal. Chem.* **467** (1999) 291.
[44]J. Lipkowski, Z. Shi, A. Chen and B. Pettinger, *Electrochim Acta* **43** (1998) 2875.
[45]N. S. Marinkovic, J. X. Wang, J. S. Marinkovic and R. R. Adzic, *J. Phys. Chem. B* **103** (1999) 139.
[46]Z. Shi, J. Lipkowski, M. Gamboa, P. Zelenay and A. Wieckowski, *J. Electroanal. Chem.* **366** (1994) 317.
[47]V. B. Paulissen and C. Korzeniewski, *J. Electroanal. Chem.* **351** (1993) 329.
[48]R. M. Hammaker, A. Francis, R. P. Eischens, *Spectrochim. Acta* **21** (1965) 1295.
[49]K. Ashley, S. Pons, *Chem. Rev.* **88** (1988) 673.
[50]S. D. Ross, *Inorganic Infrared and Raman Spectra*, McGraw-Hill, Maidenhead, UK 1972.
[51]J. K. Foley, C. Korzeniewski, J. L. Daschbach and S. Pons in: *Electroanalytical Chemistry*, Vol 14, Ed. by A.J. Bard, Marcel Dekker, New York 1986.
[52]Y. Shingaya and M. Ito, *J. Electroanal. Chem.* **467** (1999) 299.
[53]Y. Sawatari, J. Inukai and M. Ito, *J. Electron Spectrosc. Related Phenomena* **64-65** (1993) 515.
[54]T. Iwasita, F. C. Nart in: *Advances in Electrochemical Science and Engineering*, Vol 4, Ed. by H. Gerisher, C. Tobias, Verlag, Weinheim (1995).
[55]H. Over, *Prog. Surf. Sci.* **58** (1998) 249.
[56]H. Over, *Appl. Phys. A* **75** (2002), 37.
[57]W. Chrzanowski and A. Wieckowski, *Langmuir* **13** (1997) 5974.
[58]J. K. Norskov, T. Bligaard, A. Logadottir, S. Bahn, L. B. Hansen, M. Bollinger, H. Bengaard, B. Hammer, Z. Šljivančanin, M. Mavrikakis, Y. Xu, S. Dahl, C.J.H. Jacobsen, *J. Catal.* **209** (2002) 275–278.
[59]I. J. Malik and J. Hrbek, *J. Phys. Chem.* **95**, (1991) 10188.
[60]A. Bottcher and H. Niehus, *J. Chem. Phys.* **110**, (1999) 3186.
[61]C. Stampfl, S. Schwegmann, H. Over, M. Scheffler, and G. Ertl, *Phys. Rev. Lett.* **77** (1996) 3371.
[62]A. Bottcher, H. Niehus, S. Schwegmann, H. Over and G. Ertl, *J. Phys. Chem. B* **101** (1997) 11185.
[63]A. Bottcher, M. Rogazia, H. Niehus, H. Over, G. Ertl, *J. Phys. Chem.* **103** (1999) 6267.
[64]H. Over, Y. D. Kim, A. P. Seitsonen, S. Wendt, E. Lundgren, M. Schmid, P. Varga, A. Morgante and G. Ertl, *Science* **287** (2000) 1474.
[65]H. Madhavaram, H. Idriss, S. Wendt, Y. D. Kim, M. Knapp, H. Over, J. Aßmann, E. Loffler, and M. Muhler, *J. Catal.* **202** (2001) 296.
[66]Lj. Atanasoska, W. E. O'Grady, R. T. Atanasoski, F. H. Pollak, *Surf. Sci.* **202** (1988) 142.

[67]H. Over, A. P.Seitsonen, E. Lundgren, M. Schmid and P. Varga, *Surf. Sci.* **515** (2002) 143.
[68]M. Pourbaix (Ed.), *Atlas of Electrochemical Equilibria in Aqueous Solutions*, NACE, Houston, TX, 1974.
[69]F. Ludwig, R.K. Sen and E.B. Yeager, *Elektrokhymiya* **13** (1977) 847.
[70]B. E. Conway and L. Bai, *J. Electroanal. Chem.* **198** (1986) 149.
[71]H. Gasteiger, N. M. Markovic and P. N. Ross, *J. Phys. Chem.* **99** (1995) 8290.
[72]J. Wang, S. R. Brankovic, Y. Zhu, J. C. Hanson and R. R. Adzic, *J. Electrochem. Soc.* **150** (2003) A1108.
[73]H. A. Gasteiger, N. M. Markovic, P. N. Ross and E. J. Cairns, *J .Phys .Chem.* **97** (1993) 12020.
[74]H. Over and M. Muhler, *Prog.Surf.Sci.* **72** (2003) 3.
[75]R. L. C. Wang, H. J. Kreuzer, P. Jakob and D. J. Menzel, *J. Chem. Phys.* **111** (1999) 2115.
[76]P. He, H. Dietrich and K. Jacobi, *Surf. Sci.* **345** (1996) 241.
[77]C. H. F. Peden and D. W. Goodman, *J. Phys. Chem.* **90**, (1986) 1360.
[78]A. Botcher and H. Conrad, *J. Chem. Phys.* **112** (2000) 4779.
[79]A. Bottcher and H. Niehus, *Phys. Rev. B* **60** (1999) 14396.
[80]H. A. Gasteiger, P. N. Ross and E. Cairns, *J. Surf. Sci* **293**, (1993) 67.
[81]W. F. Lin, M. S. Zei, M. Eiswirth, G. Ertl, T. Iwasita and W. Vielstich, *J. Phys. Chem. B* **103** (1999) 6968.
[82]S. R. Brankovic, N. S. Marinkovic, J. X. Wang and R. R. Adzic, *J. Electroanal. Chem.* **532** (2002) 57.
[83]M. B.Vukmirovic, unpublished results.
[84]J. M. Jin, W. F. Lin and P. A. Christensen, *J. Electroanal. Chem.* **563** (2004) 71.
[85]O. A. Petrii, *Doklady AN SSSR* **160** (1965) 871.
[86]J. O'M. Bockris, H. Wroblowa, *J. Electroanal. Chem.*, **7**, (1964) 428 .
[87]M. Watanabe, S. Motoo, *J. Electroanal. Chem.* **60** (1975) 267.
[88]M. Watanabe, M. Uchida, S. Motoo, *J. Electroanal. Chem.* **229** (1987) 395.
[89]A. N. Haner, P. N. Ross Jr., *J.Phys.Chem.* **95** (1991) 3740.
[90]B. N. Grgur, N. M. Markovic, P. N. Ross, *J. Electrochem. Soc.* **146** (1999) 1613.
[91]M. Krausa, W. Vielstich, *J. Electroanal. Chem.* **379** (1994) 307.
[92]J. McBreen, S. Mukerjee, *J. Electrochem. Soc.* **142** (1995) 3399.
[93]S. R. Brankovic, J. X. Wang and R. R. Adzic, *Electrochem. Solid State Lett.* **4** (2001) A217.
[94]S. R. Brankovic, J. McBreen, R. R. Adzic, *J. Electroanal.Chem.* **503** (2001) 99.
[95]J. Clavilier, J. M. Feliu, A. Aldaz, *J.Electroanal.Chem.* **243** (1988) 419.
[96]S. R. Brankovic, J. McBreen, R. R. Adzic, *Surf.Sci.* **479** (2001) L363.
[97]W. Chrzanowski, H. Kim, A. Wieckowski, *Catal.Lett.* **50** (1998) 9.
[98]G. Tremiliosi-Fuho, H. Kim, W. Chrzanowski, A. Wieckowski, B. Grzybowska, P. Kulesza, *J. Electroanal. Chem.* **467** (1999) 143.
[99]G. Blyholder, *J.Phys.Chem.* **68** (1964) 2772.
[100]M-S. Liao, C. R. Cabrera and Y. Ishikawa, *Surf. Sci.* **445** (2000) 267.
[101]M. T. M. Kopper, T. E.Shubina and R. A. vanSanten, *J. Phys. Chem. B* **106** (2002) 686.
[102]B. Hammer and J. K. Norskov, *Adv .Catal.* **45** (2000) 71.
[103]E. Christoffersen, P. Liu, A. Ruban, H. L. Skiver and J. K. Nørskov, *J.Catal.* **199** (2001) 123.
[104]F. Buatier de Mongeot, M. Scherer, B. Gleich, E. Kopatzki, and R. J. Behm, *Surf. Sci.* **411**, (1998) 249.

[105]R. E. Benfield, *J. Chem. Soc. Faraday Trans.* **88** (1992) 1107.
[106]K. Sasaki, Y. Mo, J. X. Wang, M. Balasubramanian, F. Uribe, J. McBreen and R. R. Adzic, *Electrochim .Acta* **48** (2003) 3841.
[107]K. Sasaki, J. X. Wang, M. Balasubramanian, J. McBreen, F. Uribe and R. R. Adzic, *Electrochim. Acta* **49** (2004) 3873.
[108]A. N. Mansour, J. W. Cook, D. E. Sayers, *J. Phys. Chem.* **88** (1984) 2330.
[109]K. Kinoshita, in: *Modern Aspects of Electrochemistry*, Vol. 14, Ed. by J. O'M. Bockris, B .E. Conway, R. E. White, Plenum, New York, 1982.
[110]A. V. Ruban, H. L. Skriver and J. K. Nørskov, *Phys. Rev B* **59** (1999) 15990.
[111]K. Sasaki, M. Shao, M. B. Vukmirovic and R. R. Adzic, *ECS Meeting Abstracts*, Denver, CO, May 7-11, 2006, p. 1103.
[112]B. Beden, C. Lamy, A. Bewick, K. Kunimatsu, *J. Electroanal. Chem.* **121** (1981) 343.
[113]Y. X. Chen, A. Miki, S. Ye, H. Sakai, M. Osawa, *J. Am. Chem. Soc.* **125** (2003) 3680.
[114]T. Yajima, H. Uchida, M. Watanabe, *J.Phys.Chem. B* **108** (2004) 2654.
[115]F.Mallard, G.-Q. Lu, A.Wieckowski and U.Stimming, *J.Phys.Chem. B* **109** (2005) 16230.
[116]E. Herrero, K. Franszczuk and A. Wieckowski, *J. Electroanal. Chem.* **361** (1993) 269.
[117]S. Cramm, K. Fiedrich, K. P. Geyzers, U. Stimming and R. Vogel, *J. Anal. Chem.* **358** (1997) 189.
[118]M. M. P. Jansen and J. Moolhusen, *Electrochim. Acta* **21** (1976) 861.
[119]H. Hoster, T. Iwasita, H. Baumgartner and W. Vielstich, *J. Electrochem. Soc.* **148** (2001) A496.
[120]C. E Lee and S. H. Bergens, *J. Phys .Chem. B* **102** (1998) 193.
[121]K. A. Friedrich, K. P. Geyzers, U. Linke, U. Stimming and J. Stumper, *J. Electroanal. Chem.* **402** (1996) 123.
[122]S. Z. Zou, I. Villegas, C. Stuhlmann and M. J. Weaver, *Electrochim.Acta* **43** (1998) 2811.
[123]A. Crown, C. Johnson and A. Wieckowski, *Surf. Sci. Lett.* **506** (2002) L268.
[124]G.Tremiliosi-Filho, H. Kim, , W. Xhrzanowski, A. Wietckowski, B. Grzybowska and P. Kulezsa, *J. Electroanal. Chem.* **467** (1999) 143.
[125]E. Herrero, J. M. Feliu and A. Wietzkowski, *Langmuir* **15** (1999) 4944.

High-Performance Electrodes for Medium-Temperature Solid Oxide Fuel Cells

Hiroyuki Uchida* and Masahiro Watanabe**

*Interdisciplinary Graduate School of Medicine and Engineering and
**Clean Energy Research Center, University of Yamanashi, Kofu 400-8511, Japan

I. INTRODUCTION

1. Characteristics of SOFCs

Solid oxide fuel cells (SOFCs) are expected to provide high energy conversion efficiencies because of enhanced electrode performance due to high-temperature operation and of the efficient recovery of waste heat by additional means such as a bottoming cycle (gas turbine and steam turbine). The principle of the SOFC is shown in Fig. 1. Oxygen is reduced at the porous cathode to oxide ions (O^{2-}). The oxide ions migrate across the solid oxide electrolyte to react with fuel gas at the porous anode. The cells can use a variety of fuels, which can be reformed internally to reactive fuels such as hydrogen or carbon monoxide on the anode using the waste heat energy at the high operating temperature of SOFCs. The operating temperature of the current first-generation SOFCs is restricted to very high values of about 1000°C due to insufficient performance of the state-of-the-art electrolyte (yttria stabilized zirconia, YSZ) and

Modern Aspects of Electrochemistry, Number 42, edited by C. Vayenas et al., Springer, New York, 2008.

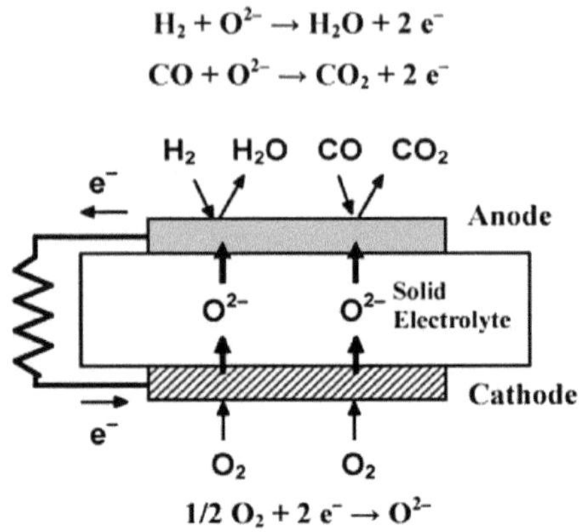

Figure 1. Principle of a solid oxide fuel cell.

electrodes (Ni-YSZ cermet anode and Sr-doped $LaMnO_3$ cathode) at lower temperatures.

2. Development of Medium-Temperature SOFCs

It is desirable to operate SOFCs at a medium temperature (700~800 °C) to overcome problems such as degradation of the construction materials, sintering of electrodes, and a limited choice of materials. For example, alloys can be used as the interconnect or as the heat-exchange material in place of conventional ceramics (doped $LaCrO_3$). Extensive challenges to develop the medium-temperature SOFCs started in the 1990's. Figure 2 shows such a trend manifested by the number of papers presented at the international symposium on SOFCs from 1995 (SOFC IV) to 2005 (SOFC IX). The total number of papers increased, especially since 2003, showing an acceleration of the R&D on SOFCs. In 1995, the number of papers concerning the reduced temperature operation was only 13 (*ca.* 10 %) and two of them were our contribution. However, this percentage has grown steeply up to about 72 % of the total in 2005, reflecting world-wide efforts for reduction of the operating temperature of SOFCs.

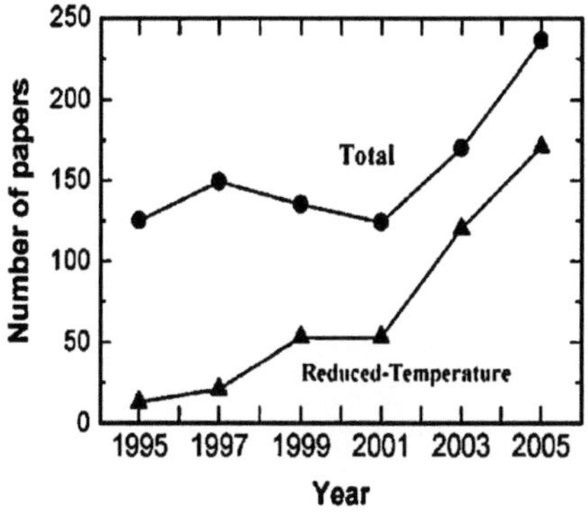

Figure 2. Variation in the number of papers presented at the international symposium on SOFCs from 1995 (SOFC IV, Yokohama, Japan) to 2005 (SOFC IX, Quebec, Canada).

In order to operate SOFCs at medium temperatures, two major obstacles must be addressed. The first is to reduce the ohmic losses in the solid electrolyte. There has been progress in the development of very thin film zirconia electrolyte[1-7] and in introducing novel solid electrolytes with higher ionic conductivity.[8-11] It was reported that the ohmic loss of 5 μm-YSZ electrolyte was estimated to be only ca. 2 % of the total voltage loss at 800°C.[4] The overpotentials of the conventional anode and cathode become a large fraction of the voltage loss with lowering operating temperature. Hence, the development of high-performance electrodes is very important, besides the reduction of the ohmic resistance.

3. Design Concept of Catalyzed Reaction Layer for Medium-Temperature SOFC

It is very important for a high-performance electrode to have both a highly active electrocatalytic reaction zone and a sufficient gas-supply network in its microstructure. The conventional anode material used so far is Ni-YSZ cermet prepared from μm-sized NiO and YSZ particles as shown in Fig. 3A. Because all reactants (fuel gas, electrons, and oxide ions) must meet together at the reaction sites, the so-called "three phase boundary (TPB) zone" is the effective

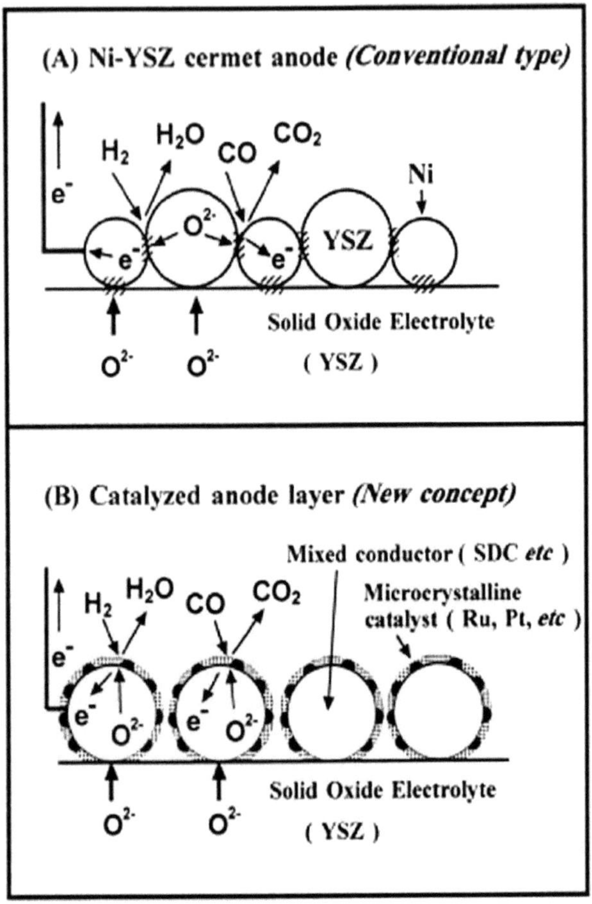

Figure 3. Schematics of anode reaction in SOFC using a conventional type Ni-YSZ cermet anode (A), and catalyzed anode layer (B). Reproduced from Ref. 14, Copyright (1996), by permission from The Electrochemical Society.

reaction zone (ERZ: hatched zone in Fig. 3A). It is intrinsically difficult to increase significantly the ERZ, since the surface areas of the contacting interfaces between the particles of μm-sized Ni (electronic conductor) and YSZ (pure ionic conductor) are limited. Also, the low specific surface area of Ni cannot result in high catalytic activity for the internal reforming reactions of fuel gases. There have been papers reporting an improvement of the anode performance at medium temperatures by optimizing the microstructure of such a conventional Ni-YSZ cermet.[3-7] For example, when a thin YSZ electrolyte film was prepared on a porous anode support,[6,7] the YSZ film penetrated into the pores, resulting in good adhesion and a large electrolyte/anode interfacial area, which contributes to high activity. In this case, however, the ERZ is enlarged two-dimensionally. To promote both the anode reaction and the internal reforming reaction, the use of highly-dispersed catalysts is desirable. The dispersion of catalyst particles on the YSZ (pure ionic conductor) surface, however, may lead to insufficient electric conductivity through the contacts between the particles.

We have proposed a new design concept of a catalyzed-reaction layer for SOFC anodes as illustrated in Fig. 3B.[12-14] Microcrystalline catalysts are highly dispersed on particles of mixed ionic-electronic conducting oxides (MIECO) such as doped ceria. This figure assumes that the catalyst microcrystals are of spherical or hemispherical shape with a mean diameter d and are embedded in hemispherical shape on the MIECO support, i.e., as ultra-fine domes on the support. We can expect the following effects. Since the specific surface area is inversely proportional to d, a high rate of fuel-reforming is obtainable on the microcrystals as the particle size decreases; the activity of 10 nm-size catalyst is expected to be 2 orders of magnitude higher than that of conventional μm-sized particles. The use of MIECO anode results in an enlarged ERZ beyond the physical TPB.[15] When active electrocatalysts are loaded on the MIECO surface, they certainly activate the anodic process at the boundary. The circumference length L, the parts of catalyst domes contacting with the MIECO, is a measure of such a highly activated boundary. Since L is proportional to d^{-2}, the size-effect of the catalysts must appear dramatically on the electrochemical reaction rate when the rate determining step occurs in the boundary region. Thus, only a small amount of microcrystalline catalysts on the MIECO may promote both the chemical (reforming) and electrochemical (electrode) reactions.

For the SOFC cathode, mixed conducting perovskite-type oxides based on $LaMnO_3$ have been widely applied. Although they exhibited a high performance at operating temperatures $T_{cell} = 1000$ °C, the cathodic overpotential becomes significant with lowering operating temperature.[16] The cathode can also be activated by highly dispersed catalyst microcrystals in a similar manner as mentioned for the catalyzed anode.

II. ACTIVATION OF MIXED-CONDUCTING CERIA-BASED ANODE

1. Effect of Various Metal Catalysts Dispersed on Samaria-Doped Ceria

An 8-mol% YSZ disk (diameter: 13 mm, thickness: 1 mm) was used as the solid electrolyte in the test cell. The IR-free polarization characteristics (*I* - *E* curves) of various electrodes were measured by a current-interruption method in a three-electrode configuration at temperatures of 800 to 1000°C. A mixed conducting samaria-doped ceria [$(CeO_2)_{0.8}(SmO_{1.5})_{0.2}$, denoted as SDC] anode without any metal catalyst exhibited a fairly small polarization at 1000 °C; the overpotential (IR-free), η, at the current density, j, of 0.6 A/cm^2 was only about 50 mV,[13] consistent with that reported by Takahashi and Iwahara for La- or Y-doped ceria.[15] Such high performance of pure SDC can be explained by its high ionic and electronic conductivity at high temperature in hydrogen atmosphere,[17] resulting in an enhanced ERZ. However, at reduced temperature of $T_{cell} = 800$ °C, an appreciable polarization is seen for the pure SDC as shown by a dashed line in Fig. 4. The depolarizing effect of the nm-size metal catalyst such as Ru, Rh, Ir, and Pt is remarkable. This effect of catalyst-loading was prominent with lowering the T_{cell}.[13] The Ru-SDC anode was found to exhibit an excellent performance in a synthetic gas ($4\ H_2 + CO_2$), which could be formed by reforming methane.[14] The Ru- or Ir-dispersed SDC powders also exhibited high catalytic activities for steam reforming of methane,[18] suggesting a potential ability for the application in internal reforming type SOFCs.

2. Effect of the Composition and Microstructure on the Performance of SDC Anodes

If the SDC layer itself can be optimized, the nm-size metal catalysts loaded on the SDC should work more effectively. Here, we demonstrate that the performances of the SDC anode dispersed with nm-sized Ru-catalysts are greatly improved by controlling the composition and the microstructure.[19,20]

First, we examined the effect of SDC composition on the performance. Among the SDCs with a composition of $(CeO_2)_{1-X}(SmO_{1.5})_X$ ($0 \leq X \leq 0.4$) examined, an SDC anode with X = 0.2 was found to exhibit the maximum current density at a given η at T_{cell} = 800 to 1000 °C, when operated under hydrogen atmosphere. This high current density was attributed to the large ERZ resulting from improved conductivities of both oxide ions (σ_{ion}) and electrons (σ_e).[19] The ceria-based materials, however, may have the disad-

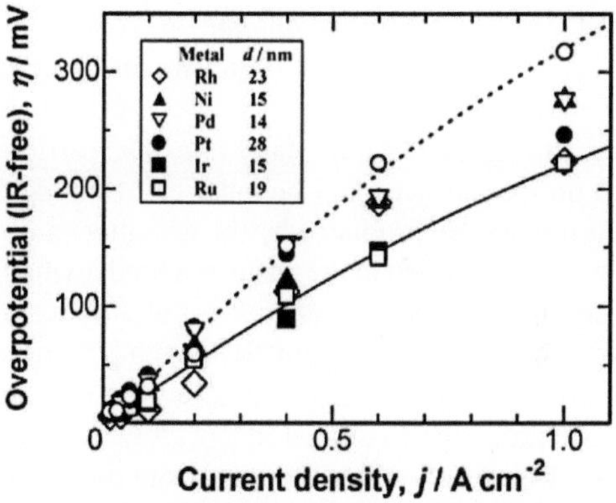

Figure 4. Polarization curves (IR-free) of catalyzed SDC anode layer (slurry coated on YSZ) in humidified H_2 ($p[H_2O]$ = 0.04 atm) at T_{cell} = 800°C. Dashed line is for SDC without metal catalyst, and solid line is the regression line for Ru-SDC. SDC particle size=1.6 μm, the amount of each metal catalyst: 0.1 mg/cm^2 (*ca.* 2.5 wt%). These metal catalysts were loaded on the SDC by impregnating 2 μL of metal salt solution containing each metal ion at 12.5 g/L, followed by thermal treatments. The particle size d estimated from XRD are shown in the table. Reproduced from Ref. 13, Copyright (1994), by permission from The Electrochemical Society.

vantage of a high isothermal lattice expansion upon reduction.[11,21] However, this problem can be overcome by applying a buffer layer with a ceria-zirconia composition gradient on the YSZ electrolyte surface.[12] An anchoring layer of YSZ particles was also found to be effective to improve the adhesion of the porous ceria anode.[22] Indeed, we found that attaching a very thin film of SDC (ca. 1 μm) onto the YSZ electrolyte before coating the SDC anode is an efficient method for significantly reducing η.[19]

Next, we examined the effects of control of the microstructure such as pore-size distribution (or porosity) and the ohmic resistance of SDC layers on the improvement of the electrocatalytic activity of the SDC anode with and without nm-sized Ru metal catalyst loading. Fine polymer beads (cross-linked polystyrene, $d = 1.2$ μm) were added to the SDC paste, resulting in the formation of μm-sized pores after sintering, which can improve the gas-diffusion rates in the electrode.

When the SDC paste (without adding polymer beads) was screen-printed on the YSZ and sintered at 1050°C, a relatively dense layer with a thickness of *ca.* 20 μm was formed as shown in Fig. 5A. Original large SDC particles (d = 1.6 μm) in the paste agglomerated into *ca.* 5 to 10 μm. Many fine SDC particles (≤1 μm) were formed on the surfaces and boundaries of the agglomerates. These fine particles were produced by the decomposition of cerium and samarium nitrates added in the paste as a binder which enhances the ionic and electronic conduction without disturbing the gas-diffusion.[19] In contrast, when the SDC with polymer beads was sintered, large-sized pores ranging from a few μm to several ten μm were uniformly produced as shown in Fig. 5B. Thus, SEM observation clearly demonstrates the change in microstructure brought about by the addition of polymer beads in the SDC paste.

Figure 6 shows integration curves of pore volumes per unit weight of various SDC layers attached on the YSZ electrolyte as a function of pore diameter, d, measured by mercury porosimetry. Two different types of micropores were found in all SDC layers, i.e., primary pores with a pore diameter between 0.2 and 0.5 μm and secondary pores ($d > 10$ μm). It is considered that the primary pores and the secondary pores correspond to a space within the SDC agglomerates and a space between the agglomerates, respectively (see the illustration in Fig. 6). Then, the volume of the primary pores, V_{prim} ($d \leq$ μm) and that of secondary pores, V_{sec} ($d > 1$ μm)

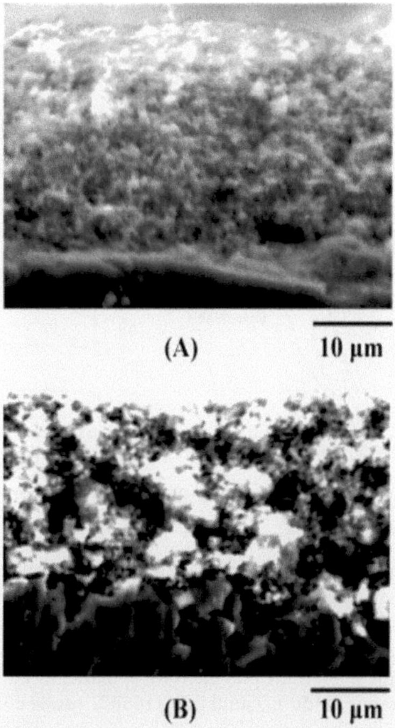

Figure 5. SEM of SDC anode cross-sections. (A) prepared from the screen-printed SDC paste without polymer beads and sintered at 1050 °C for 4 h, (B) prepared from the screen-printed SDC paste with 1.0 wt% polymer beads and sintered at 1150 °C for 4 h. Reproduced from Ref. 20, © 1999, by permission from The Electrochemical Society.

were calculated for each SDC layer. As shown in the inset of Fig. 6, the addition of polymer beads (0.5 to 1 wt%), followed by sintering, preferentially increases V_{sec}. By adding 1 wt% polymer, the value of V_{sec} increased by 1.5 times (sintered at 1050 °C for 4 h) and by 1.8 times (sintered at 1150 °C for 4 h) compared with the corresponding value without polymer. Because the value of V_{prim} was almost constant irrespective of the amount of polymer added, the increase in apparent porosity was not so marked. Therefore, in considering a change in the electrochemical properties of the SDC layers ex-

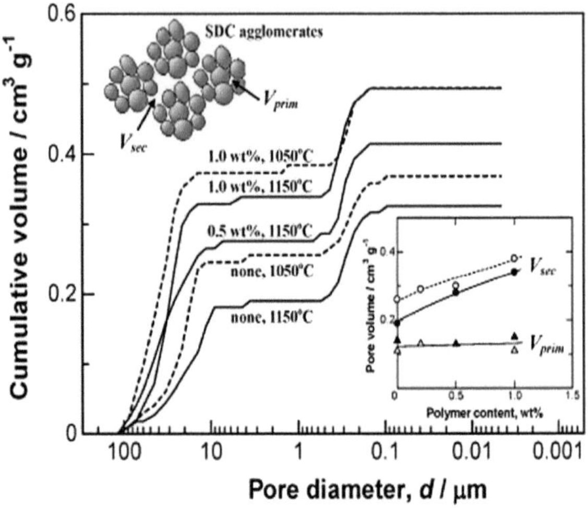

Figure 6. Integration curves of pore volume per unit weight of various SDC layers attached on YSZ electrolyte as a function of pore diameter, d. V_{sec} = secondary pore volume ($d > 1$ µm). V_{prim} = primary pore volume ($d \leq 1$ µm). Inset shows the relation between each pore volume and the amount of polymer (wt%) added to the paste. Sintering temperature: 1150°C (solid line), 1050°C (dashed line).

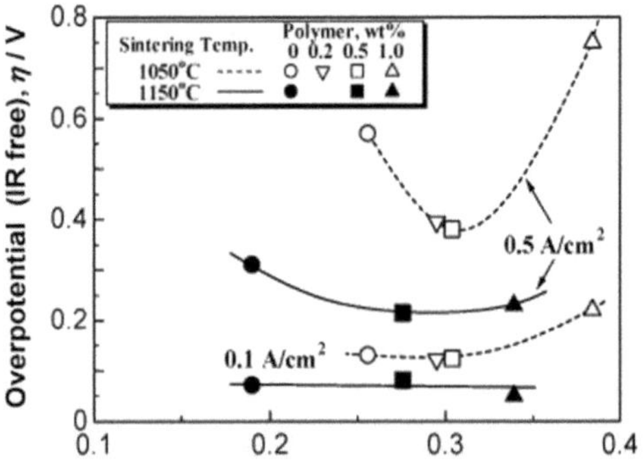

Figure 7. Dependence of IR-free overpotential η at different current densities of 0.1 A/cm² and 0.5 A/cm² as a function of secondary pore volume of V_{sec} for various SDC anodes at T_{cell} = 800°C. Reproduced from Ref. 20 Copyright (1999), by permission from The Electrochemical Society.

amined, the value of V_{sec} is a more essential factor than the apparent porosity.

Figure 7 shows the variation of η at $j = 0.1$ A/cm^2 and 0.5 A/cm^2 as a function of V_{sec} for various SDC anodes at T_{cell} = 800°C. At $j = 0.1$ A/cm^2, the dependence of η on V_{sec} is small for the SDC sintered at 1050°C and 1150°C because a delay in gas-diffusion is not so significant at low current densities. In contrast, the overpotential η at high current densities ($j = 0.5$ A/cm^2) shows a minimum in the case of 0.5 wt% polymer addition but increases by increasing the amount of polymer to 1.0 wt%, especially for the anode sintered at 1050°C. It was found that this was caused by the increase in the ohmic resistance of the SDC layer, i.e., the reduction of ERZ due to insufficient σ_{ion} and σ_e outweighs the performance improvement due to enhanced gas-diffusion rates. The highest performance was obtained for the SDC sintered at 1150°C with 0.5 wt% polymer (V_{sec} = 0.3 cm^3/g), where the best balance exists between high gas-diffusion rates and low ohmic resistance.

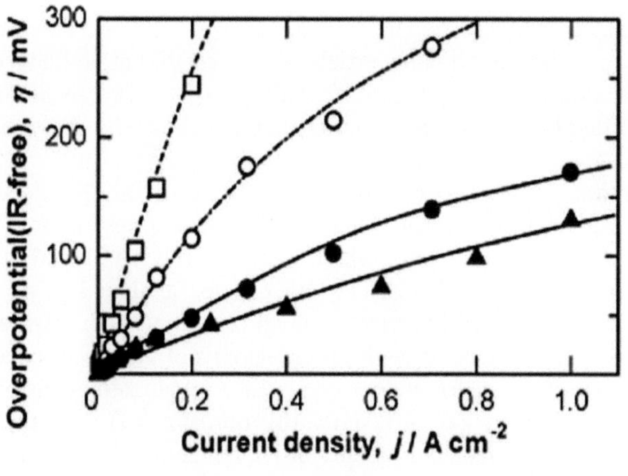

Figure 8. Polarization curves (IR-free, measured in humidified H_2) at various SDC anodes at T_{cell} = 800°C.□: original one (sintered at 1050 °C, without polymer); O: the optimized microstructure (sintered at 1150°C with 0.5 wt% polymer); ●: Ru-catalyzed (0.1 mg-Ru/cm^2 loaded on the optimized one); ▲: Ni-catalyzed [0.75 mg-Ni/cm^2 (8 vol%) loaded on the optimized one, see Section II.3).

In order to activate the SDC anode with optimized microstructure, nm-sized Ru catalysts were highly dispersed on the surface (0.1 mg/cm^2). Figure 8 shows the IR-free polarization curve for the Ru-SDC (by • symbol) in comparison with those for the original (□: sintered at 1050 °C, without polymer) and the optimized SDC anodes (○: sintered at 1150 °C with 0.5 wt% polymer). At η = 0.1 V and T_{cell} = 800 °C, the value of j on the optimized SDC was about 2 times higher than that of the original one. The Ru loading significantly enhances the performance of the optimized anode. The current density at η = 0.1 V was 0.5 A/cm^2 at T_{cell} = 800 °C, which is 2.5 times higher than that without Ru catalyst.

3. Activation of SDC Anode with Highly-Dispersed Ni Electrocatalysts

We have examined the possibility to enhance the anode performance by dispersing inexpensive metal catalysts on the SDC. As shown in Fig. 4, the SDC was not sufficiently activated by nm-sized Ni catalysts with an amount of 0.1 mg/cm^2 (*ca.* 1 vol%). However, higher Ni loading can be allowed to improve the performance, because Ni is much cheaper than Ru by *ca.* 1/40. We have found that 6 to 8 vol% of Ni nanoparticles (d_{Ni} = *ca.* 20 nm) enhanced the SDC anode performance significantly at T_{cell} = 700 to 800°C in humidified H_2.[23-25] A typical performance is shown by ▲symbol in Fig. 8.

It was also found that the performance of this Ni-dispersed SDC was much higher than that of Ni-SDC cermet anodes (8-70 vol% Ni, d_{Ni} = 1 to 3 μm), which were recently employed in medium-temperature SOFCs.[26-29] We have analyzed why the dispersion-type and the cermet-type anodes show such a large difference in the performances from viewpoints of the electrocatalytic activity (exchange current density, j_0), ERZ (evaluated by the ohmic resistance R), and gas diffusivity (by porosimetry).[24] In the cermet-type, the j_0 increased steeply at Ni content > 50 vol% and the value of R decreased with increasing Ni, reaching nearly zero at Ni content > 60 vol% at T_{cell} = 700 to 900 °C. Thus, both the ohmic loss and the polarization loss decrease with increasing the Ni content in the cermet-type. The contacting areas between the μm-sized particles of Ni and SDC are limited, mostly to the outside of the SDC

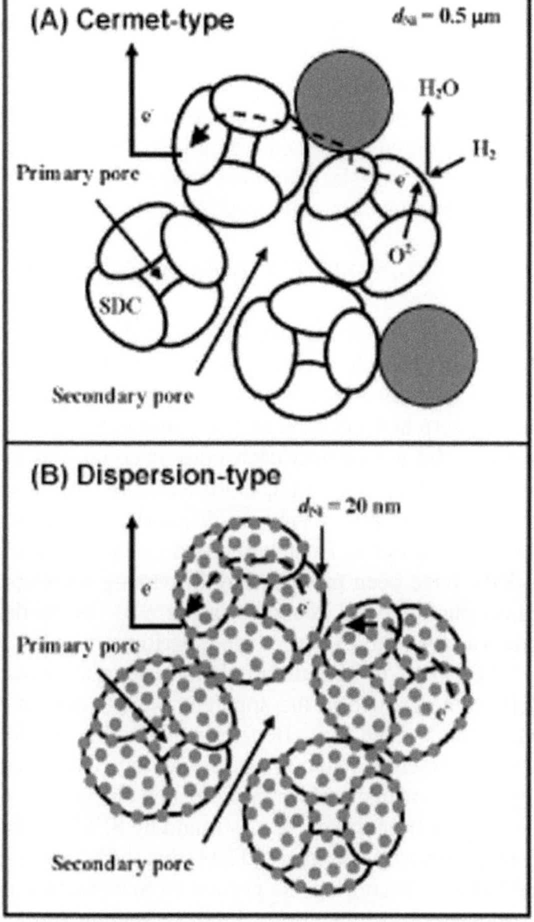

Figure 9. Schematic illustration of anode reaction zones in Ni cermet-type (A) and Ni dispersion-type (B) SDC layers. Reproduced from Ref. 24, Copyright (2005), by permission from The Electrochemical Society of Japan.

agglomerates, requiring a large Ni content to reduce the polarization loss. The electrons formed at such contacts must travel a long distance along the SDC networks when the Ni loading is lower than the percolation limit, as shown schematically in Fig. 9A. In contrast, the dispersion-type anodes of only 6 to 8 vol% Ni exhibited much higher j_0 values than those of any cermet-type anodes at all cell temperatures. Since the average circumference length L (the parts of Ni catalyst in the dome shape contacting with the SDC support) is proportional to d_{Ni}^{-2}, nm-sized Ni must promote the anode reaction (Eq. 1) more effectively than μm-sized Ni.

$$O^{2-}\,[\text{Ni-SDC}] + H_2[\text{Ni-SDC}] \rightarrow H_2O(\text{gas}) + 2e^{-}[\text{Ni-SDC}] \qquad (1)$$

where [Ni-SDC] is the active site formed at the boundary of Ni catalysts and SDC particle surfaces. Another noteworthy point for the dispersion-type electrode was that the ohmic resistance R reached nearly zero at 8 vol% Ni at T_{cell} = 800 to 900 °C, although the Ni particles did not contact each other according to the SEM observation. Since both σ_{ion} and σ_e in the SDC are sufficiently high at high temperatures (> 800 °C) in humidified H_2, negligibly small R can be obtained at 8 vol%. Thus, both good electronic network and sufficient ERZ have been produced by dispersing nm-sized Ni on the mixed conducting SDC as shown in Fig. 9B. The Ni-dispersed SDC anode was found to exhibit a stable performance at T_{cell} = 800 °C and j = 0.6 A/cm^2 in humidified H_2 for a long term over 1100 h.[24,25] While some growth in the apparent average size of Ni particles was found, both the IR-free polarization performance (reflecting the effective reaction area) and the ohmic resistance (reflecting the electronic network) were not changed noticeably during the long-term operation. It was found by SEM and scanning transmission electron microscope (STEM) that the Ni particles were rather stabilized by anchoring the portion to SDC surface presumably due to a strong interaction.[25] Such a structure must be also advantageous to maintain the ERZ against Ni sintering, compared with that of the conventional cermet type catalyst layer.

III. ACTIVATION OF MIXED-CONDUCTING PEROVSKITE-TYPE OXIDE CATHODES

1. La(Sr)MnO_3 Cathode with Highly Dispersed Pt Catalysts

We have examined the performance of conventional $(La_{0.85}Sr_{0.15})_{0.9}MnO_{3-\delta}$ (denoted as LSM), where δ is the number of oxygen deficiencies per unit cell, with and without loading micro-crystalline Pt catalysts.[13, 30] As is well known,[16] LSM exhibits excellent electrocatalytic activity for oxygen reduction at 1000 °C. However, as shown in Fig. 10, a significant η is seen for LSM without catalysts at T_{cell} = 800 °C. Then, Pt catalysts (0.1 mg/cm^2 or 0.5 mg/cm^2) were dispersed on the LSM surface by impregnating an aqueous solution of H_2PtCl_6 [anion-type Pt complex, denoted as Pt(A)] or $[Pt(NH_3)_4]Cl_2$ [cation-type complex, denoted as Pt(C)], followed by heating at 1000 °C for 1 h in air. The cation-type complex, Pt(C), probably has a strong interaction with OH groups at the oxide surface and, hence, is widely used in the preparation of

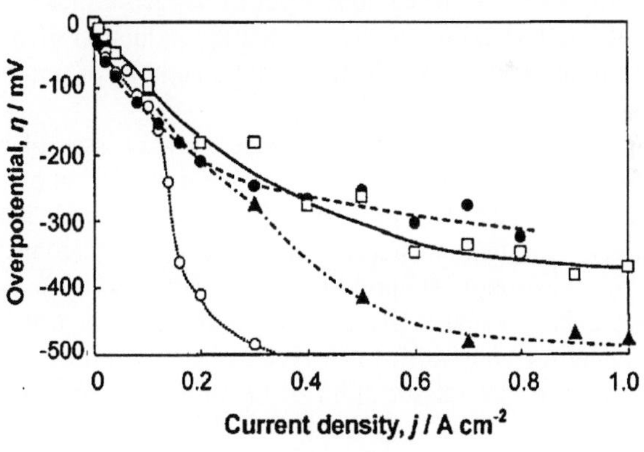

Figure 10. Polarization curves (IR-free) for oxygen reduction at LSM cathode without (○) and with (●▲□) loading Pt catalysts in O_2 at T_{cell} = 800 °C. ●: Pt(A) 0.5 mg/cm^2; ▲: Pt(A) 0.1 mg/cm^2; □: Pt(C) 0.1 mg/cm^2. Reproduced from Ref. 30, Copyright (1996), by permission from The Electrochemical Society of Japan.

conventional Pt-loaded oxide catalysts. As shown in Fig. 10, Pt catalysts dispersed on the LSM exhibit a significant depolarizing effect especially in the high current region. The performance is improved with increasing amount of Pt from 0.1 to 0.5 mg/cm^2 in the case of Pt(A). However, the LSM cathode loaded with 0.1 mg-Pt/cm^2 (1.0 wt%) from Pt(C) exhibits higher performance than that with the same Pt-loading from Pt(A), especially at high current density region, which is comparable to that with 0.5 mg-Pt/cm^2 from Pt(A). It was found that the average diameter of Pt particles, d_{Pt}, in the former was *ca.* 30 nm, smaller than that of the latter (40 nm), indicating that the use of $[Pt(NH_3)_4]Cl_2$ was effective to suppress the aggregation of Pt particles on the LSM. Thus, by decreasing the d_{Pt}, the Pt loading amount on LSM required for almost the same performance is reduced down to 1/5. Judging from the extent of the activation, much smaller Pt particles, which could not be detected by XRD, might be produced on the LSM with the use of Pt(C).

Assuming that spherical Pt particles were uniformly dispersed on spherical LSM particles (d_{LSM} = 3.8 μm), the mean coverage of Pt with d_{Pt} = 30 nm on the LSM was calculated to be only 10 %. This coverage corresponds to 0.3 vol% of Pt. Since the electrical conductivity of the LSM cathode layer in oxygen atmosphere was unchanged by Pt-loading, it is clear that the reduction of η can be brought about by the high electrocatalytic activity of nm-sized Pt particles.

It has been reported that the rate-determining step of oxygen reduction on LSM cathode lies either in the dissociation of oxygen molecules on the surface,[16] or in the diffusion of the oxide ion in the LSM layer,[31] or in the charge transfer to oxygen atoms.[32] The present results strongly support our consideration on the size-effect of loaded catalysts that activate the cathodic reaction at the boundary between Pt and LSM. Such reaction step(s) promoted by Pt catalysts may be the dissociation of O_2 molecules and/or the sequential exchange reaction, because such processes must be in quasi-equilibrium at low current density but become significant at high current density. The cathode reaction mechanism will be discussed in later Section.

2. La(Sr)CoO_3 Cathode with Ceria-Interlayer on Zirconia Electrolyte

It has been known that La(Sr)CoO_3 (denoted as LSC) exhibits higher cathodic performance than LSM,[2, 33, 34] *e.g.*, the polarization at an LSC cathode sputtered on YSZ is very small even at 800 °C.[34] However, LSC tends to react more readily with the YSZ electrolyte than LSM at high temperatures (≥ 1000 °C),[2,34] and the resulting $La_2Zr_2O_7$ or $SrZrO_3$ compounds have high ohmic resistance. There have been no practical and convenient methods to prepare LSC cathodes on YSZ electrolyte without forming such solid by-products. Although the LSC cathode works well in SOFCs using ceria-based electrolyte,[35] which does not react with LSC, the suppression of the electronic conduction in the ceria-based electrolytes remains a difficult problem to be overcome. In this Section, we describe the usefulness of samaria-doped ceria (SDC) interlayer for the prevention of unfavorable solid-state reactions between LSC cathode and YSZ electrolyte.[36, 37]

A thin film of SDC interlayer was formed on one side of the YSZ surface before coating the LSC cathode layer. The SDC thin film was prepared by screen-printing a mixed solution of cerium and samarium nitrates (Ce : Sm = 8 : 2) with a thickener, followed by heat treatment at 400 °C for 0.5 h.[19,20] Then, the same solution was screen-printed, followed by a second heat treatment at 400 °C or 1150 °C for 4 h. These SDC thin films, thus prepared on the YSZ surface are denoted as SDC400 and SDC1150, respectively. Under the atmosphere of the SOFC cathode, SDC is expected to act as a pure oxide ionic conductor with higher conductivity than that of YSZ. Onto the SDC interlayer, porous LSC cathodes were prepared by screen-printing a paste of LSC particles ($La_{0.6}Sr_{0.4}CoO_3$, mean diameter = 0.7 μm) dispersed in n-pentanol, followed by firing at 1050 °C for 1 h (LSC1050).

Figure 11 shows the polarization curves (measured in O_2) with the LSC cathodes prepared on SDC interlayers. For LSC1050/SDC1150, fairly small η and a very low ohmic resistance (R_{LSC}) were found over the entire T_{cell} region between 1000 and 800 °C. Reaction products such as $SrZrO_3$ or $La_2Zr_2O_7$ were not detected by XRD. It was found by SEM equipped with energy dispersive X-ray analyzer (EDX) that SDC1150 formed a uniform and relatively dense film with a thickness of *ca.* 1 μm. Hence, YSZ electrolyte coated with SDC1150 interlayer was protected against unfavorable solid-state reactions. In contrast, SDC400 did not work as a good interlayer, because it was not sintered well and gave large values of η and R_{LSC} by the formation of $SrZrO_3$. A disadvantage of the LSC cathode is the fairly large thermal expansion coefficient (TEC), i.e., 23.7 ppm,[2] compared with those of SDC and YSZ, i.e., 11 and 10 ppm, respectively. The mismatch, however, can be overcome by mixing SDC powder in the LSC cathode layer. The SDC interlayer can be applied to other combinations of cathodes and electrolytes, *e.g.*, $LaGaO_3$-based electrolytes[10] (and, of course, YSZ or Sc-SZ) can be protected from the solid-phase reactions with $La_{1-x}Sr_xCo_{1-y}Fe_yO_3$ (LSCF) cathodes,[38,39] which have a TEC more compatible to the SDC interlayer than that of LSC.

3. Control of Microstructure of LSC Cathodes

In this Section, we demonstrate the effects of controlling the microstructure, and thus pore-size distribution and ohmic resistance, of LSC layers on the enhancement of the electrocatalytic activity.[40]

Onto the SDC1150 interlayer described above, porous LSC1050 cathodes were prepared by screen-printing the LSC paste containing fine polymer beads as a pore-former (0 to 1.5 wt%, d = 1.2 μm), followed by firing at 1050 °C for 1 h. SEM observation clearly showed that large-sized pores ranging from a few μm to several ten μm were additionally formed when the LSC with polymer beads was sintered. Similar to the case of SDC anode in Section II.2, two different types of micropores were found in all LSC layers by mercury porosimetry, i.e., primary pores with a pore diameter between 0.2 and 1 μm and secondary pores (d > 10 μm). Figure 12A shows values of V_{prim} and V_{sec} as a function of the amount of polymer added to the paste. It is seen that the addition of polymer (0.5 to 1.5 wt%), followed by sintering, increases both V_{prim} and V_{sec}, *e.g.*, by 1.7 times at the V_{prim} and 1.5 times at the V_{sec} by adding 1.5 wt% polymer, respectively. Thus, an increase in the oxygen diffusion rate in the layer can be expected.

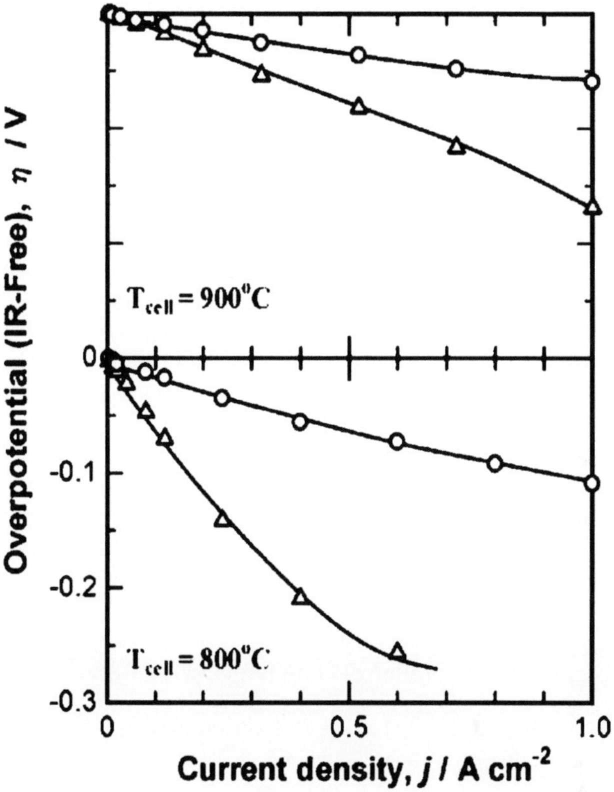

Figure 11. Polarization curves (IR-free, measured in O_2) for oxygen reduction at LSC cathodes. Δ: LSC1050/SDC400, O: LSC1050/ SDC1150.

Figure 12B shows the IR-free overpotential η at $j = 0.8$ A/cm^2 in air as a function of the polymer content in the paste for various LSC cathodes. At lower T_{cell}, the dependence of η on the polymer content becomes remarkable. At T_{cell} = 800 and 900 °C, the overpotential η first decreases with the polymer addition and shows a minimum at 1.0 wt%. It is distinctive that η increases steeply by the further addition of polymer. This suggests that the change in the polarization behavior is not simply due to the change in the gas-diffusion rates in the layer.

The values of R_{LSC} are shown in Fig. 12C. While the values of R_{LSC} are negligibly small for all specimens at T_{cell} = 1000 °C, those at T_{cell} = 900 and 800 °C are almost constant up to 1.0 wt% polymer and increase significantly at 1.5 wt% addition. The steep increase in R_{LSC} with increasing pore volume indicates that the electrical contact among LSC particles became worse, which can be correlated to the steep increase in the overpotential η.

It is worth discussing the polarization behavior at the LSC cathodes in more detail. Because LSC is a good mixed-conductor at high temperatures,[2,38] the ERZ is expected to be enlarged over the entire LSC-O_2(gas) interfacial area in addition to the TPB as schematically shown in Fig. 13. The oxygen reduction reaction at the LSC cathode can be envisioned to consist of the following elementary steps,

Diffusion of O_2(gas) in gas phase through porous LSC layer (2)

Dissociative adsorption of O_2 to form O_{ad} on LSC :

$$O_2(\text{gas}) \rightarrow 2\, O_{ad}(\text{LSC}) \tag{3}$$

Dissociative adsorption of O_2 to form O_{ad} at ERZ :

$$O_2(\text{gas}) \rightarrow 2\, O_{ad}(\text{LSC*}) \tag{3'}$$

Surface diffusion of O_{ad} to ERZ :

$$O_{ad}(\text{LSC}) \rightarrow O_{ad}(\text{LSC*}) \tag{4}$$

Charge transfer at ERZ :

$$O_{ad}(\text{LSC*}) + 2\, e^-(\text{LSC*}) \rightarrow O^{2-}(\text{LSC}) \tag{5}$$

Ionic transfer of O^{2-} from LSC into SDC interlayer :

$$O^{2-}(\text{LSC}) \rightarrow O^{2-}(\text{SDC}) \tag{6}$$

Ionic transfer of O^{2-} from SDC interlayer into YSZ:

$$O^{2-}(\text{SDC}) \rightarrow O^{2-}(\text{YSZ}) \tag{7}$$

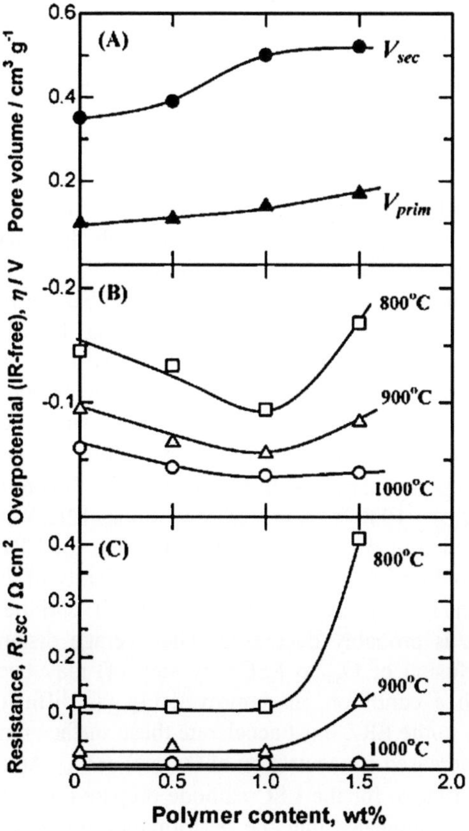

Figure 12. (A) Effect of the polymer content added in LSC cathodes as a pore-former on the primary pore volume (V_{prim}) and the secondary pore volume (V_{sec}); (B) Effect of the polymer content added in LSC cathodes on the IR-free overpotential η at 0.8 A/cm^2 operating at 800 to 1000°C in air; (C) Effect of the polymer content added in LSC cathodes on the area-specific ohmic resistances of LSC layers (R_{LSC}), which were determined by subtracting the resistance of YSZ from that of the cell LSC/SDC/YSZ/Pt. The resistances of 1 mm thick YSZ (= the resistance of the cell, Pt/YSZ/Pt) were 0.67 Ω cm^2, 0.93 Ω cm^2 and 1.74 Ω cm^2 at 1000, 900 and 800°C, respectively. Reproduced from Ref. 40, Copyright (2002), by permission from The Electrochemical Society.

where $O_{ad}(LSC^*)$ and $e^-(LSC^*)$ are adsorbed oxygen atoms and electrons at the active site on the LSC surface in the ERZ, respectively. $O^{2-}(LSC)$, $O^{2-}(SDC)$ and $O^{2-}(YSZ)$ denote the oxide ion in the LSC particles, SDC interlayer and YSZ electrolyte, respectively.

The surface active sites in the ERZ (denoted as LSC*) play a very important role in the dissociation of O_2 molecules to form O_{ad} (step 3') and/or in the sequential charge transfer (step 5). The number of LSC* sites and their activity strongly depend on the T_{cell}, gas-diffusion rate, electrode composition, etc.

At the high T_{cell} of 1000 °C, σ_{ion} and σ_e in LSC are very high, as actually seen in the negligibly small R_{LSC} in Fig. 12C. Because the number of LSC* sites is, therefore, considered to be sufficient, a large fraction of the LSC particle surface in the cathode can work well as the ERZ as illustrated in Fig. 13A. In such a case, an increase in the gas-diffusion rate, if any, could barely affect the polarization performance; a slight dependence of η on the polymer content at T_{cell} = 1000 °C is indeed seen in Fig. 12B. With lowering T_{cell}, a reduction of σ_{ion} in the LSC may restricts the ERZ in a position close to the SDC interface due to the difficulty of the transport of oxide ions by steps (6) (see Fig. 13C and D) and the number of LSC* sites is probably decreased. The average distance for the surface diffusion of O_{ad} to LSC* by step (4) may become long. Under such a condition, an improved O_2 gas diffusion through micropores to the ERZ must accelerate these surface reaction steps due to an increased concentration of O_2(gas) or O_{ad}. An appreciable reduction of the η for the LSC cathode prepared with 0.5 and 1.0 wt% polymer indicates that the performance improvement is ascribed to enhanced gas-diffusion rates with increasing V_{prim} and/or V_{sec} without a noticeable increase in R_{LSC}, as depicted in Fig. 13D. However, for the LSC with 1.5 wt% polymer, the performance improvement due to enhanced gas-diffusion rate seemed to be cancelled by the performance decrease by an increase of R_{LSC}. Hence, it is clear that the essential factors for enhancing the performance of LSC cathode are a high gas-diffusion rate as well as a low ohmic resistance, because both factors greatly contribute to an enhanced effective reaction zone, especially at low T_{cell}.

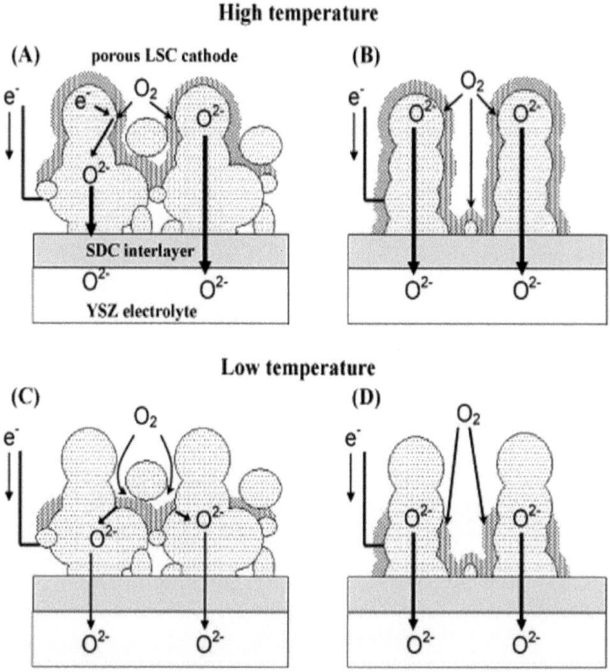

Figure 13. Schematic illustration of the cathode reactions on LSC layer with low (A, C) and high (B, D) porosity at high and low operating temperatures. The LSC cathode layer is prepared on SDC interlayer. Hatched zone is the effective reaction zone (ERZ). Reprinted from Ref. 40, Copyright (2002), by permission from The Electrochemical Society.

4. Activation of the Optimized LSC Cathode by Loading nm-Sized Pt Catalysts

In order to enhance the steps (3)-(5) on the LSC cathode with the optimized microstructure (sintered at 1050 °C with 1.0 wt% polymer), nm-sized Pt particles were dispersed on the LSC surface.[40] In this case, the following reaction steps are expected to be activated greatly by the Pt-loading and high O_2 gas-diffusion rates in the LSC layer,

$$O_2(\text{gas}) \rightarrow 2\ O_{ad}(\text{Pt-LSC}) \tag{8}$$

$$O_{ad}(\text{Pt-LSC}) + 2\ e^-(\text{Pt-LSC}) \rightarrow O^{2-}(\text{LSC}) \tag{9}$$

where the active site, (Pt-LSC), is formed at the boundary of the Pt catalysts and the LSC surface. Figure 14 shows a IR-free polarization curve for the Pt-LSC (● symbol) measured in air in comparison with those for the original (□: sintered at 1050 °C, without polymer) and the optimized LSC (○) cathodes. At $\eta = -0.05$ V and $T_{cell} = 800$ °C, the j at the optimized LSC is about 1.6 times higher than that of the original one. The Pt loading enhances the performance of the optimized cathode further especially at low T_{cell}. The current density of 1 A/cm^2 at $\eta = -0.05$ V was achieved at 800 °C in air, which is 2.5 times higher than that without Pt catalysts.

Based on these data described above, the IR-free terminal voltage (E_{cell}) for the SOFC with the Ni-SDC anode and Pt-LSC cathode was calculated to be 0.96 V at $j = 0.50$ A/cm^2 and $T_{cell} = 800$ °C, operated with humidified H_2 and air. The value of E_{cell} can

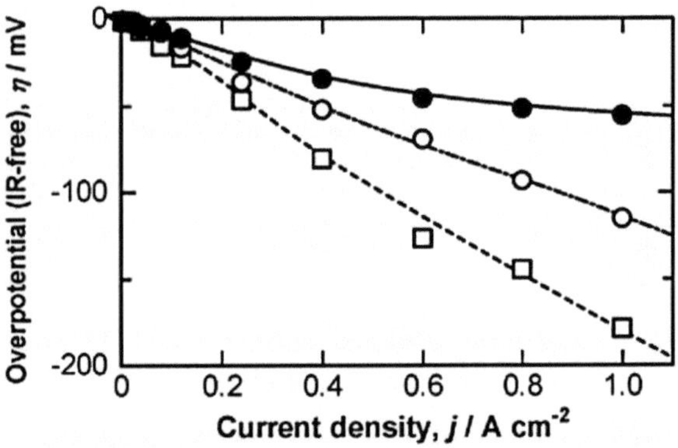

Figure 14. Polarization curves (IR-free, measured in air) at various LSC cathodes at $T_{cell} = 800$ °C. □: original one (sintered at 1050 °C without polymer), O: the optimized microstructure (sintered at 1050 °C with 1.0 wt% polymer), ●: Pt-catalyzed (0.1 mg-Pt/cm^2 loaded on the optimized one).

still be as high as 0.93 V when using the LSC cathode without any metal catalyst. The total ohmic resistance of Ni-SDC and LSC (or Pt-LSC) layers was about 0.1 Ω cm^2, resulting in a IR-loss of 0.05 V. If a thin YSZ electrolyte with the thickness of 25 μm were employed in our cell, one may achieve E_{cell} = 0.88 V at j = 0.50 A/cm^2 for the cell, Ni-SDC/YSZ(25 μm)/Pt-LSC, including 0.03 V of the electrolyte IR-loss. Such performances evaluated for a small test cell under low fuel utilization (3.2% at 0.5 A/cm^2) are important, because they provide a clear ultimate target.

IV. EFFECTS OF IONIC CONDUCTIVITY OF ZIRCONIA ELECTROLYTES ON THE POLARIZATION PROPERTIES OF VARIOUS ELECTRODES IN SOFCS

It is essential to investigate in detail the factors that control the polarization properties of electrodes in SOFCs. Many researchers have investigated electrochemical reaction kinetics or mechanisms at some metal or oxide electrodes interfaced with zirconia electrolytes. Although it has been demonstrated that the polarization loss at the electrodes can be reduced by using mixed conducting solid electrolytes,[15] or by introducing mixed conducting layers on zirconia surfaces,[41-43] no reports have been available on the effects of the purely-ionic conductivity of the electrolyte on the electrode polarization. The higher the ionic conductivity of the electrolyte, of course, the lower is the ohmic loss. We, for the first time, have found that the IR-free polarization of a platinum anode is greatly influenced by the ionic conductivity σ_{ion} of the zirconia electrolyte.[44] In the following Sections, we present the experimental results and our analysis of the mechanism at various electrodes/zirconia (doped with Y_2O_3, Yb_2O_3 or Sc_2O_3) interfaces at 800 to 1000°C.

1. Effect of σ_{ion} on the Hydrogen Oxidation Reaction Rate at Porous Pt Anode

The solid electrolytes employed were zirconia doped with yttria $(ZrO_2)_{1-X}(Y_2O_3)_X$ (X = 0.03, 0.04, 0.08; denoted as 3Y, 4Y, 8Y) or ytterbia $(ZrO_2)_{0.92}(Yb_2O_3)_{0.08}$ (denoted as 8Yb). These were confirmed to be dense (gas-tight) and purely ionic conductors. The σ_{ion} increases in the order 3Y < 4Y < 8Y < 8Yb.

Figure 15 shows IR-free polarization curves for H_2 oxidation at Pt anodes on various zirconia electrolytes. At T_{cell} between 1000 and 800°C, the η at a given current density increases with decreasing σ_{ion} of the electrolyte; the Pt anode on the 3Y-electrolyte with the lowest σ_{ion} exhibited the largest polarization.

Since linear relationships are seen between log j and η ranging from 0.1 V to 0.3 V in Fig. 15, each exchange current density j_0 is obtained by extrapolating the Tafel plot to $\eta = 0$.

$$\log (j / j_0) = \alpha_a\, n\, F\, \eta \,/\, 2.303\, RT \tag{10}$$

where j, j_0, α_a, and n are a current density at an overpotential η, the exchange current density, the anodic transfer coefficient, and the number of electrons transferred, respectively. F, R, and T have their usual meanings. At a given T_{cell}, all the Tafel lines with the various electrolytes have approximately the same slope (= 2.303 $RT/\alpha_a nF$), i.e., 125 mV/decade at 1000°C, 118 mV/decade at 900°C, and 104 mV/decade at 800°C, respectively. The values of log j_0 are plotted as a function of logarithm of the σ_{ion} of the electrolyte disks in Fig. 16. It must be emphasized that these plots correspond to the data obtained at different T_{cell} and different electrolytes. The solid line represents the least square fitting for all the data. The correlation factor of the line was 0.95 and the slope was 1.97. Therefore, it becomes clear that the exchange current density at Pt anode increases linearly with the square of the σ_{ion} for all of the zirconia electrolytes.

In order to clarify the effect of the σ_{ion} on the j_0 at the Pt anode, another experiment was carried out. We measured the IR-free polarization curves at Pt anodes on 8Y-electrolytes with different thicknesses of 0.5, 1.0 and 1.5 mm, where σ_{ion} is same but the conductances are different. The values of j_0 showed fairly good agreement with each other at a given T_{cell}. Hence, the essential factor controlling the j_0 is the σ_{ion} and not the total conductance along the thickness of the electrolytes after the pure ohmic drop of the electrolyte is corrected. In contrast, the value of $n\alpha_a$ calculated from the Tafel slope was approximately two for all the cells, suggesting that the rate-determining step is independent of the σ_{ion} or T_{cell}.

These observations have been well explained kinetically,[44] i.e., the anode performance is controlled by the rate of O^{2-} supply to the anode layer via the electrolyte, because the transfer rate of O^{2-} at the electrode/zirconia interface is proportional to the σ_{ion} of zirconia electrolyte as described later. We will discuss the mechanism in the next Section together with that of various electrodes.

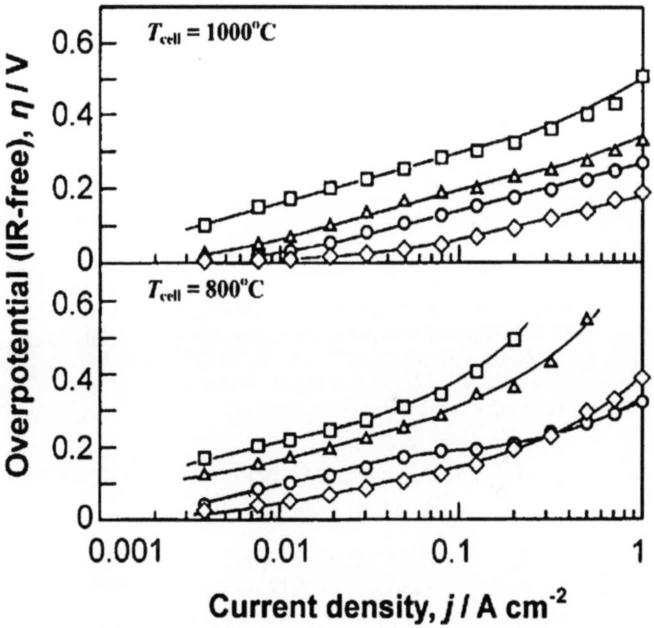

Figure 15. Polarization curves (IR-free) for hydrogen oxidation at Pt anodes attached to various zirconia solid electrolytes measured in humidified hydrogen ($p[H_2O]$ = 0.04 atm). Zirconia electrolyte; ◊: 8Yb, O: 8Y, Δ: 4 Y, □ : 3Y. Reprinted from Ref. 44, Copyright (1995), American Chemical Society.

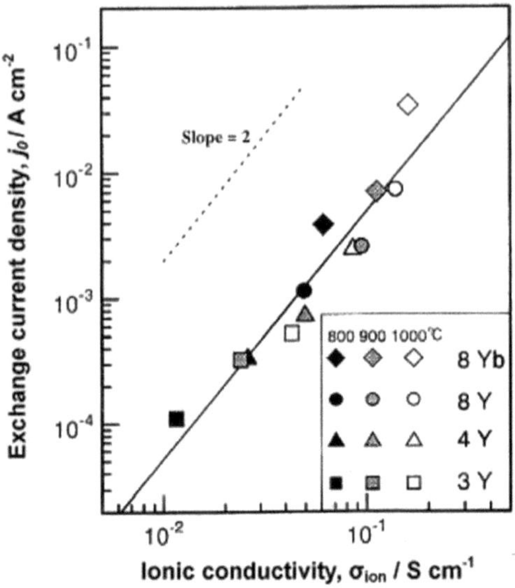

Figure 16. Plots of the exchange current density j_0 for hydrogen oxidation (in humidified H_2, $p[H_2O] = 0.04$ atm) at Pt anode against ionic conductivity σ_{ion} of zirconia solid electrolyte measured. Solid line is the least square fitting for all the data. Reprinted from Ref. 44, Copyright (1995) American Chemical Society.

2. Effect of σ_{ion} on Activities of Various Electrodes and the Reaction Mechanism

We have also found that the catalytic activities j_0 at various anodes and cathodes depend on the σ_{ion} of ZrO_2-electrolytes.[45, 46] Figure 17 shows such dependencies of j_0 at various electrodes on the σ_{ion} of ZrO_2-electrolytes (doped with Y_2O_3, Yb_2O_3 or Sc_2O_3). Results are summarized as follows;

1. As shown in Fig. 17A, the value of the exchange current density j_0 of SDC anodes was not influenced by the σ_{ion} at 900 and 1000 °C, whereas the j_0 increased proportionally to σ_{ion} at a lower T_{cell} of 800 °C. However, when the SDC was activated by the dispersion of nm-sized Ru catalysts, the j_0

increased proportionally to σ_{ion} in the whole T_{cell} range between 800 and 1000 °C.[45]

2. A platinum cathode exhibited very similar dependence of j_0 on the σ_{ion} to that of SDC anode without any catalyst. The j_0 was not influenced by the σ_{ion} at 900 and 1000 °C but increased proportionally to σ_{ion} only at the lower T_{cell} of 800 °C (Fig. 17B).[46]
3. The j_0 at LSM cathodes increased with the σ_{ion} at T_{cell} = 800 to 1000 °C (Fig. 17C). The dispersion of nm-sized Pt catalysts on LSM particles greatly enhances the performance, the magnitude of which depends on T_{cell}, the σ_{ion}, and the microstructure of LSM.[46]

In order to explain the experimental results for the effect of σ_{ion} on the electrocatalytic activity, we discuss the reaction mechanism. As a typical case, we consider the oxygen reduction reaction at the Pt cathode. By means of various experimental techniques, it has been well recognized that ERZ for the Pt/zirconia interface is restricted to the portion around the physical TPB. The cathodic reaction consists of the following elementary steps:

Diffusion of O_2(gas) in gas phase through porous layer (11)

Dissociative adsorption of O_2 to form O_{ad} on Pt :

$$O_2(\text{gas}) \rightarrow 2\, O_{ad}(\text{Pt}) \tag{12}$$

Surface diffusion of O_{ad} to TPB :

$$O_{ad}(\text{Pt}) \rightarrow O_{ad}(\text{TPB}) \tag{13}$$

Charge transfer at TPB :

$$O_{ad}(\text{TPB}) + V_o(\text{zirconia}) + 2\, e^-(\text{Pt}) \rightarrow O^{2-}(\text{zirconia}) \tag{14}$$

where V_o(zirconia) is an oxygen vacancy mobile in zirconia electrolyte. Note that we do not use Kröger-Vink notation, i.e., an oxide ion has two negative charges (O^{2-}) and an oxygen vacancy V_o is regarded as a neutral vacant site without having any electric charge.

The j_0 for the reaction (14) is written as,

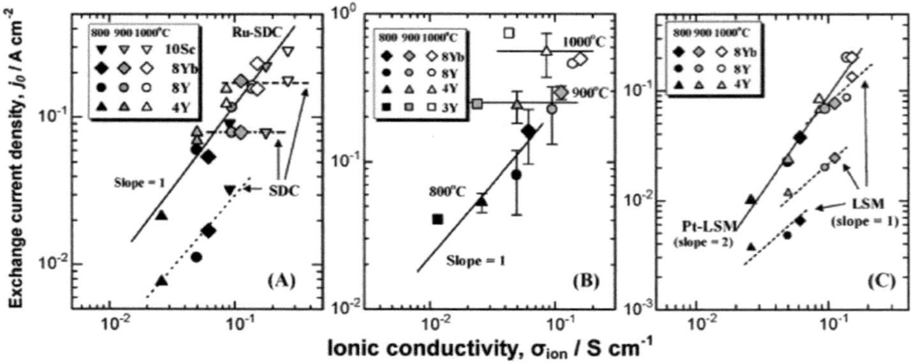

Figure 17. Dependencies of exchange current density j_0 at various electrodes on the ionic conductivity σ_{ion} of various zirconia solid electrolytes; (A) SDC (dashed, dash-dotted, and dotted lines) and Ru (0.5 mg/cm^2)-dispersed SDC anodes in humidified H_2. Reprinted from Ref. 45, Copyright (1997), with permission from The Electrochemical Society, (B) Pt cathodes in O_2, (C) $La_{0.85}Sr_{0.15}MnO_3$ (LSM, dashed lines) and Pt (0.1 mg/cm^2)-dispersed LSM cathodes in O_2. Each j_0 value was calculated from the polarization resistance (R_p, Ω cm^2), since linear relationships were observed between η and j for $\eta < 0.1$ V at all the electrodes and T_{cell} between 800 and 1000°C: $j_0 = (RT/nF)\,R_p^{-1}$. Reproduced by Ref. 46, Copyright (1999), by permission from The Electrochemical Society.

$$j_0 = 2\,F\,k^{\mathrm{o}}\,[\mathrm{O_{ad}}]^{\alpha}[\mathrm{V_o}]^{\beta} \tag{15}$$

where k^{o} is the standard rate constant, and $[O_{ad}]$ and $[V_o]$ are the activities of O_{ad} and V_o, respectively. α and β are the reaction order for the corresponding species and presumably take a value of zero or unity. On the other hand, the σ_{ion} of zirconia electrolyte is expressed as,

$$\sigma_{\mathrm{ion}} = 2\,F\,\mu\,N(\mathrm{V_o}) \tag{16}$$

where μ is the ionic mobility and $N(V_o)$ is the concentration of oxygen vacancy in the electrolyte. It must be noted that $[V_o]$ in Eq. (15) is not equal to $N(V_o)$ in Eq. (16). The value of $N(V_o)$ is usually calculated from the dopant concentration with a simple defect equilibrium in the solid. For example, $N(V_o)$ in 8Y (8 mol% Y_2O_3-doped zirconia) is equal to that in 8Yb, i.e., $N(V_o) = 8$ mol%. However, σ_{ion} of 8Yb is higher than 8Y at all temperatures. This is explained as the difference in μ. We denote by $[V_o]$ the effective activity of V_o, which takes part in the reaction (14) and also includes a factor of μ. Thus, σ_{ion} is proportional to $[V_o]$. The last term $[V_o]^{\beta}$ in Eq. (15) has never been taken into account except in our series of studies. The present result on Pt cathode indicates that β was zero at high T_{cell} of 900 and 1000 °C. If step (14) is fast compared to the preceding steps, the j_0 at the Pt cathode must be independent of σ_{ion} or the transfer rate of O^{2-}. It has been reported that the rate determining step (*rds*) at the Pt cathode is either step (12) or (13).[47-51] Therefore, our result at T_{cell} = 900 and 1000 °C is consistent with that mechanism in previous literature. In contrast, at T_{cell} = 800 °C, β was found to be unity although the value of α is not known because $[O_{ad}]$ could be kept constant at a constant temperature for a given Pt electrode material in oxygen atmosphere. Thus, the transport rate of O^{2-} at the interface controls the exchange current density j_0.

At the mixed conducting LSM cathode, the O_{ad} can be ionized at the LSM surface besides the TPB, i.e., mixed-conducting path and TPB path in parallel. The contributions of the two pathways to the overall reaction might depend on the number of active sites on the LSM or on the microstructure. Mizusaki et al.[52] reported that the reaction involving O_{ad} occurred dominantly around TPB in porous $La(Ca)MnO_3$ (LCM) cathode. This explanation seems to be valid, because the ionic conductivity in bulk LSM is considerably lower than the electronic conductivity.[53,54] However, as shown in Fig. 10 and Fig. 17C, the electrocatalytic activity of LSM is significantly enhanced by loading a small amount of Pt catalysts. As discussed in Section III.1 or III.4, nm-sized Pt particles presumably promote the following cathodic reaction step(s),

$$O_2(\text{gas}) \rightarrow 2\ O_{ad}(\text{Pt-LSM}) \tag{17}$$

$$O_{ad}(\text{Pt-LSM}) + 2\ e^-\ (\text{Pt-LSM}) \rightarrow O^{2-}(\text{LSM}) \tag{18}$$

where (Pt-LSM) denotes an active site formed at the boundary of the Pt catalyst and the LSM surface. It is noticed that the activation of step (17) and/or step (18) certainly accelerates both rates of TPB path and of mixed-conducting path. Thus, the cathode performance is controlled by the transport rate of O^{2-} at the LSM/zirconia interface when the surface reaction rate is sufficiently high. A similar explanation is possible for the change in the dependence of j_0 on the σ_{ion} by dispersing Ru catalysts on the SDC anode (Fig. 17A).

In summary, it has been shown by the work surveyed here that the high $\sigma_{\rm ion}$ in the electrolyte reduces not only the ohmic loss but also the polarization losses at the electrodes. Consequently, the combination of a solid electrolyte having high $\sigma_{\rm ion}$ and of electrodes consisting of good mixed conductors loaded with ultra-small catalysts is very important to achieve high performance SOFCs which can operate at low temperatures.

V. CONCLUSION

Catalyzed-reaction layers based on our new design concepts have worked effectively in the medium temperature operating SOFCs with yttria stabilized zirconia solid electrolytes. The use of mixed ionic and electronic conducting oxides having high $\sigma_{\rm ion}$ and $\sigma_{\rm e}$ as the porous electrode material is important for effective activation by highly dispersed electrocatalysts on the surface. The control of microstructure of the porous mixed conducting electrode layer was also found to be very important to enhance the performance, especially at low operating temperatures. The essential factors are a high gas-diffusion rate in the porous electrodes as well as a low ohmic resistance because both factors greatly contribute to an enhanced effective reaction zone.

It is very striking that the high σ_{ion} in the electrolyte reduces not only the ohmic loss but also the polarization losses at the electrodes. This gives an important strategy to achieve high performance SOFCs that can operate at low temperatures, i.e., the combination of a solid electrolyte having high σ_{ion} and a high performance electrode.

Our research results indeed contribute to the development of medium temperature operating SOFCs. For example, doped ceria has been employed in place of YSZ in the cermet-type anodes by many research groups.[27-29] Effects of σ_{ion} on the activities of some electrodes have also been supported.[55,56]

The catalyzed reaction layers can be, of course, applied to other solid electrolyte devices such as steam electrolysis, gas separation, electrosynthesis and gas sensors. Recently, we have found that a solid oxide electrolysis cell (SOEC) using the same component materials as the SOFC (Ni-dispersed SDC cathode, LSC anode with SDC interlayer, and YSZ electrolyte) produced hydrogen with a small IR-free cell voltage = 1.20 V at 0.50 A/cm^2 and 900 °C under an atmosphere of $H_2 + H_2O$ ($P[H_2] = 0.6$ atm, $P[H_2O] = 0.4$ atm) and O_2 (1 atm).[57] The performance of the SOEC was improved further by employing Sc-SZ electrolyte with higher σ_{ion}, similar to the case of SOFC electrodes.[58] Such a "reversible" cell can be regarded as a reciprocating direct energy converter between hydrogen and electricity with high conversion efficiency.

REFERENCES

1. T. Ishihara, K. Sato, Y. Mizuhara, and Y. Takita, *Chem. Lett.* 942 (1992).
2. N. Q. Minh, *J. Am. Ceram. Soc.* **76** (1993) 563.
3. M. Suzuki, H. Sasaki, S. Otoshi, A. Kajimura, N. Sugiura, and M. Ippommatsu, *J. Electrochem. Soc.* **141** (1992) 1928.
4. N. Q. Minh, in *Proc. 4th Intern. Sympo. Solid Oxide Fuel Cells* (June 1995, Yokohama, Japan), M. Dokiya, O. Yamamoto, and S. C. Singhal, Editors, PV **95-1**, p. 138, The Electrochemical Society Proceedings Series, Pennington, NJ (1995).
5. T. Tsai and S. A. Barnett, *Solid State Ionics* **93** (1997) 207.
6. S. Souza, S. J. Visco, and L. C. De Jonghe, *J. Electrochem. Soc.* **144** (1997) L35.

[7]J-W. Kim, A. V. Virkar, K-Z. Fung, K. Mehta, and S. C. Singhal, *J. Electrochem. Soc.* **146** (1999) 69.
[8]Y. Mizutani, M. Tamura, M. Kawai, and O. Yamamoto, *Solid State Ionics* **72** (1994) 271.
[9]T. Ishii and Y. Tajima, *J. Electrochem. Soc.* **141** (1994) 3450
[10]T. Ishihara, H. Matsuda, and Y. Takita, *J. Am. Chem. Soc.* **116** (1994) 3801.
[11]H. Yokokawa, N. Sakai, T. Horita, K. Yamaji, and M. E. Brito, *Electrochemistry* **73** (2005) 20.
[12]M. Watanabe, Japan Patent 4-253163 (1992).
[13]M. Watanabe, H. Uchida, M. Shibata, M. Mochizuki, and K. Amikura, *J. Electrochem. Soc.* **141** (1994) 342.
[14]H. Uchida, N. Mochizuki, and M. Watanabe, *J. Electrochem. Soc.* **143** (1996) 1700.
[15]T. Takahashi, H. Iwahara and I. Ito, *Denki Kagaku* (Presently *Electrochemistry*) **38** (1970) 509.
[16]Y. Takeda, R. Kanno, M. Noda, Y. Tomida, and O. Yamamoto, *J. Electrochem. Soc.* **134** (1987) 2656.
[17]H. Yahiro, Y. Eguchi, K. Eguchi, and H. Arai, *J. Appl. Electrochem.* **18** (1988) 527.
[18]M. J. Saeki, H. Uchida, and M. Watanabe, *Catal. Lett.* **26** (1994) 149.
[19]H. Uchida, H. Suzuki, and M. Watanabe, *J. Electrochem. Soc.* **145** (1998) 615.
[20]H. Uchida, T. Osuga, and M. Watanabe, *J. Electrochem. Soc.* **146** (1999) 1677.
[21]M. Mogensen, T. Lindegaard, U. R. Hansen, and G. Mogensen, *J. Electrochem. Soc.* **141** (1994) 2122.
[22]O. A. Marina, C. Bagger, S. Primdahl, M. Mogensen, *Solid State Ionics* **123** (1999) 199.
[23]H. Uchida, S. Suzuki, M. Watanabe, *Electrochem. Solid-State Lett.* **6** (2003) A174.
[24]S. Suzuki, H. Uchida, and M. Watanabe, *Electrochemistry* **73** (2005) 128.
[25]S. Suzuki, H. Uchida, and M. Watanabe, *Solid State Ionics* **177** (2006) 359.
[26]T. Setoguchi, K. Okamoto, K. Eguchi, and H. Arai, *J. Electrochem. Soc.* **139** (1992) 2875.
[27]R. Maric, S. Ohara, T. Fukui, T. Inagaki, and K. Miura, *Electrochem. Solid-State Lett.* **1** (1998) 201.
[28]T. Ishihara, T. Shibayama, H. Nishiguchi, and Y. Takita, *Solid State Ionics* **132** (2000) 209.
[29]J. Liu, B. D. Madsen, Z. Ji, and S. A. Barnett, *Electrochem. Solid-State Lett.* **5** (2002) A122.
[30]H. Uchida, A. Tsuno, M. Watanabe, *Denki Kagaku* (in English, presently *Electrochemistry*) **64** (1996) 686.
[31]J. Van herle, A. J. McEvoy, and K. R. Thampi, *Electrochim. Acta* **39** (1994) 1675.
[32]M. Liu and A. Khandkar, *Solid State Ionics* **52** (1992) 3.
[33]Y. Ohno, S. Nagata, and H. Sato, *Solid State Ionics* **3/4** (1981) 439.
[34]O. Yamamoto, Y. Takeda, R. Kanno, and M. Noda, *Solid State Ionics* **22** (1987) 241.
[35]M. Godickemeier and L. J. Gauckler, *J. Electrochem. Soc.* **145** (1998) 414.
[36]H. Uchida, S. Arisaka, and M. Watanabe, *Electrochem. Solid-State Lett.* **2** (1999) 428.
[37]H. Uchida, S. Arisaka, M. Watanabe, *Solid State Ionics* **135** (2000) 347.
[38]Y.Teraoka, H-M. Zhang, S. Fukuoka, and A. Yamazoe, *Chem. Lett.* (1985) 1743.
[39]L-W. Tai, M. M. Narallah, H. U. Anderson, D. M. Sparlin, and S. R. Sehlin, *Solid State Ionics* **76** (1995) 261.
[40]H. Uchida, S. Arisaka, M. Watanabe, *J. Electrochem. Soc.* **149** (2002) A13.

[41]B. C. Nguyen, T. A. Liu, and D. M. Mason, *J. Electrochem. Soc.* **133** (1986) 1807.
[42]E. J. L. Shouler and M. Kleitz, *J. Electrochem. Soc.* **134** (1987) 1045.
[43]M. P. Van Duk, K. J. De Vries, and A. J. Burggraaf, *Solid State Ionics* **21** (1986) 73.
[44]H. Uchida, M. Yoshida, and M. Watanabe, *J. Phys. Chem.* **99** (1995) 3282.
[45]M. Watanabe, H. Uchida, and M. Yoshida, *J. Electrochem. Soc.* **144** (1997) 1739.
[46]H. Uchida, M. Yoshida, and M. Watanabe, *J. Electrochem. Soc.* **146** (1999) 1.
[47]T. M. Gür, J. D. Raistrick, and R. A. Huggins, *J. Electrochem. Soc.* **127** (1980) 2620.
[48]J. Sasaki, J. Mizusaki, S. Yamauchi, and K. Fueki, *Bull. Chem. Soc. Jpn.* **54** (1981) 1688.
[49]M. J. Verkerk, M. W. J. Hammink, and A. J. Burggraaf, *J. Electrochem. Soc.* **130** (1983) 70.
[50]H. Okamoto, G. Kawamura, and T. Kudo, *Electrochim. Acta* **28** (1983) 379.
[51]J. Mizusaki, K. Amano, S. Yamauchi, and K. Fueki, *Solid State Ionics* **22** (1987) 313.
[52]J. Mizusaki, H. Tagawa, K. Tsuneyoshi, and A. Sawata, *J. Electrochem. Soc.* **138** (1991) 1867.
[53]S. Carter, A. Selcuk, R. J. Chater, J. Kajda, J. A. Kilner, and B. C. H. Steele, *Solid State Ionics* **53-56** (1992) 597.
[54]J. Mizusaki, T. Saito, and H. Tagawa, *J. Electrochem. Soc.* **143** (1996) 3065.
[55]N. Nakagawa, K. Nakajima, M. Sato, and K. Kato, *J. Electrochem. Soc.* **146** (1999) 1290.
[56]T. Kenjo and Y. Kanehira, *Solid State Ionics* **148** (2002) 1.
[57]H. Uchida, N. Osada, and M. Watanabe, *Electrochem. Solid-State Lett.* **7** (2004) A500.
[58]N. Osada, H. Uchida, and M. Watanabe, *J. Electrochem. Soc.* **153** (2006) A816.

3

Electrochemical CO_2 Reduction on Metal Electrodes

Y. Hori

Faculty of Engineering, Chiba University, Chiba 263-8522, Japan

I. INTRODUCTION

Carbon dioxide is a potential carbon resource abundant on earth. It is also a green house gas with a rapidly increasing atmospheric concentration during the last two centuries. Chemical fixation of CO_2 is an attractive technique for utilization of carbon resources, as well as for the reduction of the atmospheric concentration of CO_2. Nevertheless, CO_2 is the stablest among carbon based substances under the environmental conditions. It has not been incorporated as a major industrial material.

Carbon dioxide can be electrochemically reduced to useful products under mild conditions. However, the energy conversion efficiency, defined as the ratio of the free energy of the products obtained in electrochemical CO_2 reduction and that consumed in the reduction, would be roughly 30 to 40%.[1, 2] Such a low efficiency may discourage practical application of CO_2 reduction in the very near future. However, the significance of the CO_2 reduc-

Modern Aspects of Electrochemistry, Number 42, edited by C. Vayenas et al., Springer, New York, 2008.

tion will be reevaluated in various aspects of science in future, since many interesting facts in basic electrochemistry have been revealed by recent studies. At the moment, electrochemical reduction of CO_2 is an interesting as well as an important topic in chemistry.

Fundamental studies of the electrochemical reduction of CO_2 started as early as the 19th century, and have been further promoted by many workers in recent years. Numerous reports have been published in the last two decades. At the same time, confusions and contradictions are found in this field.

This article attempts to provide a review of CO_2 reduction at metal electrodes. Several review papers have already appeared, which contain the history and classical works of CO_2 reduction.[3-13] Thus this paper will not repeat the historical aspects of CO_2 reduction. This chapter will review recent progress, putting emphasis on basic problems and particularly on electrocatalytic aspects.

II. FUNDAMENTAL PROBLEMS

1. Reactions Related with CO_2 Reduction

(*i*) *Electrochemical Equilibria*

The chemical reactivity of CO_2 is low. However, the equilibrium potentials of CO_2 reduction are not very negative as compared with that of the hydrogen evolution reaction (HER) in aqueous electrolyte solutions. For example, electrochemical reduction of CO_2 to $HCOO^-$ in aqueous solution is given below together with the standard electrode potential at pH 7.0 at 25°C with respect to the standard hydrogen electrode (SHE).

$$CO_2 + H_2O + 2e^- = HCOO^- + OH^- \quad (-0.43 \text{ V vs. SHE at pH 7.0}) \tag{1}$$

The standard electrode potential of HER at pH 7.0 is –0.414 V vs. SHE at 25°C. The CO_2 reduction in aqueous solution involves H_2O and OH^-, and the equilibrium potential varies in accordance

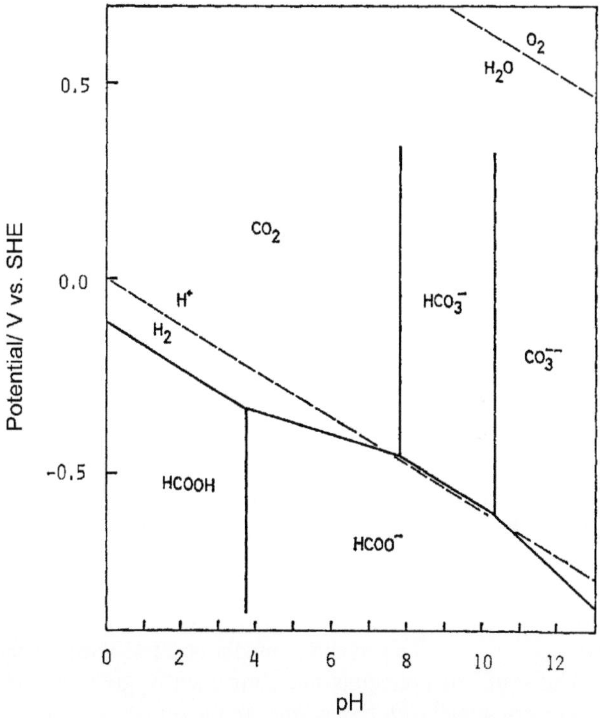

Figure 1. pH potential diagram of CO_2 and its related substances. pH potential relations for water are shown in broken lines. Reprinted with permission from Ref. 15, Copyright (1982) Chemical Society of Japan.

with the pH of the electrolyte. Equilibrium potential vs. pH relations (Pourbaix diagram)[14] can be constructed on the basis of thermodynamic data, as shown in Fig. 1 as an example for formation of HCOOH from CO_2 at 25°C.[15] Thermodynamically stable regions of CO_2 and HCOOH related species are shown with respect to pH and potential. The boundaries of regions are determined with the activity of each species unity, or activity ratio such as HCO_3^-/CO_3^{2-} unity and so on. Thus $HCOO^-$ is predominantly produced in neutral pH region instead of HCOOH from CO_2. Equilibrium relationships of H_2O and H^+ or O_2 are also given with the partial pressures of H_2 and O_2 at 1 atm; the equilibrium potential of CO_2 reduction is in the same range as HER in aqueous media.

Nevertheless, CO_2 reduction does not take place easily, and the actual electrolysis potentials for CO_2 reduction are much more negative in most cases than the equilibrium ones. The reason is that the intermediate species $CO_2\cdot^-$, formed by an electron transfer to a CO_2 molecule, proceeds as the first step at highly negative potential, such as –2.21 V vs. saturated calomel electrode (SCE) measured in dimethyl formamide (DMF), as discussed later in detail.

The standard potentials for CO_2 reduction and HER, estimated from thermodynamic data, decrease similarly with pH as is apparent in Fig. 1. However, the actual potential of CO_2 reduction does not depend on pH as described later, whereas that of HER is proportional to proton activity. Thus HER prevails over CO_2 reduction in acidic solutions, and most studies of CO_2 reduction in aqueous media have been made in neutral pH region.

Such pH potential diagrams may be given for all the products of CO_2 reduction. Only the standard electrode potentials of CO_2 reduction are given below for formation of CO, CH_4, C_2H_4, C_2H_5OH, and C_3H_7OH for brevity. The values are estimated from thermodynamic data,[16] in aqueous media at 25°C with respect to SHE. The standard potentials are conveniently given at pH 7.0, where most of actual CO_2 reductions are measured.

$$CO_2 + H_2O + 2e^- = CO + 2OH^- \qquad (-0.52\text{ V}) \tag{2}$$

$$CO_2 + 6H_2O + 8e^- = CH_4 + 8OH^- \qquad (-0.25\text{ V}) \tag{3}$$

$$2CO_2 + 8H_2O + 12e^- = C_2H_4 + 12OH^- \qquad (-0.34\text{ V}) \tag{4}$$

$$2CO_2 + 9H_2O + 12e^- = C_2H_5OH + 12OH^- \qquad (-0.33\text{ V}) \tag{5}$$

$$3CO_2 + 13H_2O + 18e^- = C_3H_7OH + 18OH^- \qquad (-0.32\text{ V}) \tag{6}$$

(*ii*) Equilibria of CO_2 Related Species in Aqueous Solution

Thermodynamic equilibria of CO_2 related species in aqueous media are important in CO_2 chemistry, and are summarized briefly as follows. An equilibrium of gaseous CO_2 with HCO_3^- is

$$CO_2 + H_2O = HCO_3^- + H^+ \tag{7}$$

The pH is related to activities of other substances, $[HCO_3^-]$ and $[CO_2]$,

$$\mathrm{pH} = \mathrm{p}K_{a1} + \log [HCO_3^-] - \log [CO_2] \qquad \mathrm{p}K_{a1} = 6.35\ (25°C) \tag{8}$$

where K_{a1} is the equilibrium constant of the reaction (7). Since $[CO_2]$ is related to the partial pressure of CO_2 with the Henry constant, h ($=3.38 \times 10^{-2}$ mol l^{-1} atm^{-1}, at 25°C),

$$\begin{aligned}\mathrm{pH} &= \mathrm{p}K_{a1} - \log h + \log [HCO_3^-] - \log P(CO_2) \\ &= 7.82 + \log [HCO_3^-] - \log P(CO_2)\end{aligned} \tag{9}$$

Another equilibrium in the aqueous solution is,

$$HCO_3^- = CO_3^{2-} + H^+ \tag{10}$$

The pH is determined by $[CO_3^{2-}]/[HCO_3^-]$ with K_{a2} the equilibrium constant of the reaction (10),

$$\mathrm{pH} = \mathrm{p}K_{a2} + \log \{[CO_3^{2-}]/[HCO_3^-]\} \qquad \mathrm{p}K_{a2} = 10.33\ (25°C) \tag{11}$$

$P(CO_2)$ is also given in combination of Eqs. (9) and (11),

$$\begin{aligned}\log P(CO_2) &= \mathrm{p}K_{a1} - \mathrm{p}K_{a2} - \log h \\ &\quad + \log [HCO_3^-] - \log \{[CO_3^{2-}]/[HCO_3^-]\} \\ &= -2.51 + \log [HCO_3^-] - \log \{[CO_3^{2-}]/[HCO_3^-]\}\end{aligned} \tag{12}$$

The pH values for equilibrium (7) in given $[HCO_3^-]$ solutions under $P(CO_2)$ 1 atm are presented in Table 1 as estimated from Eq. (9). Table 2 tabulates pHs and partial pressures of CO_2 for solutions in equilibrium (10), calculated from Eqs. (11) and (12) for given $[CO_3^{2-}]/[HCO_3^-]$ and $[HCO_3^-]$. Activity coefficient of

Table 1
Values of pH of Solutions of Given Activity [HCO_3^-] under $P(CO_2)$ = 1 atm

[HCO_3^-]/ mol dm^{-3}	1	0.5	0.1	0.05	0.01
pH of the solution	7.82	7.52	6.82	6.52	5.82

HCO_3^- $\gamma(HCO_3^-)$ is related with the activity [HCO_3^-] and the concentration $C(HCO_3^-)$,

$$[HCO_3^-] = \gamma(HCO_3^-) \cdot C(HCO_3^-) \tag{13}$$

Average activity coefficients for HCO_3^- and Na^+ in mixed solutions composed of $NaHCO_3$ and Na_2CO_3, reported by Han and

Table 2
Values of $P(CO_2)$ of Solutions of Given Activity [HCO_3^-] and Given [CO_3^{2-}]/[HCO_3^-] Ratio

[HCO_3^-] = 1.0					
	[CO_3^{2-}]/[HCO_3^-]				
	1	0.5	0.1	0.05	0.01
pH of solution	10.33	10.03	9.33	9.03	8.33
$P(CO_2)$/ atm	0.0031	0.0062	0.031	0.062	0.31

[HCO_3^-] = 0.5					
	[CO_3^{2-}]/[HCO_3^-]				
	1	0.5	0.1	0.05	0.01
pH of solution	10.33	10.03	9.33	9.03	8.33
$P(CO_2)$/ atm	0.0015	0.0031	0.015	0.031	0.155

[HCO_3^-] = 0.1					
	[CO_3^{2-}]/[HCO_3^-]				
	1	0.5	0.1	0.05	0.01
pH of solution	10.33	10.03	9.33	9.03	8.33
$P(CO_2)$/ atm	0.0003	0.0006	0.0031	0.0062	0.031

Bernardin, are 0.568 for a mixture of 0.5-M $NaHCO_3$ and 0.1 M Na_2CO_3 (denoted as 0.5-0.1), and 0.585 for (0.5-0.05) at 25°C for example.[17]

Various electrolyte solutions are employed for CO_2 electrolysis, not only HCO_3^- solutions. Since CO_2 cannot be present in basic aqueous solutions, CO_2 reduction has been studied with neutral to acidic solutions. It is apparent that aqueous 0.1 M KOH is converted to 0.1 M $KHCO_3$ after saturation and equilibration with CO_2 gas, and that aqueous 0.1 M K_2CO_3 to 0.2 M $KHCO_3$. If a solution with buffer ability, such as phosphate buffer solution, is employed, the problem is more complicated. The pH of a phosphate buffer solution changes from the value of the originally prepared one by equilibration with CO_2, forming HCO_3^-. In addition to the equilibrium between CO_2 related species, another buffer reaction must be taken into account.

$$HPO_4^{2-} + CO_2 = H_2PO_4^- + HCO_3^- \tag{14}$$

The pH is determined by the following two equilibria.

$$CO_2 + H_2O = HCO_3^- + H^+ \tag{15}$$

and

$$H_2PO_4^- = HPO_4^{2-} + H^+ \tag{16}$$

The equilibrium concentrations of respective ions are readily calculated in accordance with the equations above and pK_a values.[15]

(*iii*) Variation of pH at the Electrode During CO_2 Reduction

Equilibrium values of pH and CO_2 concentration in the electrolyte are estimated in accordance with the procedure described in Section II.1(*ii*). Nevertheless, the electrolyte solution close to the electrode is not necessarily in equilibrium during electrolysis.

OH^- is generated at the electrode, when H_2O molecule is involved with a cathodic reduction. For example,

$$CO_2 + 6H_2O + 8e^- = CH_4 + 8OH^- \tag{17}$$

The rate of neutralization between OH^- and CO_2 is slow in aqueous solution at the ambient temperature, as is well known.[18] Hence the pH adjacent to the electrode becomes higher than that of the bulk solution. CO_2 molecules are present in more or less nonequilibrium high pH region at the electrode/electrolyte interface. The enhancement of pH is more significant in solutions, such as K_2SO_4, KCl and $NaClO_4$, which do not release protons. Anions with buffer ability can neutralize OH^-, mitigating the pH enhancement at the electrode.

$$OH^- + HCO_3^- = H_2O + CO_3^{2-} \tag{18}$$

$$OH^- + HPO_4^{2-} = H_2O + PO_4^{3-} \tag{19}$$

If one compares two electrolyte solutions, a phosphate buffer solution and a KCl one with the same pH in the bulk solution, the pH at the electrode/ electrolyte interface in a phosphate buffer solution is actually lower than in a KCl one. Such a difference in the pH enhancement, depending on anions, leads to the anionic effect in CO_2 reduction,[19] as discussed in Section VIII.3.

The problem is quantitatively solved by analysis of the transport process in the electroreduction of CO_2 in an electrolyte solution, e. g. aqueous $KHCO_3$.[20] When the electrolyte solution is steadily stirred, a diffusion layer is formed close to the electrode. No convective flux is present within the diffusion layer. CO_2 diffuses to the electrode across the diffusion layer, whereas OH^-, HCO_3^- and CO_3^{2-} will move oppositely from the electrode to the bulk of the solution. The concentrations of the chemical species will be uniform and constant outside the the diffusion layer.

The following chemical reactions proceed in the diffusion layer:

$$CO_2 + H_2O = H_2CO_3 \quad \text{(slow)} \tag{20}$$

$$CO_2 + OH^- = HCO_3^- \quad \text{(slow)} \tag{21}$$

$$H_2CO_3 + OH^- = HCO_3^- + H_2O \quad \text{(instantaneous)} \tag{22}$$

$$HCO_3^- + OH^- = CO_3^{2-} + H_2O \text{ (instantaneous)} \tag{23}$$

One can denote k_{1r} and k_{1d} for the rate constants of the forward and backward reactions of (20), and k_{2r} and k_{2d} for those of reaction (21), respectively. Reactions (22) and (23) are instantaneous, and in equilibria. The equilibrium constant K_1 is for reaction (22), and K_2 for reaction (23), where Cs' are the concentrations of the chemical species,

$$K_1 = \frac{C(HCO_3^-)}{C(H_2CO_3)C(OH^-)} \tag{24}$$

$$K_2 = \frac{C(CO_3^{2-})}{C(HCO_3^-)C(OH^-)} \tag{25}$$

A set of Fick's diffusional equations are applied to the case, linear diffusion normal to the electrode with simultaneous chemical reactions,

$$\frac{\partial C(m)}{\partial t} = D_i \frac{\partial^2 C(m)}{\partial x^2} + V(m) \tag{26}$$

where $C(m)$s are concentrations with m = CO_2, HCO_3^-, CO_3^{2-}, or OH^-, respectively; $x = 0$ at the electrode surface. $V(m)$s denote the rates of formation of each chemical species as above. $V(CO_2)$ is given below.

$$V(CO_2) = k_{1d}C(H_2CO_3) + k_{2d}C(HCO_3^-) - k_{1r}C(CO_2) - k_{2r}C(CO_2)C(OH^-) \tag{27}$$

The Eq. (26) is solved for the steady state,

$$\frac{\partial C(m)}{\partial t} = 0 \tag{28}$$

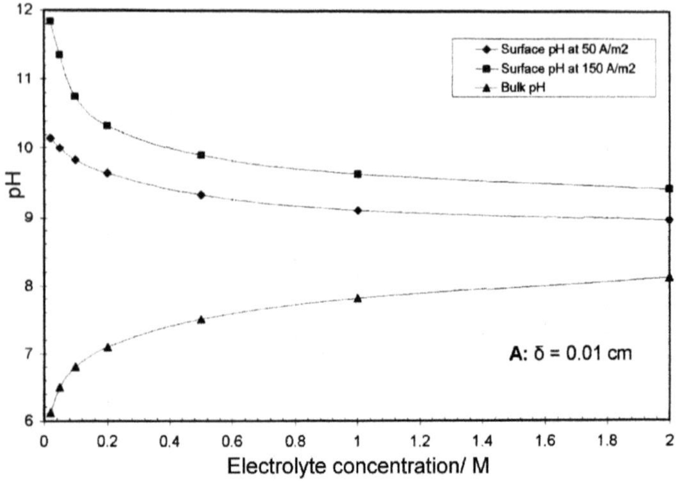

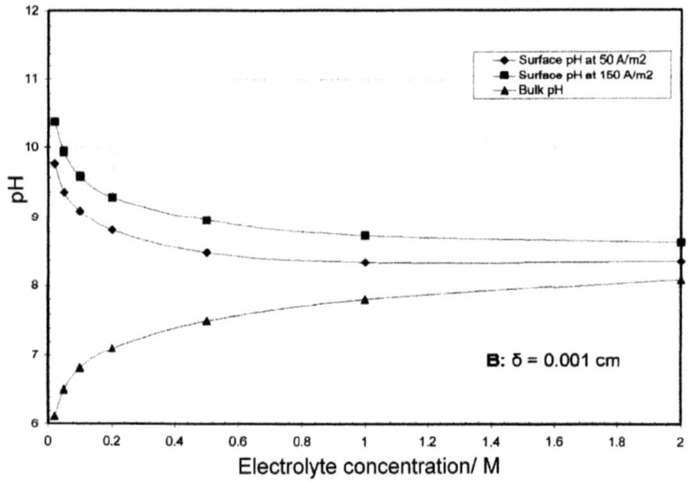

Figure 2. Calculated pH at cathode vs. electrolyte HCO_3^- concentration with the diffusion layer thickness $\delta = 0.01$ cm and 0.001 cm. Two different current densities are assumed. The pH of the bulk solution is also indicated. Reproduced from Ref. 21, Copyright (2005) with kind permission of Springer Science and Business Media.

The set of equations are solved with relevant boundary conditions, such as *C(m)*s at the outside surface of the diffusion layer, $x = \delta$, are equal to the values in the bulk solution in equibrium. Since the equations cannot be solved analytically, numerical solution was made by calculus of finite differences.[20] Gattrell et al. treated the problem more rigorously, and recently published the results.[21] Figure 2 shows the variation of pH at the electrode surface with the concentration of the electrolyte HCO_3^-; the pH values are given for two different thicknesses of the diffusion layer and two constant current electrolysis at 5 and 15 mA cm^{-2}.

2. Problems Related with Experimental Procedures and Data Analysis

(*i*) *Difference Current Obtained from Voltammetric Measurements*

Eyring and his coworkers studied the kinetic process of CO_2 reduction at a Hg electrode mainly on the basis of their current potential measurements. Since the cathodic current corresponds accurately to CO_2 reduction in the neutral pH region, the reaction kinetics can be discussed on the basis of the current potential relation.[22]

Many subsequent workers studied CO_2 reduction at various metal electrodes, employing voltammetric or polarization measurements. They judged that the electrochemical reduction of CO_2 proceeded when the cathodic current under CO_2 atmosphere was higher than that under inert atmosphere. The net CO_2 reduction current was often determined as the difference between the two cathodic currents with and without CO_2. The assumptions made in this procedure were that the current in CO_2 saturated solution is a sum of CO_2 reduction current and HER current, and that HER current in CO_2 saturated solution takes the same value as that in inert gas saturated solution.

Two problems are pointed out for this procedure. One is that the two cathodic currents, HER and CO_2 reduction, are sometimes involved with each other, when CO_2 reduction actually takes place. For example, a Cu electrode is covered with adsorbed CO during the reduction of CO_2 to CH_4 or other hydrocarbons in an aqueous media, and the current density is usually lower than in argon saturated solution. The second problem is connected with the change of pH of the solution. In CO_2 saturated solution, pH decreases owing to the dissociation of CO_2 in aqueous solution, and the HER naturally takes place at a higher rate than in an inert gas saturated solution. Thus, the fact that a higher current is observed in CO_2 saturated solution does not guarantee that CO_2 reduction proceeds. It is meaningless to discuss any current potential relationship, without any evidence that the current corresponds to a single electrochemical reaction with a faradaic efficiency of 100%. Accordingly, one has to carry out appropriate chemical analysis of the products in order to conclude whether or not the CO_2 reduction actually takes place. And one has to have faradaic balance, sometimes time consuming and painstaking work, in order to reveal what is the real chemical process at the electrode.

Thus papers which report CO_2 reduction only on the basis of the difference current will not be discussed any more.

(*ii*) *Purity of the Electrolyte Solution*

It is well known that trace amount of impurities will interfere with surface process in electrochemical reactions on the electrodes as well as any heterogeneous interfacial reactions. The electrochemical reduction of CO_2 is also a surface process, and is naturally sensitive to the cleanliness of the electrode. Any surface contamination will lead to variation of electrocatalytic property of the electrode. The product selectivity of metal electrode is often severely affected by the presence of extremely small amount of adatoms on the surface.[23] The surface contamination is sometimes a source of many controversial experimental results.

Many workers studied HER on various metal electrodes since Tafel's work. They attempted to obtain the correlation between the hydrogen overvoltage and the properties of the electrode material. During the course of these works, widely scattered voltage current relations were published by different workers. Kuhn et al. reviewed the previous works, and decided to lay down criteria to select the experimental data for their discussions.[24] They chose the data by only those workers who had used preelectrolysis for the purification of the electrolyte solution. Bockris described the importance of preelectrolysis of the electrolyte solutions in the study of electrode processes in detail on the basis of his own studies for long years as well as on the basis of preceding studies.[25,26] The preelectrolysis is to scavenge the solution with a cathodically polarized large area electrode, usually platinum black.

The situation of electrochemical reduction of CO_2 must be similar with the hydrogen electrode process. Any chemical reagents contain impurities of ppm level according to the manufacturers. For example, $KHCO_3$ of the best quality contains 5 ppm of heavy metals as Pb and 5 ppm of Fe, respectively as maximum. 0.1 M $KHCO_3$, prepared from this reagent, will contain ca. 1 μM of heavy metals or Fe ion. These metal ions are reduced to metals and deposited on the electrode surface with one monolayer or more in 10 min or less during the CO_2 reduction under most experimental electrolysis conditions. The deposited metal will affect the electrocatalytic property of the electrode. Hydrogen overpotential will drop, leading to HER prevalent in the total current. The problem is discussed in Section V in detail. Thus, an effective purification of the electrolyte solution such as preelectrolysis is required to obtain reliable results of the electrocatalytic property of metal electrodes. Eyring and his coworkers purified their electrolyte solutions by pre-electrolysis, and they obtained stabler current potential relationships than otherwise.[22]

The recrystallization technique is also effective for sparingly soluble salts e.g. $KClO_4$, when the solubility of the salt greatly changes with temperature. However, highly soluble salts such as $KHCO_3$ cannot be effectively purified by recrystallization.

III. OVERVIEWS OF ELECTROCHEMICAL REDUCTION OF CO_2 AT METAL ELECTRODES

The product distribution in CO_2 reduction varies widely, depending primarily on the electrode metals and the electrolyte solutions used for the reaction. Various reactions proceed simultaneously in parallel on the electrode surface. The electrode provides the site of the reaction, and the product selectivity in CO_2 reduction is affected by whether or not the reactants and other related species are adsorbed. The selectivity also depends on the strength of the adsorption, if any species is adsorbed on the surface. The electrolyte solution determines the concentration and the stability of the reactants and intermediate species. This Section overviews representative results of electrochemical reduction of CO_2.

Before the results are presented, some important points should be mentioned in connection with the reports of CO_2 reduction. Some papers give only the values of the faradaic efficiency of the products of CO_2 reduction in constant potential electrolysis without showing the values of the current density. Other ones do not present electrode potential in constant current electrolysis. It is impossible to evaluate the rate of the reaction as well as the quality of the measurements in the electrolysis measurements without the current density and the electrode potential. Thus these electrochemical parameters should be presented in any publications. Additionally, it is also important to state whether or not the electrolyte solution used for the work was purified by any means in the experiment. The features of reaction change to a great extent, when the electrode surface is contaminated with extremely small amount of impurities possibly contained in chemical reagents of the highest quality.

1. Aqueous Solutions

HER easily takes place in aqueous electrolytes by cathodic polarization, usually competing with CO_2 reduction. The rate is proportional to the proton activity in the electrolyte at a constant potential at various metal electrodes.[27] HER is prevalent particularly in acidic solutions,[22] whereas CO_2 molecules do not exist in a basic solution as descibed in Section II. Thus most of CO_2 reduction studies were done with neutral electrolyte solutions.

Many studies of electrochemical reduction of CO_2 in early years were carried out in aqueous media with metal electrodes of high hydrogen overvoltage such as mercury and lead, aiming at suppression of HER. Eyring and his coworkers studied CO_2 reduction at a Hg electrode in detail; they showed that $HCOO^-$ is exclusively produced with the faradaic efficiency 100% in neutral aqueous electrolytes. Hori and Suzuki revealed that the partial current of $HCOO^-$ formation at a Hg electrode does not depend on pH at a constant potential, whereas HER is proportional to proton activity.[15]

Ito et al. studied CO_2 reduction at Zn, Sn, In, Cd and Pb electrodes in hydrogen carbonate solutions with various alkali metal cations. They analyzed the products by infrared spectroscopy, and reported that formic acid is the only product from these metal electrodes. A combination of an In electrode with 0.2 M $LiHCO_3$ gave the highest current yield for HCOOH formation of 92% at 3.9 mA cm^{-2} at ca. −1.4 V vs. SHE.[28] Formic acid was the only major product in electrochemical CO_2 reduction in aqueous media until 1985 with a few exceptions that reported detections of methanol and methane with extremely low current density.

Hori and his coworkers carried out CO_2 reduction at various metal electrodes in constant current electrolysis at 5 mA cm^{-2} in 0.5 M $KHCO_3$ aqueous solution purified with preelectrolysis. They applied full chemical analysis of the products and studied the faradaic balance. They revealed that CO_2 reduction in aqueous media yields measurable amount of CO, CH_4 and other hydrocarbons as well as formic acid at ambient temperature and pressure in a reproducible way, and the product selectivity depends greatly on the metal electrodes.[29] The product distribution is tabulated in terms of faradaic efficiency with the current densities in Table 3, which contains the results revised in their later publications. The product selectivity is greatly affected by the purity of the electrode metals as well as that of the electrolyte solution. The results above were confirmed later by other workers.[30-35]

Metal electrodes are divided into 4 groups in accordance with the product selectivity indicated in Table 3. Pb, Hg, In, Sn, Cd, Tl, and Bi give formate ion as the major product. Au, Ag, Zn, Pd, and Ga, the 2nd group metals, form CO as the major product. Cu electrode produces CH_4, C_2H_4 and alcohols in quantitatively reproducible amounts. The 4th metals, Ni, Fe, Pt, and Ti, do not practically give product from CO_2 reduction continuously, but hydrogen evolution occurs. The classification of metals appears loosely related with that in the periodic table. However, the correlation is not very strong, and the classification such as *d* metals and *sp* metals does not appear relevant. More details of the electrocatalytic properties of individual metal electrodes will be discussed later.

Sakata and his coworkers applied 32 metals to CO_2 reduction at –2.2 V vs. SCE in 0.05 M $KHCO_3$. They mostly confirmed the results given in Table 3, and showed that Ni and Pt electrodes, which scarcely give products in CO_2 reduction at ambient tempera-

Table 3
Faradaic Efficiencies of Products in CO_2 Reduction at Various Metal Electrodes. Electrolyte: 0.1 M $KHCO_3$, T = 18.5 ± 0.5°C. Reprinted from Ref. 23, Copyright (1994) with Permission from Elsevier

Electrode	Potential vs. SHE V	Current density mA cm^{-2}	Faradaic efficiency, %							
			CH_4	C_2H_4	EtOH[a]	PrOH[b]	CO	$HCOO^-$	H_2	Total
Pb	−1.63	5.0	0.0	0.0	0.0	0.0	0.0	97.4	5.0	102.4
Hg	−1.51	0.5	0.0	0.0	0.0	0.0	0.0	99.5	0.0	99.5
Tl	−1.60	5.0	0.0	0.0	0.0	0.0	0.0	95.1	6.2	101.3
In	−1.55	5.0	0.0	0.0	0.0	0.0	2.1	94.9	3.3	100.3
Sn	−1.48	5.0	0.0	0.0	0.0	0.0	7.1	88.4	4.6	100.1
Cd	−1.63	5.0	1.3	0.0	0.0	0.0	13.9	78.4	9.4	103.0
Bi[c]	−1.56	1.2	-	-	-	-	-	77	-	-
Au	−1.14	5.0	0.0	0.0	0.0	0.0	87.1	0.7	10.2	98.0
Ag	−1.37	5.0	0.0	0.0	0.0	0.0	81.5	0.8	12.4	94.6
Zn	−1.54	5.0	0.0	0.0	0.0	0.0	79.4	6.1	9.9	95.4
Pd	−1.20	5.0	2.9	0.0	0.0	0.0	28.3	2.8	26.2	60.2
Ga	−1.24	5.0	0.0	0.0	0.0	0.0	23.2	0.0	79.0	102.0
Cu	−1.44	5.0	33.3	25.5	5.7	3.0	1.3	9.4	20.5	103.5[d]
Ni	−1.48	5.0	1.8	0.1	0.0	0.0	0.0	1.4	88.9	92.4[e]
Fe	−0.91	5.0	0.0	0.0	0.0	0.0	0.0	0.0	94.8	94.8
Pt	−1.07	5.0	0.0	0.0	0.0	0.0	0.0	0.1	95.7	95.8
Ti	−1.60	5.0	0.0	0.0	0.0	0.0	tr.	0.0	99.7	99.7

[a]ethanol; [b]n-propanol; [c]the data are taken from Hori et al.[23] except Bi which is read from an illustration in a paper by Kunugi et al.[117]; [d]the total value contains C_3H_5OH(1.4%), CH_3CHO(1.1%), C_2H_5CHO(2.3%) in addition to the tabulated substances; [e]the total value contains C_2H_6(0.2%)

ture and pressure, can reduce CO_2 to CO or formic acid under elevated pressure (60 atm).[35, 36]

Table 4 gives a list of electrode metals used in research works of electrochemical reduction of CO_2 published roughly after 1970. The electrode metals appear in accordance with the product selectivity mentioned above. Metals with low activity in CO_2 reduction at ambient pressure are divided into two groups; metals with which activity appears under elevated pressure, and ones otherwise.

In addition to the major products mentioned above, formation of higher carboxylic acids in aqueous media has been disputed for long years. Bewick and Greener reported that electrochemical reduction of CO_2 produces malate from a Hg cathode, and glycolate from a Pb cathode in high yield in aqueous tetraalkyl ammonium electrolytes.[37, 45] Kaizer and Heitz examined the reaction, and concluded that Bewick and Greeners' results were not reproduced.[123] They showed that oxalic acid is reduced to glycolic, glyoxylic and maleic acids at Ni, Al, Pb and Hg electrodes in aqueous electrolytes in the presence of quarternary ammonium salts. But no oxalic acid was produced from these electrodes in aqueous electrolytes. Ito et al. detected slight amount of oxalic and propionic acids in tetraethylammonium aqueous electrolytes using Pb, Sn and In electrodes at 10 atm.[39] Wolf and Rollin confirmed Bewick and Greener's results.[46] Eggins et al. also supported Bewick and Greeners' results; they employed carbon and Hg electrodes in aqueous solutions of tetramethylammonium chloride and tetramethylammonium hydroxide (pH 9). They obtained oxalic and glyoxylic acids in addition to formic acid, as confirmed by high performance liquid chromatography (HPLC).[48]

Many papers reported that CH_4 or CH_3OH is formed from the CO_2 reduction at various metal and semiconductor electrodes near the equilibrium potential. However, the amount of CH_4 or CH_3OH appearing in these reports is extremely low, and any increase to reasonable value at higher overpotential is not shown. Reproducibility of the findings has not been confirmed by other workers. Thus Table 4 does not contain the results with the partial current of the CO_2 reduction estimated below 0.1 mA cm^{-2}.

Table 4
Metal and Selected Nonmetal Electrodes in CO_2 Reduction in Aqueous Electrolytes

Electrode	Major products at 1 atm	Comments/ Major products at 30 atm	References
(A) $HCOO^-$ formation metals			
Pb	$HCOO^-$		23,28,29,31,36–42
Hg	$HCOO^-$		15,22,23,28,31,36,40,43–48
Tl	$HCOO^-$		23, 36
In	$HCOO^-$		23,28,29,31,35,39,40
Sn	$HCOO^-$		23,28,29,31,36,39,41,49
Cd	$HCOO^-$		23,28,29,31,36,49
Bi	$HCOO^-$		117
Hg/Cu	$HCOO^-$		165
Sn-Cd, Sn-Zn	$HCOO^-$		49
(B) CO formation metals			
Au	CO		23,29,35,36,50–55, 193
Ag	CO		23,29,31,35,56–58
Zn	CO, $HCOO^-$		23,28,29,31,35,49,59,60,127
Pd	CO, $HCOO^-$		23,31,36,61–68
Ga	CO		23,31
Ni-Cd alloy	CO		69
(C) Hydrocarbon formation metals			
Cu	CH_4, C_2H_4, C_2H_5OH, $HCOO^-$		19,23,29–36,70–93
Cu-Ni, Cu-Fe alloy	CH_4, C_2H_4, CO		94
Cu-Cd alloy	CH_4, C_2H_4, CO		95
Cu-Ni, Cu-Sn, Cu-Pb, Cu-Zn, Cu-Cd alloys	$HCOO^-$, CO		96
Cu-Au alloy	CH_4, C_2H_4, $HCOO^-$, EtOH, PrOH-1		97
Cu-Ag alloy	CO, C_2H_4, CH_3CHO, C_2H_5OH		96, 98

Table 4. Continuation

Electrode	Major products at 1 atm	Comments/ Major products at 30 atm	References
(D) Activity in CO_2 reduction: low at 1 atm, but enhanced under elevated pressure			
Zr		CO, $HCOO^-$	99
Cr		CO, $HCOO^-$	31,99
W		$HCOO^-$	31,99
Fe		$HCOO^-$, CO	23,29,31,35, 99,100–102
Co		$HCOO^-$, CO	31,99,101
Ni		$HCOO^-$, CO	23,29,31,35,36,99,101,103, 104
Rh		CO, $HCOO^-$	99,105
Ir		CO, $HCOO^-$	31,99,106
Pt		$HCOO^-$, CO	23,29,31,35,36,99,101,107, 108–115
(E) Activity in CO_2 reduction: low under 1 atm and elevated pressure			
Ti			23,29,31,99
V[a]			31
Nb			31,99
Ta			31,99
Mo			31,99
Mn			99
Re[a]			31
Ru[a]			31,116
Al			31, 36, 99
As[a]			117
(F) Non-metallic electrodes			
Glassy carbon		CO, $HCOO^-$	118
n-Si		$HCOO^-$	99
RuO_2+TiO_2	CH_3OH		119,120
RuO_2/Diamond[b]	$HCOO^-$, CH_3OH		121
RuO_2/TiO_2/ CNT, CNP[c]	CH_3OH		122

Most of the electrolyte solutions are 0.05 to 0.5-M $NaHCO_3$ or $KHCO_3$; tetraalkyl ammonium salts are employed in some cases without any significant differences in major products.

[a]electrolysis data under elevated pressure are unavailable; [b]RuO_2/Diamond: RuO_2 coated diamond electrode; [c]RuO_2/TiO_2/ CNT, CNP: RuO_2 and (or) TiO_2 loaded on CNT(carbon nanotube) or CNP (carbon nanoparticle).

Table 5
Metal and Selected Nonmetal Electrodes in CO_2 Reduction in Nonaqueous Electrolytes

Electrode	Electrolyte	Major products	Comments	References
		A. Oxalate formation metals		
Pb	PC, AN, DMF, TEABr	$(COOH)_2$, CO, glyoxylate, glycolate		38,124–127
Pb	MeOH/TEAP, HA	$(COOH)_2$, glyoxylate, glycolate		128
Pb	MeOH/TEAP, KOH	HCOOH, CO		129,130
Hg	DMSO/TEAP, DMF, PC/TEAP	$(COOH)_2$, CO, HCOOH		124,125,127, 131–134
Tl	PC/0.1 M TEAP	$(COOH)_2$		127
Pt	PC/TEAP, PC/TEAH AN/0.1 M TEAP	$(COOH)_2$, CO, HCOOH		123,127,129, 135
Pt	MeOH/$TBABF_4$	HCOOH, CO		129
Ni	PC/0.1 M TEAP,	$(COOH)_2$, CO		127
Ni	MeOH/TBAP	HCOOH, CO, CH_4, C_2H_4		129
		B. Oxalate and CO formation metals		
Fe	PC/0.1 M TEAP	$(COOH)_2$, CO		127
Cr	PC/0.1 M TEAP	$(COOH)_2$		127
Mo	PC/ TEAP	$(COOH)_2$, CO		127
Pd	PC/0.1 M TEAP, TBAP	CO, $(COOH)_2$		127,129
Cd	PC/0.1 M TEAP	CO, $(COOH)_2$		127
Ti	PC/0.1 M TEAP MeOH/KOH	CO, $(COOH)_2$, HCOOH		127,137
Nb	PC/0.1 M TEAP	CO, $(COOH)_2$, HCOOH		127
Cr-Ni-Mo Steel	AN/TEAH PC/TEAH	HCOOH, $(COOH)_2$, glyoxylate, glycolate		123

Table 5. Continuation

Electrode	Electrolyte	Major products	Comments	References
		C. CO formation metals		
Cu	PC/0.1-M TEAP	CO		127
Cu	MeOH/BC	CO, CH_4, C_2H_4	0 to −15°C	138
Cu	MeOH/TEAP, $TBABF_4$	CO, HCOOH, CH_4, C_2H_4	40 to 60 atm	129,139–141
Cu	MeOH/LiCl, LiBr, etc.	CH_4, C_2H_4, CO		142
Cu	MeOH/$TBABF_4$	CO	Near critical condition	143
Cu	EtOH	CH_3OH	95 atm	144
Sn	PC/0.1-M TEAP MeOH/BC, $TBABF_4$	CO		127,129,145
Ag	PC/TEAP, MeOH/KOH, BC, $TBABF_4$	CO, HCOOH		127,129,145, 146
Zn	PC/0.1-M TEAP, MeOH/BC, MeOH/KOH, $TBABF_4$	CO		127,145 129,147
In	PC/0.1-M TEAP	CO		127
In	MeOH/KOH	CO, HCOOH		148
Au	DMSO/TEAP, PC/0.1-M TEAP, MeOH/KOH, BC	CO, HCOOH		127,131,145, 149
		D. Composite electrodes		
Re/Au	MeOH/$LiClO_4$	CO, CH_4		150
Ppy-Re/Au[a]	MeOH/$LiClO_4$	CO, CH_4		151
Ppy-Re, Cu/Au[b]	MeOH/$LiClO_4$	CO, CH_4		151
Ppy/Pt	MeOH/$LiClO_4$	HCHO, HCOOH, HOAc	20 bar	152
PAn	MeOH/$LiClO_4$, H_2SO_4	HCHO, HCOOH, HOAc	20 bar	153,154

Typical data are presented for the major products. BC: Benzalconium chloride, HA: Hydroxylamin, TBATF: Tetrabutyl ammonium tetrafluoroborate, TEAB: Tetraethyl ammonium bromide, TEAH: Tetraethyl ammonium hydroxide, TEAP: Tetraethyl ammonium perchlorate. Ppy: Polypyrrole, PAn: Polyaniline.

[a]Rhenium dispersed on polypyrrole film supported on Au.

[b]Rhenium-copper alloy dispersed on polypyrrole film supported on Au.

2. Nonaqueous Solutions

Electrochemical reduction of CO_2 in nonaqueous solutions is significant from the following viewpoints.[4] Firstly, hydrogen evolution reaction can be suppressed. Secondly, the concentration of water as a reagent can be accurately regulated and the reaction mechanism may be more easily studied. Thirdly, the solubility of CO_2 in organic solvents is much higher than in water. Various metal electrodes have been tested for CO_2 reduction in some nonaqueous solvents, such as propylene carbonate (PC), acetonitrile (AN), DMF, and dimethyl sulfoxide (DMSO), as tabulated in Table 5. Methanol is also used for CO_2 reduction, and mentioned in the next Section.

Haynes and Sawyer conducted a chronopotentiometric study of CO_2 reduction at Au and Hg electrodes in DMSO. They showed that CO and formate are the reaction products from CO_2.[131] Tyssee et al. reported that oxalic acid is exclusively formed from CO_2 reduction at Hg in DMF.[133]

Kaizer and Heitz employed Cr-Mo-Ni steel as a cathode in AN and PC electrolytes with tetraalkyl ammonium salt as the electrolyte.[123] They showed that oxalic acid is the main product. Addition of slight amount of water to the electrolyte gives formate as the main product.

Savéant and his coworkers studied the electrochemical reduction of CO_2 at Pb and Hg in H_2O-DMF solutions. The main products were CO, HCOOH and $(COOH)_2$, the product distribution depending upon the operational conditions.[124] They discussed the reaction schemes with an assumption that a presumed intermediate $CO_2^{\cdot-}$ is adsorbed on neither Pb nor Hg electrode and that no specific interaction exists between intermediates or products and the electrodes.[125, 134] The details are discussed later.

Ito and his coworkers employed various electrode metals for constant potential electrolysis of CO_2 reduction at –2.8 V vs. Ag/AgCl in PC with tetraethylammonium perchlorate (TEAP) as the electrolyte with water content 300 ppm. Table 6 shows their results of the electrolysis of the electric charge 100 C in terms of the faradaic efficiency.[127] Electrode metals are classified into three groups. Pb, Hg, and Tl give $C_2O_4^{2-}$ as the main product. Oxalic acid is partly reduced to glycolic acid and glyoxylic acid in the presence of slight amount of water in the electrolyte solution. Cu,

Ag, Au, Zn, In, Sn, Ni, and Pt predominantly yield CO from CO_2. Fe, Cr, Mo, Pd, and Cd form both $C_2O_4^{2-}$ and CO in comparable yields. The product selectivity by Ito et al. is mostly confirmed by other workers' results[136] with exceptions mentioned below.

Tomita et al. pointed out that Ag^+ ion easily leaks from an Ag/AgCl reference electrode employed in the electrolysis cell with AN electrolyte. Ag^+ contaminates the Pt electrode, and CO is consequently the major product from the contaminated Pt electrode. A Pt electrode without Ag contamination gives $(COOH)_2$ as the major product. When polarized cathodically in CO_2 saturated electrolyte, Pt is readily covered with strongly adsorbed CO as evidenced by IR spectroscopy, and is chemically inert like Hg and Pb,

Table 6
Faradaic Efficiencies of the Products in CO_2 Reduction in 0.1-M TEAP in PC at Various Metal Electrodes at −2.8 V vs. Ag/AgCl/(0.01 M LiCl + 0.1 M TEAP)/PC. Quantity of the Electric Charge: 100 C.

Electrode	Faradaic efficiency(%)					
	$(COOH)_2$	OHCCOOH	HCOOH	CO	H_2	Total
Pb	76.6	2.9	2.5	10.4	0.0	92.4
Hg	60.2	1.0	1.2	17.8	0.0	80.2
Tl	70.4	1.0	2.1	3.1	0.0	76.6
Fe	46.1		1.4	15.6	22.2	85.3
Cr	31.8	1.0	4.4	11.2	36.5	84.9
Mo	29.5		5.4	13.9	40.9	89.7
Pd	21.8		3.1	51.9	0.0	76.8
Cd	21.6	3.9	1.5	63.1	0.0	90.1
Ti	19.9	1.2	4.2	20.1	45.4	90.8
Nb	7.5	0.8	8.1	6.8	49.4	72.6
Cu	0.6			74.9	0.0	75.5
Sn	1.9	1.0	3.5	81.8	0.0	88.2
Ag	1.4		2.1	77.4	1.9	82.8
Zn	0.4			89.7	0.0	90.1
In	0.2		2.3	89.2	0.0	91.7
Au	0.2		3.7	83.2	13.9	100.9
Ni	0.2		8.3	45.7	44.0	98.2
Pt	1.0		7.7	66.6	17.4	92.7

Faradaic efficiencies were calculated from the data published by Ito et al.[127]

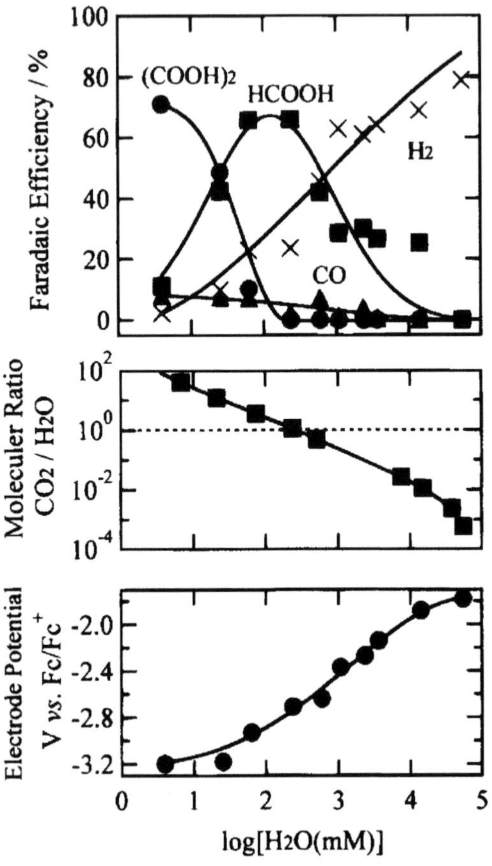

Figure 3. Constant current electroreduction of CO_2 at a Pt electrode in 0.1 M TEAP in AN-H_2O mixtures. Current density: 5 mA cm^{-2}. Reproduced from Ref. 135, Copyright (2000) with permission from the Electrochemical Society.

yielding $(COOH)_2$ as the major product. A Pt electrode modified deliberately with Ag adatoms prevalently produces CO in AN with low concentration of H_2O.[135] Their product distribution obtained from constant current electrolysis at 5 mA cm^{-2} is given as a function of water concentration in Fig. 3.

Oxalic acid is formed as a major product in nonaqueous electrolytes, and is further reduced to higher carboxylic acids (glyoxylic acid, glycolic acid etc.) at Cr-Ni-Mo steel electrodes,[123] Pb and Hg electrodes.[124,126-128]

$$2CO_2 + 2e^- \rightarrow (COO)_2^{2-} \tag{29}$$

$$(COO)_2^{2-} + 2H_2O + 2e^- \rightarrow HCO\text{–}COO^- + 3OH^- \tag{30}$$

$$HCO\text{–}COO^- + 2\,H_2O + 2e^- \rightarrow H_2C(OH)\text{–}COO^- + 2OH^- \tag{31}$$

Formation of glyoxylic acid was reported with metal electrodes: Ti, Nb, Cr, Mo, Ag, Cd, Hg, Tl, Sn, Pb in a PC/TEAP electrolyte solution.[127]

Neither CH_4 nor C_2H_4 was produced from CO_2 at a Cu electrode in PC with TEAP according to Ikeda et al.,[127] as confirmed by the present author using AN with TEAP.[136]

Bockris and his coworkers studied photoelectrochemical reduction of CO_2 in DMF based electrolyte solution, and proposed a concept of tetraalkyl amonium cation as a mediator in electron transfer.[132, 155] Saeki et al. showed that CO_2 reduction in a methanol based electrolyte proceeds with tetraalkyl ammonium radical working as an electron mediator. The CO_2 reduction does not take place when LiCl is used as the supporting electrolyte.[156] Tomita et al. also indicated that CO_2 reduction at a Pt electrode in AN requires some tetraalkyl ammonium salts.[135] The signigicance of electron mediator in CO_2 reduction was stressed by Gennaro et al. as well.[134, 157]

3. Methanol, another Nonaqueous Solution

Methanol is an attractive solvent for nonaqueous electrolyte solutions: inexpensive, lowly toxic, and produced in an industrially large scale. Methanol dissolves large amount of CO_2 particularly under elevated pressure. Thus many workers studied CO_2 reduction at metal electrodes in methanol-based electrolytes (Table 5). In addition to metal electrodes, polyaniline and polypyrole were also investigated as electrode materials.[152-154]

Ohta and his coworkers reduced CO_2 at a Cu electrode in 8 and 80 mM benzalkonium chloride in methanol at 0 to –15°C, and produced CH_4, C_2H_4 as well as CO in high faradaic efficiencies in their electrolysis of the electric charge 30 to 50 C.[138] Hydrocarbon formation at Cu electrode was not observed with PC or AN based electrolytes with small amount of H_2O.[127,136] Thus, one would naturally expect that the proton source of the hydrocarbons formed at a Cu electrodes would be hydrogen of the hydroxyl group attached to a methanol molecule; the hydrogen at this position is, as is well known, reactive and readily released as hydrogen gas by cathodic polarization[158]. However, Ohta et al. argued that the proton in the products is derived from residual water contained in methanol (less than 0.09%) on the basis of their electrolytic measurements with deuterated methanol;[138] any relevant experimental data has not been presented.

Fujishima and his coworkers employed methanol-CO_2 mixture with tetraalkyl ammonium salts as a supporting electrolyte under elevated pressure, aiming at high concentration of CO_2 in the electrolyte solution. $HCOCH_3$ and CO are the major products from a Cu electrode at the current density of 200 mA cm^{-2} with less faradaic efficiencies of CH_4 and C_2H_6 in tetrabutyl ammonium tetrafluoroborate solution ($TBABF_4$)[129, 139]. $HCOCH_3$ may be formed by esterification of HCOOH with CH_3OH. They demonstrated that the carbon atoms come from the CO_2 molecule on the basis of 13C labeled CO_2 experiments. They showed that CH_4 and C_2H_4 were mainly produced in TEAP solution, in stead of CO and $HCOCH_3$.[140] They presumed that the proton source of methyl formate and hydrocarbons is alcoholic hydrogen, since the high partial current densities for production of these substances cannot be supported if otherwise.

Methanol is a nonaqueous solvent, but normally classified as a protic solvent with many similarities with water. The product selectivity of CO_2 reduction depends on the electrode metal in methanol based electrolytes as well.[130, 145, 148] The features of the selectivity obtained from methanol base electrolytes are similar to those presented in Table 3 for aqueous electrolytes. Formate is favorably produced instead of oxalate in methanol based electrolytes. Taking into account of the reaction scheme discussed in Section VI, it might not be appropriate to classify methanol simply as a nonaqueous electrolyte. However, Table 5 lists papers in accordance with the selectivity in other nonaqueous media for convenience.

4. Electrochemical Reduction of CO_2 in High Concentration

(*i*) *CO_2 Reduction under Elevated Pressures*

CO_2 reduction under elevated pressures has been studied for an attempt of enhancing the concentration of CO_2 in the electrolyte and expecting higher transport rate. This topic will be referred to in Sect. IX.1, and only the electrocatalytic aspects are briefly discussed here.

Ni and Pt electrodes do not yield products continuously in CO_2 reduction in aqueous solutions under 1 atm, as shown in Table 3. Platinum and nickel reduce CO_2 to CO strongly adsorbed on the electrode surface, as confirmed by spectroscopic measurements.[159, 160] The adsorbed CO totally covers the electrode surface, preventing further reduction of CO_2. Only HER can proceed at such an electrode surface under atmospheric pressure in aqueous electrolytes. Sakata and his coworkers revealed that these metal electrodes can reduce CO_2 in 0.1 M $KHCO_3$ under elevated pressures.[101] Table 7 shows their results that Fe, Co, Ni, and Pt electrodes yield CO and $HCOO^-$ at 50 or 60 atm.

The heat of adsorption of CO on Ni or Pt is 40 kcal mol^{-1} or higher.[161] The increment of free energy $RT \ln P(CO)$ due to the increase of CO_2 partial pressure to 50 to 60 atm is 2 to 3 kcal mol^{-1} at most, much smaller than the heat of adsorption of CO. Thus the elevation of CO_2 pressure will not energetically affect the adsorption of CO on the electrode surface to a great extent. The apparent enhancement of the electrocatalytic activity in CO_2 reduction is derived from the increase of the CO_2 concentration in the electrolyte solution.

In addition to the enhancement of the CO_2 reduction rate, product selectivity can be changed by elevation of the CO_2 pres sure. Pb and In electrodes, forming $HCOO^-$ exclusively at 1 atm, yield CO preferentially at a low current density (1 mA cm^{-2}) in 0.5-M $KHCO_3$ at 20 atm.[40] The reaction scheme will be discussed in Section VI.5.

Table 7
Faradaic Efficiencies of the Products in CO_2 Reduction in 0.1-M $KHCO_3$ at Various Metal Electrodes Under Elevated Pressures at –1.8 V vs. Ag/AgCl. Quantity of the Electric Charge: 2000–3000 C.Reprinted from Ref. 101, Copyright (1991) with permission from Elsevier.

Electrode	Pressure of CO_2 (atm)	Faradaic efficiency (%)					
		CO	$HCOO^-$	CH_4	C_2H_4	C_2H_6	Other hydrocarbons
Fe	1	0	0	0	0	0	0
Fe	50	6.7	3.7	1.61	0.05	1.32	2.73
Co	1	1.2	t[a]	0.31	t	0.2	0.28
Co	60	14.4	4.1	0.93	0.13	0.45	0.81
Ni	1	0	0.1	0.62	0.63	0	0.06
Ni	60	10.4	23.2	1.78	0.44	0.88	0.77
Pd	1	5.3	4.4	0	0	0	0
Pd	50	57.9	4.4	0	0	0	0
Pt	1	0	0	0.02	0	0	0
Pt	60	9.3	24.1	0.31	t	0.08	0.08

[a]t = trace.
For the measurements under 1-atm CO_2 pressure, Ar was introduced to balance the total pressure of 50 or 60 atm. Temperature: 30°C. The original table was rearranged for the sake of simplicity.

(ii) CO_2 Reduction Promoted by High Concentration

Table 6 and the discussion above showed that Ni, Fe, Pt, and Ti electrodes, inert to CO_2 reduction in aqueous media at 1 atm, yield products in nonaqueous electrolyte solution. HER and CO_2 reduction proceed in parallel competitively in aqueous electrolytes saturated with CO_2. Nonaqueous electrolyte solution does not contain water except small amount of residual one, and thus CO_2 reduction will take place with HER severely suppressed.

Tomita et al. studied CO_2 reduction at a Pt electrode in mixtures of H_2O and AN under 1 atm CO_2.[135] Figure 3 shows the result of their constant current electrolysis of CO_2 reduction at 5 mA cm^{-2}. The major product is $(COOH)_2$ in low water concentration region, and changes to HCOOH and H_2 with the increase of water concentration. The electrode potential is as negative as −3.2 V vs. Fc/Fc$^+$ (ca. −2.7 V vs. SHE), much more negative than the standard potential of CO_2^-/CO_2. The electrode surface is fully covered with adsorbed CO, and the electron transfer from the electrode to CO_2 has to take place across the adsorbed CO layer by tunnel effect. The faradaic efficiency of HCOOH is equal to that of H_2 at the molecular ratio CO_2 to H_2O ca. 0.25 in the electrolyte solution. HER becomes predominant in the solution with the molecular ratio below 0.25, and CO_2 reduction gradually diminishes down to zero in pure water.

One can thus understand the appearance of the electrocatalytic activity of Ni and Pt under elevated pressure in a similar way to the situation in nonaqueous solutions. Elevation of CO_2 pressure enhances the solubility of CO_2 in H_2O. The solubility at 25°C is 0.034 mol l^{-1} at 1 atm, and magnified by 36 times to 1.22 mol l^{-1} at 50 atm.[162] Such an increase of the molecular ratio will increase the rate of CO_2 reduction under elevated pressure.

Some papers report that low temperatures favor the CO_2 reduction in comparison with ambient temperature.[36,149,163] Enhancement of the CO_2 solubility in electrolyte solutions as well as favorable adsorption equilibrium of CO_2 may lead to effective reduction of CO_2.

IV. ELECTROACTIVE SPECIES IN THE ELECTROCHEMICAL REDUCTION OF CO_2

Teeter and Rysselberghe reported that CO_2 is reduced to HCOOH at a Hg cathode with the faradaic efficiency nearly 100%.[43] They showed that the polarographic wave height depends on the CO_2 concentration in the electrolyte solution, and concluded that the electroactive species of CO_2 reduction is CO_2 molecule, neither HCO_3^- nor CO_3^{2-}.[164]

Since the solubility of CO_2 is as low as 30 mM in water under 1 atm at the ambient temperature, the rate of electrochemical reduction of CO_2 in aqueous media must be limited by the transport process. It is thus important whether HCO_3^- or CO_3^{2-} can be electrochemically reduced.

Several articles were published, arguing that HCO_3^- can be reduced at a Hg coated Cu electrode,[165] Pd impregnated polymer electrodes,[166] or metallic Pd electrodes.[167] However, the partial current density of CO_2 reduction was below 0.1 mA cm^{-2},[166] or the data were not confirmed by quantitative chemical analysis.[167]

Hori and Suzuki studied the electrochemical reduction of $NaHCO_3$ on the basis of chemical analysis of the product $HCOO^-$[47]. They showed that HCO_3^- is apparently reduced to $HCOO^-$ at a Hg electrode in aqueous $NaHCO_3$ solutions. The partial current density of formate formation j_C at a constant potential depends greatly on pH. It is proportional to the concentration of $NaHCO_3$ at constant potential as shown in Fig. 4, when the CO_3^{2-} and HCO_3^- concentration ratio is kept constant at a constant pH. The partial current density increases with the increase of the negative potential, reaching a limiting value (Fig. 5). The limiting current is below 1.0 mA cm^{-2} for a mixture of 0.05 M CO_3^{2-} and 1.0 M HCO_3^-. The limiting current would be too low under usual experimental

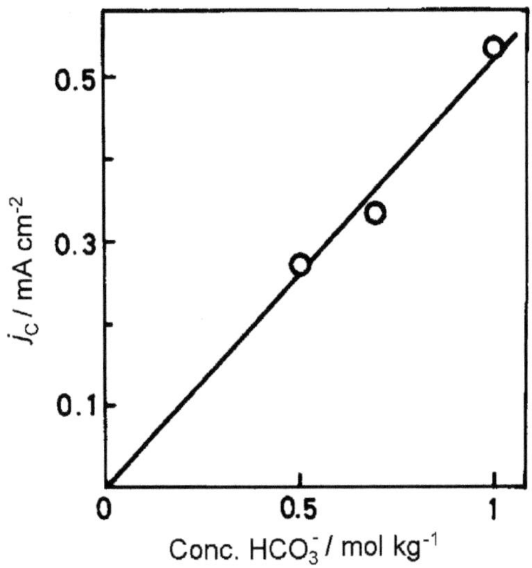

Figure 4. Partial current density of $HCOO^-$ formation j_c vs. HCO_3^- concentration from reduction of HCO_3^- with a Hg pool electrode. The concentration ratio $C(CO_3^{2-})/C(HCO_3^-)$ is kept constant at 0.05. The electrode potential is −1.813 V vs. SHE. Reprinted from Ref. 47, Copyright (1983) with permission from the Electrochemical Society.

conditions, if HCO_3^- were an electroactive species and directly reduced at the electrode.

In addition, two data are shown for inert gas bubbling through the electrolyte as well in Fig. 5. These two data are well correlated smoothly in a single curve with other electrolysis data obtained without the gas bubbling. One of the data is close to the limiting current value. Thus the convective motion of the electrolyte solution caused by the gas bubbling does not affect the rate process of the electrochemical reduction of HCO_3^-. These facts evidently show that HCO_3^- is not an electroactive species, suggesting that HCO_3^- is likely decomposed to CO_2, then reduced at the electrode.

The decomposition of HCO_3^- is a slow process in ambient temperature as is well known.

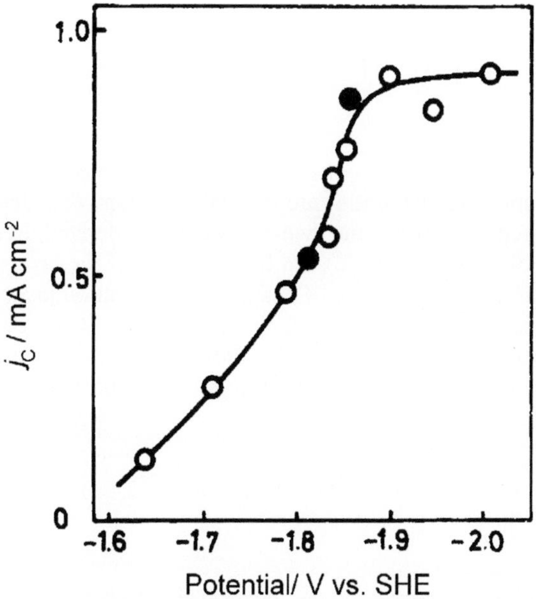

Figure 5. Partial current density of $HCOO^-$ formation j_c vs. electrode potential from reduction of HCO_3^- with a Hg pool electrode with (○) and without (•) N_2 bubbled. Electrolyte: $NaHCO_3$ (m = 1.0 mol kg^{-1})- Na_2CO_3 (m = 0.05 mol kg^{-1}). The potential axis negative to the right. Reprinted from Ref. 47, Copyright (1983) with permission from the Electrochemical Society.

$$HCO_3^- = CO_2 + OH^- \tag{32}$$

The resultant OH^- is instantaneously neutralized.

$$HCO_3^- + OH^- = CO_3^{2-} + H_2O \tag{33}$$

k_d and k_r denote the rate constants of the forward and backward reactions of reaction (32), respectively. Reaction (33) is instantaneous, remaining virtually in equilibrium. The rate of CO_2 formation in reaction (32) is

$$V(CO_2) = k_d\, C(HCO_3^-) - k_r\, C(CO_2)\, C(OH^-) \tag{34}$$

where Cs' are the concentrations of the respective species. A Fick's equation is

$$\frac{\partial C(CO_2)}{\partial t} = D\frac{\partial^2 C(CO_2)}{\partial x^2} + V \tag{35}$$

where D is the diffusion constant of CO_2 in the electrolyte solution. Equations (34) and (35) are equivalent to Eqs. (27) and (26). The difference between the two treatments is derived from the direction of the reaction (32) and the thickness of the layers, i. e., the diffusion layer in Section II and the decomposition layer in the present Section.

Equation (35) is solved under the steady state for the CO_2 concentration at the electrode surface with relevant boundary conditions.

$$\frac{\partial C(CO_2)}{\partial t} = 0 \tag{36}$$

The experimental results of the electrolysis of HCO_3^- with a Hg electrode were analyzed by the procedure above, and compared with those obtained from electrochemical reduction of gaseous CO_2 at a Hg electrode. The two sets of data, from HCO_3^- and gaseous CO_2, showed a good compatibility. The thickness of the decomposition layer in which HCO_3^- decomposition proceeds was estimated to be approximately 0.008 cm, much thinner than the diffusion layer under usual experimental conditions.[47] It is natural that moderate convective motion of the electrolyte solution does not affect the partial current of CO_2 reduction.

Thus HCO_3^- is decomposed to CO_2 in the thin decomposition layer; the resultant CO_2 diffuses to the Hg electrode, then electrochemically reduced to $HCOO^-$. The equilibrium CO_2 pressure of $NaHCO_3$-Na_2CO_3 (1.0–0.05) given in Fig. 5 is estimated at 0.03 atm from Table 2 with an assumption of the activity coefficient 0.5. The rate of decomposition of HCO_3^- limits the supply of CO_2 to the electrode, and the reduction of HCO_3^- stays at a low value as shown in Fig. 5.

An Au electrode yields CO in 0.5 M KHCO3 without supply of gaseous CO_2 from outside. The partial current of CO formation agreed with the value derived from the reduction of CO_2 formed in the dissociation of HCO3$^-$ in the solution.[50]

If the decomposition of HCO_3^- proceeded more rapidly at an electrode, reduction of HCO_3^- would be a more effective process than the direct reduction of CO_2. Eggins et al. later examined the reduction of HCO_3^- at a glassy carbon electrode by cyclic voltammetry. They applied a similar analytical procedure as above, and confirmed that the decomposition of HCO_3^- to CO_2 is the rate determining step in the reduction of HCO_3^-.[168]

V. DEACTIVATION OF ELECTROCATALYTIC ACTIVITY OF METAL ELECTRODES

Deactivation of metal electrodes in CO_2 reduction has been reported by many workers, and regarded as an unavoidable defect irrespective of the potential importance of CO_2 reduction. For example, the deactivation was described that the formation of CH_4 and C_2H_4 at Cu electrode decays rapidly and hydrogen evolution prevails in 10 to 30 minutes after the start of the CO_2 reduction. However, the features of the deactivation have been various, depending on the research groups. No agreement is found for the reason of the deactivation as well. Bard et al.[73] and Vielstich et al.[74] reported that the surface of the deactivated copper electrode is blackened after electrolysis. Bard et al. analyzed the surface by X-ray photoelectron spectroscopy (XPS), and reported that the surface is contaminated with some carbonaceous substance. They presumed that the carbonaceous substance is formed during the CO_2 reduction, and poisons the reaction. Kyriacou et al.,[77] Augustynski,[80] and Tributsch,[82] also argued that the deactivation of copper electrode is derived from some organic substance adsorbed on the electrode surface.

Nogami et al.[79] and Augustynski et al.[80] showed that the activity of copper electrode, deactivated during the CO_2 reduction, is recovered by anodic polarization of the electrode. Periodic anodic pulses are effective to maintain the electrocatalytic activity in prolonged electrolysis of CO_2 reduction.

Hori et al. pointed out that the deactivation takes place due to the presence of heavy metal impurities originally contained in chemical reagents used as the electrolytes.[169] Heavy metal ions in the electrolyte solution are cathodically reduced and deposited on the electrode surface during the CO_2 reduction, deteriorating the electrocatalytic properties of metal electrodes. They applied a classically established technique of preelectrolysis to purification of electrolyte solutions since their early works. Frese also referred to the impurity heavy metals, and mentioned the presence of Fe and Zn on the Cu electrode after electrolysis on the basis of the surface analysis by XPS.[5] The importance of the purity of the electrolyte solution was mentioned in Section II.2(*ii*) as well. The mechanism of the deactivation was recently established, and summarized below.[170]

Hori et al. carried out CO_2 reduction with a Cu plate electrode in aqueous $KHCO_3$ electrolyte solutions prepared from reagents from various sources previously used by other workers. They reproduced the deactivation, whereas the features of the deactivation varied depending on the source of the reagents. Fig. 6 gives an example of the time course of CO_2 reduction in 0.3 M $KHCO_3$ prepared from puriss p.a. ACS by Fluka in a constant current electrolysis at 5 mA cm^{-2}. CO_2 gas was supplied to the electrolysis cell and bubbled in the solution at a constant flow rate continuously, and small portions of gas were sampled from the effluent gas for gas chromatographic analysis. The yields of the products are indicated in faradaic efficiencies. The electrode rapidly loses the electrocatalytic activity; CH_4 formation diminishes in 100 min, whereas H_2 formation increases up to 90% in faradaic efficiency. C_2H_4 formation is very low. The electrocatalytic activity is recovered by an anodic stripping voltammetry followed by an anodic polarization at –0.05 V, but the activity deteriorates again in a short period. The cyclic voltammogram shown in Fig. 6 shows a broad peak between –0.3 and 0.1 V in the anodic scanning. Such deactivations were observed more or less with all the chemicals used for the electrolytes; reagents from puriss p.a. ACS by Fluka, A. C. S.

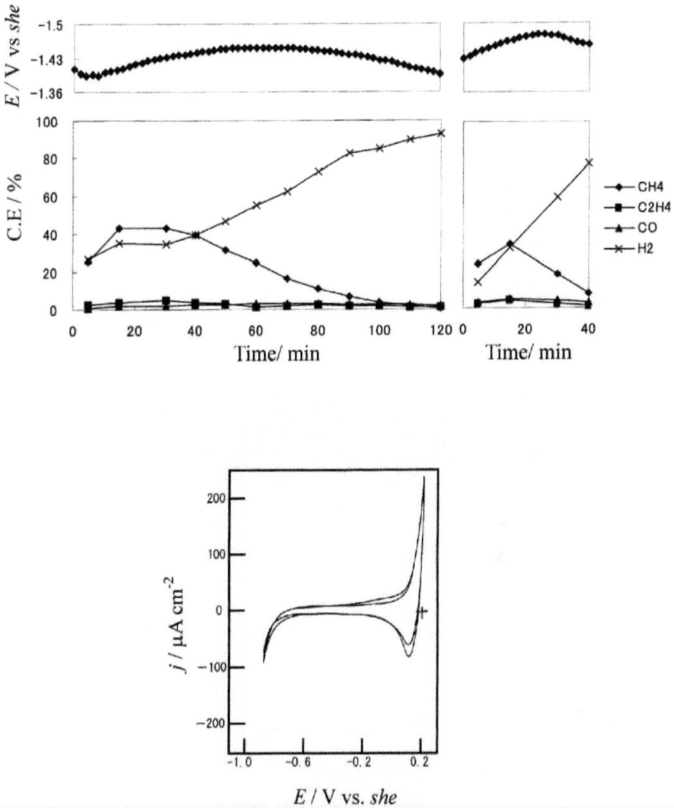

Figure 6. A time course of CO_2 reduction at a Cu electrode in an electrolyte solution 0.3 M $KHCO_3$ prepared from a chemical reagent puriss p.a. ACS by Fluka *without* preelectrolysis. The right hand side shows a time course of a repeated CO_2 reduction after an anode stripping voltammetry and an anodic polarization at –0.05 V for 5 min. Voltammetric scan rate: 100 mV s^{-1}. Reprinted from Ref. 170, Copyright (2005) with permission from Elsevier.

reagent by Aldrich, GR for analysis by Merck, GR by Wako Jun-yaku, and GR by Nacalai Tesque.

Figure 7 presents a time course with 0.3 M $KHCO_3$ prepared from a chemical reagent GR by Wako Jun-yaku purified with preelectrolysis with a Pt black cathode more than 16 h under Ar atmos-

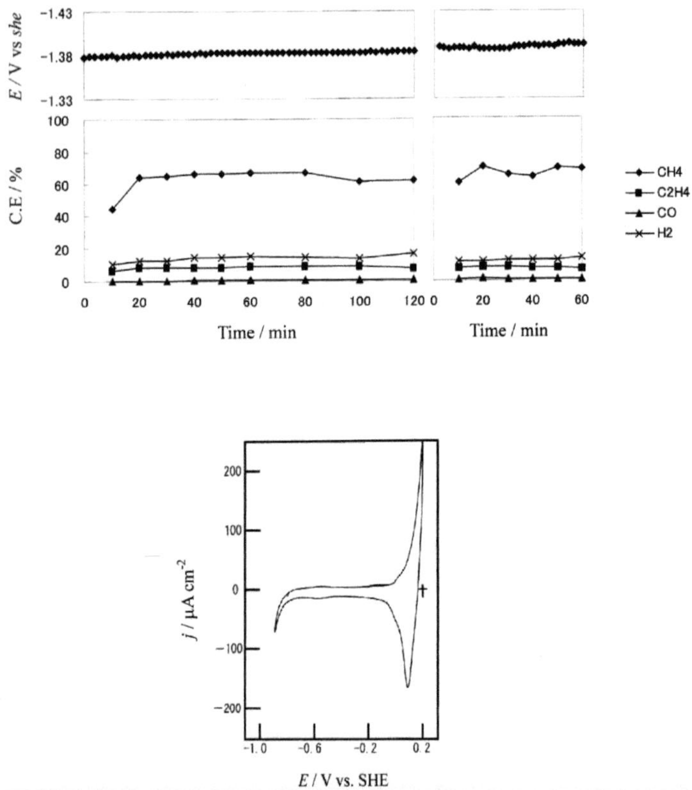

Figure 7. A time course of CO_2 reduction at a Cu electrode in an electrolyte solution of 0.3 M $KHCO_3$ prepared from a chemical reagent GR by Wako *with* preelectrolysis. The right hand side shows a time course of repeated CO_2 reduction after an anode stripping and an anodic polarization at –0.05 V for 5 min. Voltammetric scan rate: 100 mV s^{-1}. Reprinted from Ref. 170, Copyright (2005) with permission from Elsevier.

phere to remove heavy metal impurities possibly contained in the chemicals. The electrochemical measurements were conducted in the same electrolysis cell used for the preelectrolysis without transferring the electrolyte solution. No deactivation takes place during the electrolysis for 120 min. Practically no anodic peak appears in the cyclic voltammogram.

Deliberate addition of 0.1 μM Fe^{++} was examined with $KHCO_3$ electrolyte solution prepared from GR by Wako Jun-yaku purified with preelectrolysis. Figure 8 demonstrates that the deactivation is reproduced; CH_4 formation rapidly decays and HER grows simultaneously. C_2H_4 formation also diminishes. Anodic polarization after the electrolysis recovers the electrocatalytic activity. The cyclic voltammogram gives an anodic peak at –0.1 V. Identical measurements with 0.1 μM Zn^{++} addition also reproduced more gradual deactivation; an anodic peak appears at –0.6 V in an anode stripping voltammogram.

The concentration of heavy metals in the electrolyte solutions can be estimated on the basis of the electric charge of the anode peaks in the anode stripping voltammogram. The concentration originally contained in the chemical reagents was evaluated below 0.2 ppm both as Fe and Zn, mostly 0.1 ppm or less, far below the standard of the impurity levels guaranteed by the manufacturers. Thus the deactivation of Cu electrode in CO_2 reduction is not caused by adsorption of the products or the intermediates produced in CO_2 reduction.

The usefulness of the pre-electrolysis was stressed by Hori et al.[170] They showed that the concentration of Fe^{++} in $KHCO_3$ solutions is remarkably decreased by the pre-electrolysis, as evidenced by Electrothermal Atomic Absorption Spectroscopy (ETAAS). Some papers report that deactivation still takes place even with solution treated with pre-electrolysis. Nevertheless, any analytical data has not been presented with regard to the electrolyte solutions. If any preelectrolysis is conducted in a cell, and the solution is transferred to another electrolysis cell, the solution is possibly contaminated during the transfer process.

Deactivation of Cu electrode can appear in another manner. When a Cu electrode is covered with Cd in submonolayer coverage, hydrocarbon formation is suppressed and CO production increases.[95] Au and Ag were also reported to suffer from the deactivation during the electrolysis.[52, 56, 58]

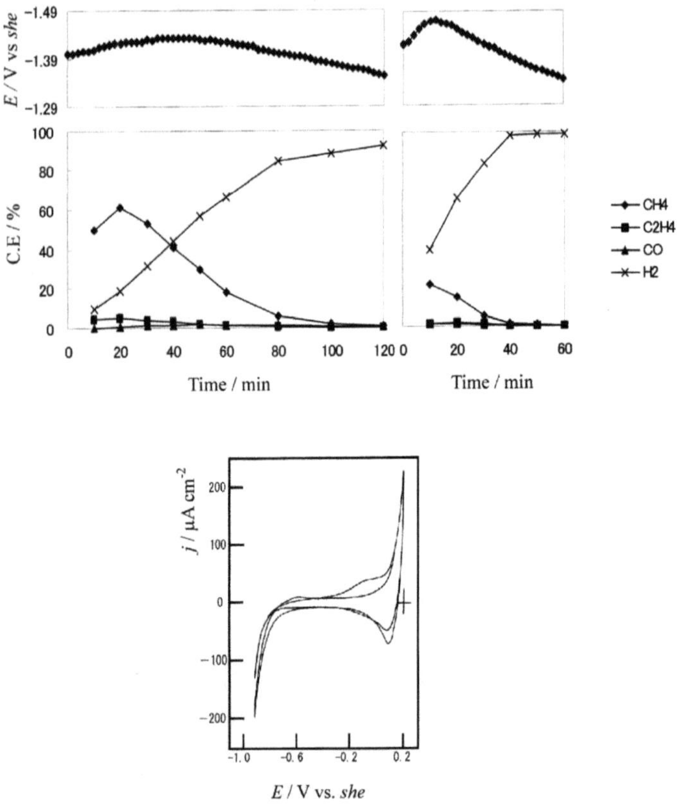

Figure 8. A time course of CO_2 reduction at a Cu electrode in an electrolyte solution of 0.3 M $KHCO_3$ prepared from a chemical reagent GR by Wako *with preelectrolysis*, deliberately added with 0.1 μM Fe^{2+}. The right hand side shows a time course of repeated CO_2 reduction after an anode stripping and an anodic polarization at –0.05 V for 5 min. Voltammetric scan rate: 100 mV s^{-1}. Reprinted from Ref. 170, Copyright (2005) with permission from Elsevier.

Some organic substances suppress CO_2 reduction at Cu electrode. Slight amount of trimethylamine, occasionally contained in anion exchange membrane, prevents CO_2 reduction with enhanced HER.[170] Addition of ca. 10% acetonitrile to 0.1 M $KHCO_3$ severely diminishes formation of CH_4 and C_2H_4 at a Cu electrode with simultaneously increased formation of CO.[171]

Another possibility for the deactivation may be derived from the water from which electrolytic solutions are prepared. Very tiny amount of organic substances, such as surface active reagent, is recently contained in water.[172] These organic substances are sometimes very hard to remove by distillation. Adsorption of surface active reagents may deteriorate electrocatalytic activity of any electrode.

VI. CLASSIFICATION OF ELECTRODE METALS AND REACTION SCHEME

1. Classification of Electrode Metals and CO Selectivity

Product selectivity in CO_2 reduction depends on many factors: concentration of the reactants, i.e. CO_2 and H_2O; electrode potential; temperature. Among these factors, the electrode metal and the electrolyte solution basically determine the product selectivity, as recently accepted by most workers. In aqueous systems, electrode metals can be divided into 4 groups in accordance with the product selectivity, discussed in Section III.1. These electrode metals may be regrouped broadly into 2 categories, i.e. CO formation metals and formate formation metals. Cu electrode yields CH_4, C_2H_4 and alcohols from CO_2, in which process CO is formed as an intermediate species as described later. Ni, Pt and other platinum family metals give CO strongly adsorbed on the electrode surface, which practically prevents further reduction of CO_2 under atmospheric pressure, forming H_2 in aqueous systems. Au, Ag, Zn, Pd, and Ga give CO as a major product in CO_2 reduction. These metals (Cu, Au, Ag, Zn, Pd, Ga, Ni and Pt) belong to CO formation metals. Formate formation metals (Pb,Hg, In, Sn, Cd, and Tl) give formate as a major product in neutral aqueous electrolytes in CO_2 reduction.

Such a classification obtained from the selectivity in aqueous systems is combined with that in nonaqueous systems shown in Table 6, resulting in Table 8. CO formation metals in aqueous electrolyte yield CO in PC. Some formate formation metals in aqueous electrolyte, Hg, Tl and Pb, give oxalic acid in PC. The intermediate metals, Cd, Sn, and In, give CO in PC solutions as the major product, whereas formic acid in aqueous solutions.

Table 8
Classification of Electrode Metals in Accordance with Product Selectivity in CO_2 Reduction in Aqueous and Nonaqueous Electrolyte (PC). Reproduced from Ref. 23, Copyright (1994) with permission from Elsevier

Nonaqueous electrolyte (PC) \ Aqueous electrolyte	CO	$HCOO^-$
CO	Au, Ag, Cu, Zn	Cd, Sn, In
COO^{2-}	–	Hg, Tl, Pb

The presence of small amount of foreign atoms on the electrode surface changes greatly the reaction selectivity of CO_2 reduction, as exemplified by metal electrodes alloyed with other metals or adatom modified electrode. Adatom modified electrodes were prepared by underpotential deposition (UPD) or overpotential deposition techniques.[23] The coverage of the adatoms ranged between 1.1 and 1.6, mostly between 1.25 and 1.35. CO selectivity r in CO_2 reduction, defined as $F(CO)/[F(CO) + F(HCOO^-)]$, was

Table 9
CO Selectivity of Adatom Modified Electrodes, *r*. Reproduced from Ref. 23, Copyright (1994) with permission from Elsevier

Substrate metal	Potential (V vs. SHE)	Without adatoms	With adatoms				
			Cd	Sn	In	Pb	Tl
Au	−1.14	0.99	0.82	0.54	0.60	0.53	0.95
Ag	−1.37	0.99	0.92	0.59	0.40	0.39	0.05
Cu	−1.44	0.69	0.82	0.36	0.20	0.00	0.00
Zn	−1.54	0.93	0.19	0.08	0.03	0.01	0.01
Cd	−1.63	0.15					
Sn	−1.48	0.07					
In	−1.55	0.02					
Pb	−1.63	0.00	0.00	0.03			
Tl	−1.68	0.00					
Hg	−1.51	0.00					

CO selectivity, $r = CO/(CO + HCOO^-)$; current density for Hg: 0.5 mA cm^{-2}, other metals: 5 mA cm^{-2}, electrolyte: 0.1 M $KHCO_3$: temperature: 18.5 ± 0.5°C.

measured with various adatom modified electrodes, where F(CO) and $F(HCOO^-)$ denote the faradaic efficiencies for CO and $HCOO^-$. F(CO) contains the faradaic efficiencies not only for CO itself but also for CO which is further reduced to hydrocarbons. Table 9 shows the value of r, obtained from CO_2 electrolysis. r of CO formation metal electrodes, Au, Ag, Cu and Zn, varies significantly by modification with formate formation metal adatoms. One may expect that the electrocatalytic properties of adatom modified electrodes are intermediate between those of substrate and modifier metals. Such analysis, not a precise one, may still provide a rough measure of electrocatalytic feature of metals. The dotted line in Table 9 indicates a border of $r = 0.5$, which divides the adatom modified electrodes of CO formation from those of $HCOO^-$ formation. Tables 3, 8, and 9 lead to the order of CO selectivity of metals; Au > Ag > Cu > Zn >> Cd > Sn > In > Pb > Tl ≈ Hg.

2. Electrode Potential of CO_2 Reduction at Various Metal Electrodes

The standard potential of the reduction of CO_2, calculated from the thermodynamic data, depends on the products, –0.52 V vs. SHE at pH 7 at 25°C for CO and so on as given in Section II.1. Table 3 shows that the reduction of CO_2 actually takes place at much more negative potentials than the values estimated from the thermodynamic data. Such high overpotentials must result from involvement of intermediate species which requires highly negative potential for the formation; $CO_2{\cdot}^-$ anion radical is the presumed intermediate species.

A correlation was investigated between the potentials of CO_2 reduction at various metal electrodes and the heat of fusion of the metals. The potentials of CO_2 reduction was determined by voltammetric measurements at 0.5 mA cm^{-2} at the scanning rate 100 mV s^{-1} in 0.1 M $KHCO_3$ solution with exceptions of Pt and Hg. The potential for Pt is the value observed for "reduced CO_2" formation on the electrode surface. The potential for Hg was determined as that of steady state electroreduction of CO_2 at 0.5 mA cm^{-2}. Figure 9 shows that the potential for CO_2 reduction is well correlated with the heat of fusion of metals.[16] CO formation takes place at less negative potentials on metal electrodes of high heat of

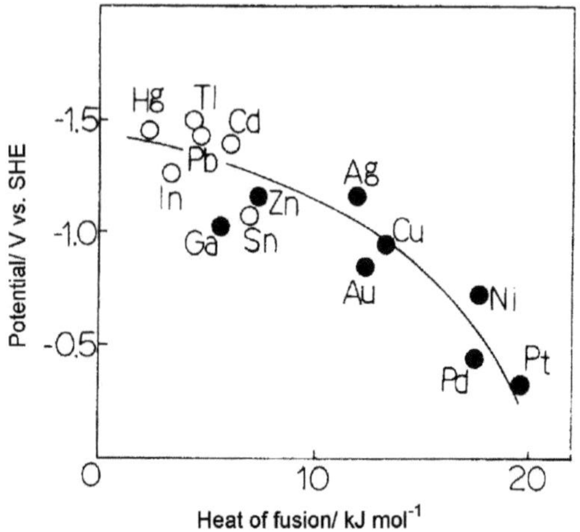

Figure 9. Correlation between the potential of CO_2 reduction and the heat of fusion of various metals. ●: CO formation metals, ○: $HCOO^-$ formation metals. Reprinted from Ref. 23, Copyright (1994) with permission from Elsevier.

fusion. It is noteworthy that CO formation metals are clearly separated from $HCOO^-$ ones in Fig. 9 with an exception of Sn. The CO_2 reduction potentials are also well correlated with hydrogen evolution overpotential in 0.1 M $KHCO_3$ solution with an exception of Sn.[23]

3. Formation of $CO_2{}^{\cdot -}$ Anion Radical and Further Reduction to $HCOO^-$

Jordan and Smith proposed formation of $CO_2{}^{\cdot -}$ anion radical by one electron transfer to CO_2 molecule as an initial step in CO_2 reduction on the basis of their polarographic study,[44]

$$CO_2 + e^- = CO_2{}^{\cdot -} \qquad (37)$$

Eyring and his coworkers analyzed polarization data of CO_2 reduction to $HCOO^-$ at a Hg electrode in aqueous electrolytes containing HCO_3^-, and discussed the reaction mechanism with $CO_2{\cdot}^-$ anion radical as the initial intermediate.[22] The resultant $CO_2{\cdot}^-$ subsequently accepts a H^+ and another electron, reduced to $HCOO^-$,

$$CO_2{\cdot}^- + H_2O = HCOO{\cdot} + OH^- \tag{38}$$

$$HCOO{\cdot} + e^- = HCOO^- \tag{39}$$

They obtained a logarithm of current and potential relationship composed of two linear parts, and rationalized the polarization data. They determined the transfer coefficient in the higher overpotential region 0.25; confirmed later by Hori and Suzuki in their measurements of the partial current densities of $HCOO^-$ formation at a Hg pool electrode.[15] This value indicates that the rate determining step is the first electron transfer to form $CO_2{\cdot}^-$ anion radical. The transfer coefficient in the lower overvoltage region is 0.67.

Pacansky et al. studied scf ab initio molecular orbital energies and atomic population analysis of $CO_2{\cdot}^-$ at the minimum energy geometry (Fig. 10). According to their analysis, the unpaired electron density at the highest occupied orbital is localized at C atom at 84%.[173] This result suggests that $CO_2{\cdot}^-$ is ready to react as a nucleophilic reactant at the carbon atom. The standard potential of $CO_2{\cdot}^-$ formation is −2.21 V vs. SCE in DMF,[174] −1.90[175] or −1.85 V vs. SHE[176] in aqueous media.

Detection of $CO_2{\cdot}^-$ anion radical was conducted with a Pb electrode in CO_2 saturated aqueous, acetonitrile and propylene carbonate electrolytes during cathodic polarization by ultraviolet (UV) spectroscopic measurements by Aylmer-Kelly et al.[38] $CO_2{\cdot}^-$ anion radical is mostly present freely in both aqueous and nonaqueous electrolyte solutions. Stabilization of $CO_2{\cdot}^-$ due to hydrogen bond formation in aqueous electrolyte solution was suggested on the basis of the red shift of the observed absorption band.

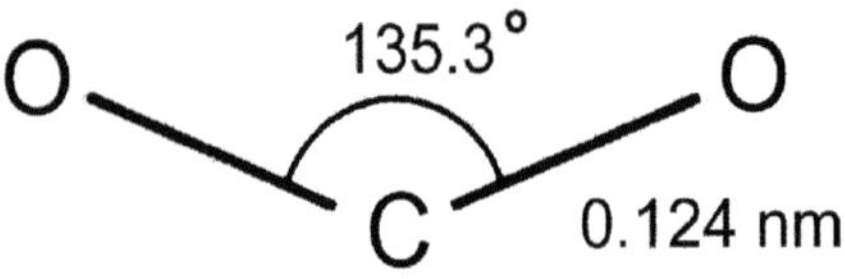

Figure 10. Atomic configuration of $CO_2^{\cdot -}$ employed in scf ab intio study by Pacansky et al.[173]

Formation of $CO_2^{\cdot -}$ anion radical and subsequent reduction to formate were studied by photoemission and polarization measurements with Hg electrodes[177] as well as the capacitance and potential decay measurements at Sn and In electrodes;[178] all these measurements agreed that very low fraction of the electrode is covered by the adsorbed species. Schiffrin investigated reduction of $CO_2^{\cdot -}$ anion radical formed in a photoemission measurement with a Hg electrode, and showed that the potential of the reduction does not depend on pH. He thus concluded that H_2O is the proton donor in the formate formation from $CO_2^{\cdot -}$,[179] as confirmed later by Hori and Suzuki.[15] They demonstrated that the electrode potential is constant in the pH range 2 to 8 at a constant partial current density of $HCOO^-$ formation 0.5 mA cm^{-2} at a Hg electrode.

Hence, the formation of $HCOO^-$ may be initiated by one electron transfer to CO_2 at the potential negative of −1.6 V vs. SHE, forming $CO_2^{\cdot -}$ present mostly freely in the solution close to the electrode. The concentration of $CO_2^{\cdot -}$ would be ca. 10^{-5} mol dm^{-3}, if one takes into account the standard potential −1.85 V or −1.90 V and the Nernst relation 59 mV decade^{-1} at 25°C. Free $CO_2^{\cdot -}$ will take a proton from a H_2O molecule at the nucleophilic carbon atom, forming $HCO_2^{\cdot}$. H^+ will not be bonded to the O atom of $CO_2^{\cdot -}$, since pK_a value of the acid-base couple ($CO_2^{\cdot -}/CO_2H$) is low (= 1.4).[180] $HCO_2^{\cdot}$ is subsequently reduced to $HCOO^-$ at the electrode in aqueous media. The reaction scheme would be valid for Hg, Tl, Pb, In, Sn, and Cd, and depicted in Fig. 11(1). The electrode potentials of Pb, Cd and Tl are −1.6 V or more negative at 5.0 mA cm^{-2} as given in Table 3. The potential of Hg is −1.51 V at

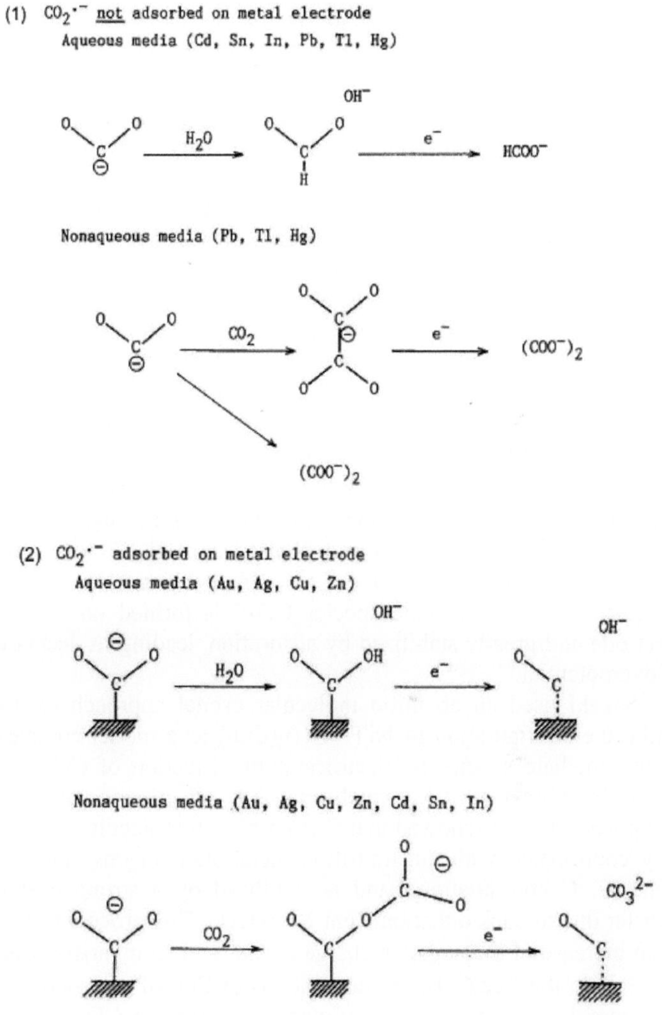

Figure 11. Reaction scheme of the electrochemical reduction of CO_2. Reprinted from Ref. 23, Copyright (1994) with permission from Elsevier.

0.5 mA cm^{-2}, which would be equivalent to −1.7 V at 5.0 mA cm^{-2} with an assumption of the transfer coefficient for $HCOO^-$ formation 0.25.[15,22,181]

4. Formation of Adsorbed $CO_2^{\cdot -}$ Leading to further Reduction to CO or $HCOO^-$

Vassiliev and his coworkers published polarization curves of various metal electrodes in CO_2 reduction in aqueous media, and showed great differences in these polarization curves, which depends on the electrode metals. They discussed their current potential relationships, and argued that the great difference of the polarization curves must be derived from adsorption of the intermediate species $CO_2^{\cdot -}$ on the metal electrode surface.[7,182]

Table 3 shows that CO formation takes place with relatively lower overpotentials than $HCOO^-$ formation. Among all the electrodes, Au electrode reduces CO_2 to CO at remarkably low cathodic potential, −1.14 V at 5 mA cm^{-2}. This fact strongly suggests that CO is produced by a mechanism different from $HCOO^-$ formation which may proceed with free $CO_2^{\cdot -}$ intervening. Hori et al. suggested that intermediate species $CO_2^{\cdot -}$ is formed on the Au electrode and greatly stabilized by adsorption, leading to decrease of overpotential.[50]

Sakaki used an ab initio molecular orbital approach to describe the configuration of $Ni^{I}F(NH_3)_4(CO_2)$ as a model complex of intermediate species in electrocatalytic reduction of CO_2 mediated by $Ni(cyclam)Cl_2$, which proceeds effectively with low overpotential.[183] He showed that C atom of CO_2 molecule is favorably coordinated with the transition metal atom in this complex (Fig. 12, C coordination), and is stabilized by a strong charge transfer due to back donation from Ni to CO_2. This process results in an increase of the negative charge (−0.58 e^- in comparison with −0.33 e^- in the free CO_2) on the O atoms of CO_2 of the coordination complex in an atomic configuration similar to $CO_2^{\cdot -}$ anion radical. The extra negative charge on O atom would facilitate protonation to the coordinated CO_2. This result supports the reaction mechanism of CO formation from CO_2 with Ni cyclam catalyst proposed by Sauvage et al. (Fig. 13).[184]

Chemisorbed $CO_2(\delta-)$ was detected at Cu surfaces in gas-solid interface by XPS at low temperature, such as 160 °K.[185 - 187] Detection of adsorbed $CO_2^{\cdot-}$ in electrochemical systems has been investigated by several workers,[188, 189] but confirmative results have not been obtained to date.

The potentials of CO_2 reduction is well correlated with the heat of fusion of the electrode metals (Fig. 9). The heat of fusion is related to the extent of *d* electron contribution to metallic bond[190], and may be taken as a measure of *d* electron avilability; *d* electron availability will affect the back donation and thus determine the extent of the stabilization of adsorbed $CO_2^{\cdot-}$. Stabilized $CO_2^{\cdot-}$ will have the extra negative charge on O atoms like CO_2 coordinating

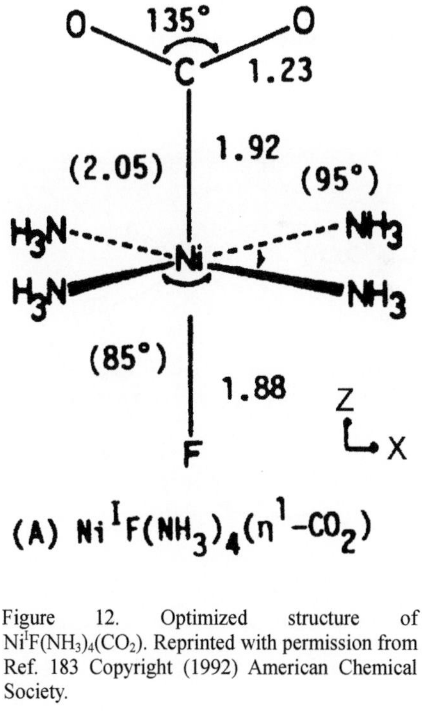

Figure 12. Optimized structure of $Ni^{I}F(NH_3)_4(CO_2)$. Reprinted with permission from Ref. 183 Copyright (1992) American Chemical Society.

with Ni(cyclam)$^{2+}$, which facilitates protonation leading to CO formation. Thus adsorbed $CO_2^{\cdot -}$ would be nucleophilic at O atoms, and a reaction similar to coordinated CO_2 will take place.

Electrophilic reagents, H_2O in aqueous solution, react with the O atom of adsorbed $CO_2^{\cdot -}$, forming CO(ad) and OH^- as depicted in Fig. 11(2). H^+ will not take part in the CO formation as discussed with regard to Au and Zn electrodes in Section VII.1, since the partial current of CO formation is independent of pH. CO(ad) is readily desorbed from the electrode as a gaseous molecule. The reaction scheme may be applied to Au, Ag, Cu and Zn electrodes in aqueous media. The sequence of CO selectivity roughly agrees with that of the electrode potentials shown in Fig. 9. This agreement verifies the above hypothesis that CO is favorably produced from the elecrode metals which stabilize $CO_2^{\cdot -}$ effectively.

The potentials of In and Sn, which are $HCOO^-$ formation metals, are much less negative than Pb, Cd and Tl. On these metal electrodes, $CO_2^{\cdot -}$ could be weakly adsorbed, or adsorbed on the electrode in O coordination with one or two O atoms adsorbed on

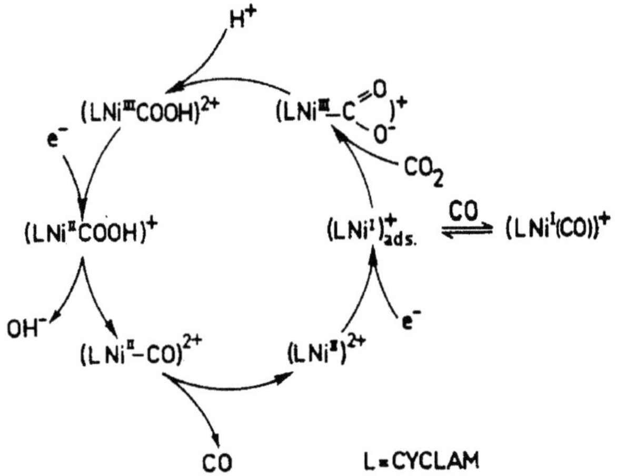

Figure 13. Postulated mechanistic cycle for the electrocatalytic reduction of CO_2 into CO by Nicyclam^{2+} in water. Reprinted with permission from Ref. 184, Copyright (1986) American Chemical Society.

the surface; C atom of the O coordinated $CO_2^{\cdot-}$ will react with H_2O, forming $HCOO^-$ subsequently. Cu produces $HCOO^-$ as well at −1.44 V in a low yield as shown in Table 3. O coordinated $CO_2^{\cdot-}$ may be present in a small fraction on Cu electrode in addition to C coordinated $CO_2^{\cdot-}$ in the most part.

5. Reaction Scheme in Nonaqueous Electrolyte

Savéant and his coworkers studied the electrochemical reduction of CO_2 at Pb and Hg in H_2O-DMF solutions.[124,125,134] The major products are $(COOH)_2$, CO and HCOOH. They proposed a reaction scheme that an electron transfer to CO_2 molecule initiates the process, forming $CO_2^{\cdot-}$. They suggested that the interactions between the electrode and the reactants, intermediates and products are negligible since the CO_2 reduction at Hg and Pb electrodes proceeds at potentials close to the standard potential of $CO_2^{\cdot-}/CO_2$ couple. Several competing homogeneous reactions sequently take place in parallel in the electrolyte solution (Fig. 14). They obtained rate constants of the homogeneous reactions in the analysis of the experimental data,[124] and rationalized that the product distribution at Pb and Hg electrodes is determined by the current density and the concentration of CO_2 and H_2O. Oxalate is formed by coupling of $CO_2^{\cdot-}$, Fig. 14(1), and CO is formed by sequential reactions (2) and (3) or (3') in Fig. 14. However, Hg and Pb electrodes produce CO with relatively low yield, and the major product is oxalate or formate depending on the concentration of H_2O. Oppositely, Table 6 shows that Au and Ag electrodes give mainly CO with scarce formation of oxalate or formate. Thus the product selectivity is primarily determined by the extent of stabilization of $CO_2^{\cdot-}$ by adsorption on electrode surface in nonaqueous electrolytes in a way similar to that in aqueous electrolyte.

As discussed in Section VI.3, $CO_2^{\cdot-}$ is not adsorbed on Pb, Hg, and Tl, and is freely present in nonaqueous electrolyte solution as well. If H_2O is present in the electrolyte, a $CO_2^{\cdot-}$ will react with a H_2O molecule at the nucleophilic carbon atom, as shown in Fig. 11(1), further reduced to formate. The H_2O molecule reacts as a Lewis acid. If H_2O molecule is not available and plenty of CO_2 is

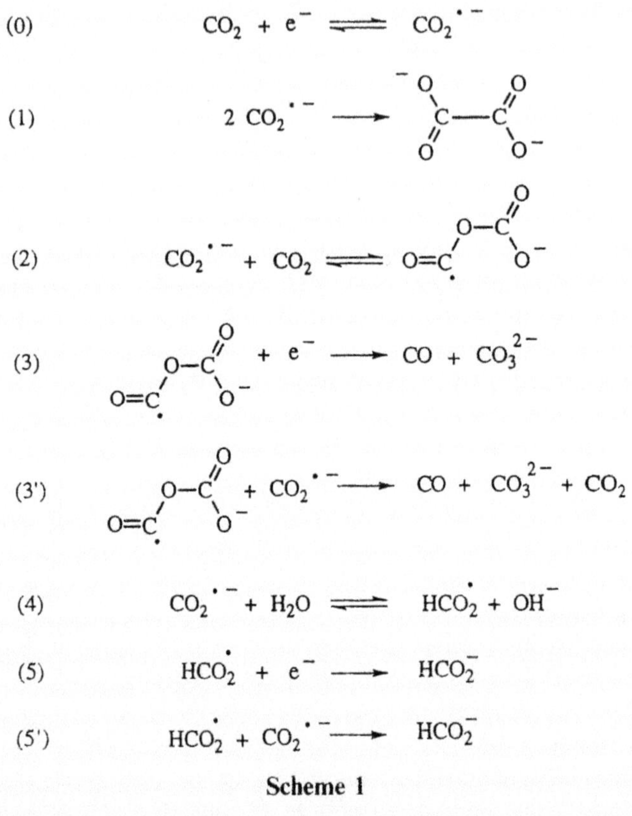

Figure 14. Reaction scheme of CO_2 reduction without specific interaction between electrode and the reactants, intermediate, and products proposed by Savéant et al. Reference 134–Reproduced by permission of The Royal Society of Chemistry.

present, a CO_2 molecule will play a role of a Lewis acid with the nucleophilic C of $CO_2^{\cdot-}$ as shown in Fig. 11(1). The coupling of $CO_2^{\cdot-}$ and CO_2 will lead to formation of an adduct $(CO_2)_2^{\cdot-}$. Aylmer-Kelly et al. observed a UV absorption band, apart from that of $CO_2^{\cdot-}$. They assigned the absorption band to an intermediate $(CO_2)_2^{\cdot-}$, and proposed that oxalate is produced from the adduct

$(CO_2)_2^{\cdot -}$.[38] Bagotzky and Osetrova suggested that this scheme is more probable than Savéant's couping mechanism.[7]

On the electrode surface of CO formation metals Au, Ag, and Zn, $CO_2^{\cdot -}$ is stabilized by adsorption both in aqueous and nonaqueous electrolytes. In nonaqueous media, a CO_2 molecule reacts as a Lewis acid with adsorbed $CO_2^{\cdot -}$, and allows a C–O bond of the $CO_2^{\cdot -}$ to be broken. This process forms CO(ad) and CO_3^{2-} as postulated by Savéant and his coworkers[191] for electrochemical reduction of CO_2 catalysed by iron porphyrins. CO(ad) thus formed is easily desorbed, as shown in Fig. 11(2).

Cd, Sn, and In, of medium CO selectivity, do not strongly adsorb $CO_2^{\cdot -}$. $CO_2^{\cdot -}$ will be mostly freely present in aqueous electrolyte owing to the hydrogen bond stabilization by water molecules;[38] high dielectric constant of water molecule will also contribute to the stabilization of $CO_2^{\cdot -}$. $CO_2^{\cdot -}$ stabilized in the electrolyte will be further reduced to $HCOO^-$. However, $CO_2^{\cdot -}$ is not sufficiently stabilized in nonaqueous electrolyte due to lack of hydrogen bond formation and low dielectric constant of the solvents. Thus $CO_2^{\cdot -}$ adsorbed on Cd, Sn, and In may be relatively stabler than $CO_2^{\cdot -}$ dissolved in the electrolyte. These metals yield CO in nonaqueous media in the same manner as Au, Ag, and Zn in Fig. 11(2). The CO selectivity mentioned above will be closely connected to the stability of adsorbed $CO_2^{\cdot -}$ on the electrode.

A similar case is likely to happen in 0.5 M $KHCO_3$ aqueous solution under elevated pressures. Sakata et al. reported that an In electrode yields CO preferentially instead of $HCOO^-$ at a low current density (1 mA cm^{-2}) under 20 atm.[40] $CO_2^{\cdot -}$, adsorbed on the electrode with a higher coverage than under 1 atm, may react with CO_2 molecules present abundantly in pressurized solution, forming CO as the major product.

Ikeda et al. showed that addition of small amount of H_2O to nonaqueous electrolyte greatly affects the product selectivity; the product distributions in 0.8 to 4.2% of H_2O (ca. 0.5 to 2.8-M H_2O) in 0.1-M TEAP/ PC become similar to those in aqueous electrolytes obtained by the same authors[127] (Table 10). A similar result was obtained with a Sn electrode in 0.1 M-TEAP/ AN-H_2O mixtures.[192] CO formation at Sn electrodes is predominant in AN based electrolytes of H_2O below 1000 mM (Fig. 15). $HCOO^-$ for-

Table 10
Variation of Product Distribution in CO_2 Reduction in Nonaqueous Electrolyte Added with H_2O. A part of this table (Pb and In) was reprinted with permission from Ref. 127, Copyright (1987) Chemical Society of Japan

Electrode metal	H_2O conc. M	Faradaic efficiency/ %			
		Oxalic acid	Glyoxylic acid	Formic acid	CO
Pb[a]	0.02	73.3	3.4	2.2	11.0
	0.5	9.8	0.0	56.9	14.4
	2.8	trace	0.0	72.8	8.6
In[a]	0.02	0.1	0.0	1.3	85.3
	0.5	0.0	0.0	33.3	57.8
	2.8	0.0	0.0	63.6	20.0
Sn[b]	0.02	0.7	n.a.[c]	28.1	73.9
	0.5	0.0	n.a.	46.2	51.5
	2.8	0.0	n.a.	60.0	33.8

[a]0.1-M TEAP in PC at −2.6 V vs. Ag/AgCl/(0.01-M LiCl + 0.1-M TEAP)/PC. Electric charge spent in the electrolysis: 100 C.[127]
[b]Data interpolated from those obtained in constant current electrolysis at 5 mA cm^{-2} in 0.1-M TEAP in acetonitrile.[192] Electrode potential: −2.46, −2.27, −2.07 V vs. Fc^+/Fc (0.1 M TEAP/AN) for H_2O concentration 0.02, 0.5, 2.8 M, respectively. 0.0 V vs. Fc^+/Fc (0.1 M TEAP/AN) is roughly 0.08 V vs. Ag/Ag^+/(0.01 M, 0.1 M TEAP).
[c]not analyzed.

mation exceeds CO at H_2O concentration 1000 mM, increasing further with the increase of H_2O concentration. Thus the stabilization of $CO_2^{\cdot -}$ due to hydrogen bond formation will become significant at H_2O 1000 mM. The data for In electrodes in Table 10 is interpreted in a similar manner. The data for Pb electrodes indicate that $CO_2^{\cdot -}$ is adsorbed at the electrode neither in aqueous nor nonaqueous electrolyte. Lewis acid, which reacts with $CO_2^{\cdot -}$ freely present in the electrolyte with a Pb electrode, is H_2O in aqueous solution, and CO_2 in nonaqueous one.

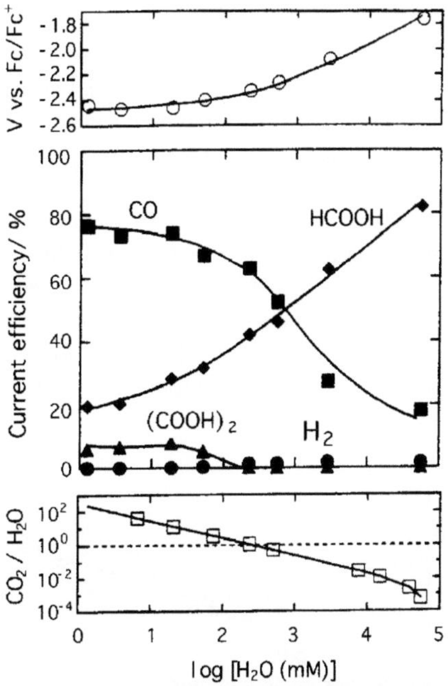

Figure 15. CO_2 reduction at a Sn electrode in 0.1 M TEAP AN-H_2O mixtures. Current density: 5 mA cm^{-2}.[192]

VII. ELECTROCHEMICAL REDUCTION OF CO_2 TO CO AT SELECTED METAL AND NONMETAL ELECTRODES

1. CO Formation at Au, Ag, and Zn

(*i*) *Au*

Gold electrodes yield CO in CO_2 reduction in aqueous media.[23,29,35,50-55,193] Some papers reported kinetic aspects of the reaction. CO formation begins at –0.8 V vs. SHE.[50] Such a low overpotential strongly suggests that the formation of CO takes place with intermediate intervention of $CO_2{\cdot}^-$ stabilized by adsorption on the electrode as is previously mentioned. Hori et al. carried out potentiostatic measurements in 0.5 M $KHCO_3$ preelectrolyzed before the electrolysis operation, and obtained a relationship between the potential and logarithm of the partial current of CO formation (i_c) determined from analytical values (Fig. 16). They showed that the Tafel slope is approximately 130 mV decade^{-1} in the lower current region where the transport process of CO_2 does not interfere with the supply of CO_2 to the electrode; the Tafel slope corresponds to the transfer coefficient 0.46. i_c at a constant potential is linear to the CO_2 partial pressure. These data suggest that the CO_2 reduction proceeds in the 1st order with respect to CO_2, and the rate determining step of the reaction is the first electron transfer to CO_2. These facts support the reaction scheme discussed in Section VI.

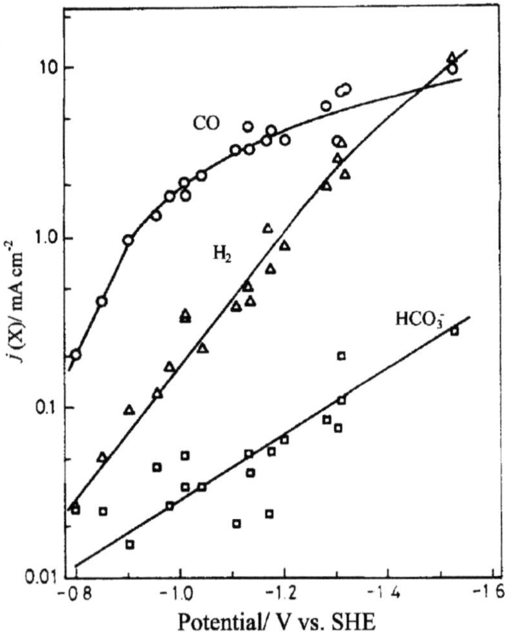

Figure 16. Partial current densities of the products j(X) vs. electrode potential in CO_2 reduction at a Au electrode. 0.5 M $KHCO_3$, 1 atm CO_2, 18 °C. The potential axis negative to the right. Reference 50–Reproduced by permission of The Royal Society of Chemistry.

Ikeda et al. later published results on the CO_2 reduction at an Au electrode in phosphate buffer solutions of pH 2 to 6.8.[53] They showed that i_c, also obtained from analytical measurements, is proportional to the pressure of CO_2. The Tafel slope is ca. 120 mV decade^{-1}. Their potential partial current relation obtained with electrolytes at pH below 4.3 agreed well with Hori et al.'s[50] obtained with 0.5 M $KHCO_3$ at pH 7.5 after the double layer correction due to the difference of the electrolyte concentration. This fact shows that CO formation at Au electrode does not depend on pH of the electrolyte, and that the proton donor is not H^+ but H_2O molecules.

(ii) Ag

Silver electrodes give CO as the primary product in CO_2 reduction in aqueous media.[23,29,31,56,58] Sakata et al. reported that lower temperature, e.g. 0°C, is effective to enhance the faradaic efficiency of CO as well as $HCOO^-$.[36] Single crystal Ag electrodes, of fcc crystal structure, were employed for CO_2 reduction in aqueous media.[57] Major products from CO_2 reduction are CO and $HCOO^-$, accompanied with hydrogen evolution of lower faradaic efficiency at three low index planes, (110), (111) and (100), same as polycrystal electrodes. Among these crystal orientations, (110) shows the highest electrocatalytic activity in CO formation, followed by (111) and (100). With regard to $HCOO^-$ formation, the activity changes slightly still with the sequence (110) > (111) > (100). It may be natural that (110), atomically roughest and abundant in active sites, provides highest catalytic activity in the reaction. The (111) surface, which is the most stable crystal orientation in fcc crystals with lowest dangling bonds, usually gives the lowest catalytic activity, whereas Ag(111) is a little more active than (100) in CO_2 reduction.

(iii) Zn

Zinc is occasionally classified as a metal that forms $HCOO^-$ as the primary product. A discrepancy has been reported with regard to the major product, CO or $HCOO^-$, between research groups.[23,36,127] The reproducibility of the product distribution is sometimes poor in the faradaic efficiency.[29] The reason of the low reproducibility was found to come from surface contamination of Zn electrode, as demonstrated by Table 9.[23] Even submonolayer coverage of Zn electrode with $HCOO^-$ formation metals strongly enhances the production of $HCOO^-$. Zn electrode, if free from the surface contamination, will yield CO as the primary product. Ikeda et al. reported that CO formation is promoted in electrolyte solutions added with Zn^{2+} up to 1 mM, whereas $HCOO^-$ formation is not affected by the presence of Zn^{2+}.[60] Such enhancement of CO formation is likely to be derived from electrodeposition of Zn^{2+} and the subsequent regeneration of Zn surface during CO_2 reduction. They also reported that CO formation does not depend on pH of the solution, and is proportional to CO_2 partial pressure.

2. Platinum Group Metals

(*i*) *Pt*

Platinum electrodes do not give products continuously in CO_2 reduction in aqueous media under 1 atm as shown in Table 3. Platinum electrodes initially reduce CO_2 to "reduced CO_2".[107,108] The entity of the "reduced CO_2" is CO strongly adsorbed on the Pt electrode, as revealed by Beden et al. by means of infrared spectroscopy.[159] This fact is later confirmed by other workers.[194, 195] In addition to linearly bonded CO as the major adsorbed species, small amounts of bridged and multibonded CO, COH and HCOO species are also detected on Pt electrode surface.[196-198] The presence of "reduced CO_2" on Pt electrode practically inhibits further reduction of CO_2 in aqueous media. The formation of "reduced CO_2" proceeds as below in the potential region in which adsorbed hydrogen is stably present.

$$CO_2 + 2H(ads) = CO(ads) + H_2O \qquad (40)$$

Formation of "reduced CO_2", or adsorbed CO, on Pt electrodes has been studied using Pt single crystal electrodes. Yeager and his coworkers reported reduction of CO_2 at Pt(111), Pt(100) and Pt(110) in $HClO_4$ solution. They showed by voltammetry and infrared spectroscopy that Pt(110) electrode is the most active in

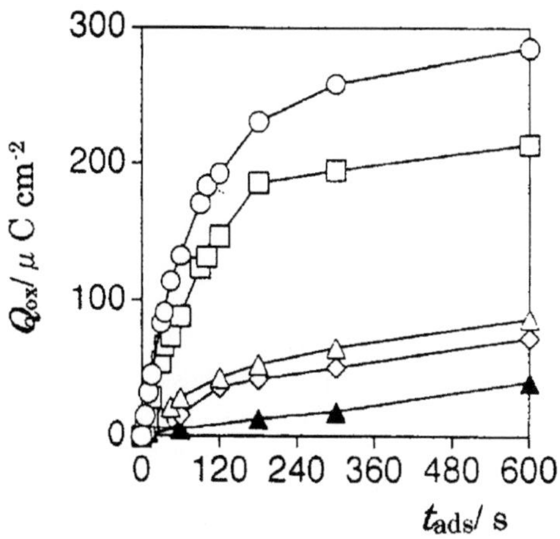

Figure 17. Formation of adsorbed CO from CO_2 on Pt(110) (○), Pt(100) (Δ) and Pt(111) (▲) electrodes at 0.10 V vs. RHE in 0.5 M H_2SO_4 solution. Amounts of adsorbed hydrogen consumption during the reaction are also shown, Pt(110) (□) and Pt(100) (◊). Reprinted from Ref. 109, Copyright (1994) with permission from Elsevier.

CO_2 reduction, whereas Pt(111) is inert. Later Aramata et al., Rodes et al. and Hoshi et al. confirmed this fact. Fig. 17 gives time courses of adsorbed CO formation on Pt(111), Pt(100) and Pt(110) in $HClO_4$ solution published by Aramata et al.[109] The amount of CO was determined by anode stripping. Rodes et al. showed in their study using a Pt(110) electrode that adsorbed hydrogen consumed is regenerated during the reaction as above.[110]

Hoshi et al. extended the formation of "reduced CO_2" at Pt single crystal electrode to those of high index crystal orientations.[113] They determined the initial rate of CO_2 reduction $V_{t=0}$ from the slope of the time course of adsorbed CO formation at $t = 0$ as shown in Fig. 17. Figure 18 shows dependences of $V_{t=0}$ on the electrode potential at a series of electrodes Pt(S)-[n(111) × (111)] in 0.1-M $HClO_4$. The electrode surfaces are composed of n atomic rows of (111) terrace and one atomic height of (111) step. Adsorbed hydrogen is present below 0.3 V vs. reversible hydrogen electrode (RHE) at this series of the electrodes, in which potential region CO_2 is reduced. $V_{t=0}$ takes higher value with the increase of step atom density at any potential, i.e., with the decrease of *n* value.

The $V_{t=0}$ for $n = 2, 3, 4$ and 6 gives a peak at 0.2 V, where the amount of adsorbed hydrogen is less than at 0.1 V. This apparently strange feature is rationalized by involvement of vacant sites of the Pt surface in the activation of CO_2 reduction.[114] Other series of Pt single crystals are studied for CO_2 reduction in addition to Pt(S)-[*n*(111) × (111)], i.e., Pt(S)-[*n*(100) × (111)], Pt(S)-[*n*(111) × (100)], and kinked stepped surfaces Pt(S)-[*n*(100) × (110)] and Pt(S)-[*n*(110) × (100)]. Among all these crystal orientations, kinked surfaces are the most active in CO_2 reduction. $V_{t=0}$ depends linearly on the density of step or kink atoms as shown in Fig. 19. This fact shows that the kink or step structure is closely connected with the active sites for the CO_2 reduction.

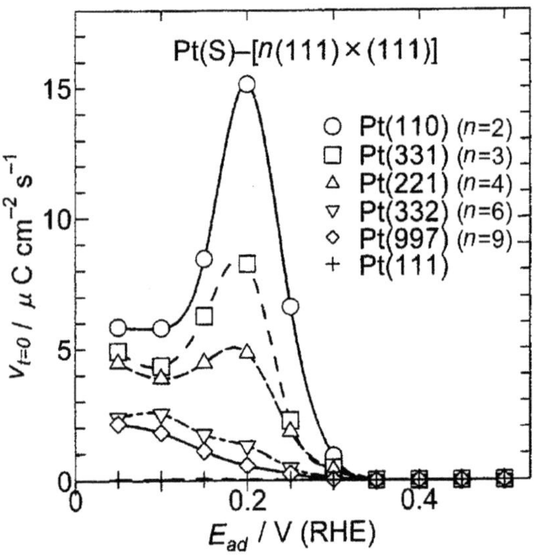

Figure 18. Initial rate of adsorbed CO formation from CO_2 at a series of Pt single crystal electrodes Pt(S)-[*n*(111) × (111)] in 0.1-M $HClO_4$. Reprinted from Ref. 114, Copyright (2000) with permission from Elsevier.

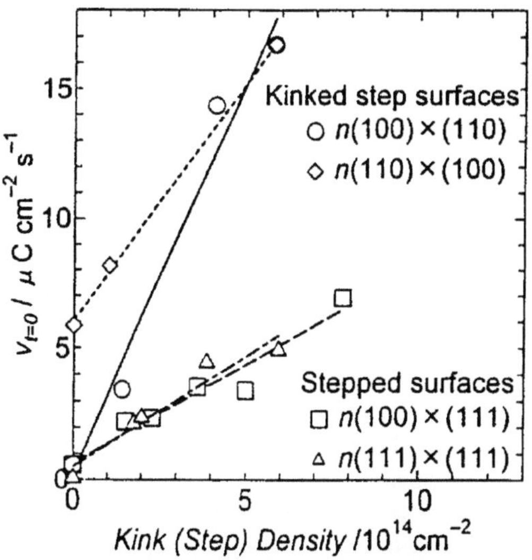

Figure 19. Initial rate of adsorbed CO formation at 0.05 V vs. RHE at various series of Pt single crystal electrodes in 0.1-M $HClO_4$ plotted against the kink and step atom density. Reprinted from Ref. 114, Copyright (2000) with permission from Elsevier.

As previously mentioned, high pressure electrolysis is effective in activation of CO_2 reduction at Pt electrode. Hara et al. reported that CO_2 can be reduced in aqueous 0.1-M $KHCO_3$ under 30 atm mainly to $HCOO^-$ and CO with the faradaic efficiency of 50.4% and 6.1% respectively at the current density 163 mA cm^{-2}.[99]

Brisard et al. studied on-line mass spectrometry of CO_2 reduction on Pt deposited on an Au substrate.[115] They demonstrated that the mass signals of methanol and formaldehyde produced in CO_2 reduction in 0.1 M $HClO_4$ are synchronized with cathodic potential steps 0.2 to –0.9 V vs. RHE applied to the electrode. Reduction of CO to methanol and formaldehyde also proceeded in the equivalent experimental conditions. If the formation of methanol and formaldehyde takes place really on Pt electrode, it must be of great interest. However, the amounts of methanol and formaldehyde formed would be very low, since these substances have not been detected in any macroelectrolytic measurements in aqueous media. In connection with this finding, Sakata et al. reported that CH_4, not methanol, is produced from CO_2 with partial current higher than 180 mA cm^{-2} in 0.5 M $KHCO_3$ aqueous solution with a Pt loaded gas diffusion electrode under elevated pressure above 5 atm.[199] Characterization of Pt catalyst was not provided. More details of gas diffusion electrode are described in Section IX.2.

(*ii*) *Pd*

The heat of adsorption of CO on Pd is moderate and intermediate between Au and Pt.[200] Hydrogen can be easily dissolved in Pd metal, and the concentration of hydrogen atoms can be controlled at the electrode surface. These properties attracted many workers' interest, and Pd has been employed as a hopeful electrode material for CO_2 reduction. However, CO and formate are the major products at Pd electrode in prolonged electrolysis in aqueous media,[63,127] as exemplified in Table 3. Reduction of HCO_3^- ions was studied with Pd electrodes,[166,167] as previously discussed in Section IV.

Ayers and Farley employed an electrolytic cell in which a Pd membrane worked as a bipolar electrode.[61] CO_2 is cathodically reduced at the front side of the Pd membrane electrode with hydrogen atoms cathodically produced and dissolved at the back side. Hydrogen atoms permeate through the Pd membrane to CO_2 reduction side. Both cathodic reactions can be regulated independently by two electrochemical systems. They reported that methanol was formed at low electrolytic current densities.

Fujishima and his coworkers studied the effect of hydrogen absorption in a Pd electrode on CO_2 reduction. They showed that Pd electrode with hydrogen absorbed provides higher activity in CO_2 reduction, giving higher partial currents for production of both CO and $HCOO^-$. They concluded that absorbed hydrogen atoms are involved with the formation of CO and $HCOO^-$.[62,64]

Yoshitake et al. studied an electrolysis device similar to Ayers and Farley's.[201] Higher supply of hydrogen atoms from the backside of Pd electrode results in enrichment of hydrogen atoms in the surface of the electrode, and enhances the faradaic yields of CO and $HCOO^-$ in CO_2 reduction at the front side of the electrode, in agreement with Fujishima et al.[62] They provided deuterium atoms from the backside of the electrode by electrolysis of heavy water. They observed that 95% of formate produced in the reaction is occupied by $HCOO^-$, not $DCOO^-$.[66] Thus the formation of formate appears to be involved with hydrogen atoms formed at the electrode electrolyte interface of CO_2 reduction side, not hydrogen atoms permeating through the electrode, whereas the formate formation is enhanced by increased supply of hydrogen atoms from the backside of the electrode. Methanol formation was not confirmed.

Iwakura et al. reduced CO_2 to formic acid at a palladized Pd/CO_2 interface in a gas compartment without any protic solvent.[68] CO_2 is reduced by atomic hydrogen permeating through the membrane, which is produced by electrolysis of aqueous KOH solution at the back side of the Pd membrane. The reaction proceeds chemically in gas solid surface, not electrochemically. The current density of the electrolysis of water was 10 mA cm^{-2}, and the faradaic efficiency of HCOOH formation corresponded to 10 to 20%.

Methanol formation at Pd electrode is reported by some workers. All these reports communicate in common that methanol is produced with extremely low current density with very low overpotential or sometimes at a potential close to the equilibrium one. However, no convincing evidence is presented; for example, higher current density of methanol formation at higher overpotential have not been accompanied and these observations are not reproduced by other workers employing the identical experimental conditions.

In addition to the continuous production of CO, Pd electrode produces adsorbed CO on the electrode surface at the potential more positive than continuous electrolysis.[202] Hoshi et al. studied the rate of formation of adsorbed CO from CO_2 at Pd single crystal electrodes with low index planes in 0.1 M $HClO_4$.[67] The amount of adsorbed CO was determined by anode stripping in a way similar to the one applied to Pt single crystals described in the previous Section. Adsorbed CO is not formed in the potential range of "adsorbed hydrogen region" where the reaction proceeds at Pt, Ir and Rh electrodes. Adsorbed CO is formed below −0.10 V vs. RHE at Pd electrode. Figure 20 presents the initial rate of CO_2 reduction $V_{t=0}$ plotted against the potential. The rate of CO_2 reduction is stri-

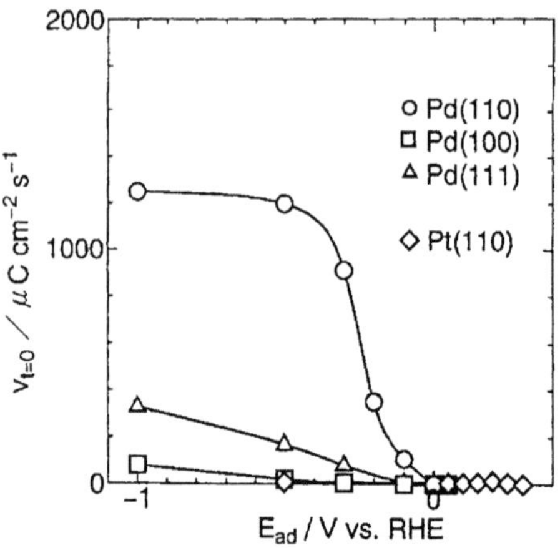

Figure 20. Initial rate of adsorbed CO formation at Pd single crystal electrodes (110), (100) and (111) in 0.1-M $HClO_4$. Values from Pt(110) is also given for comparison. Reprinted from Ref. 67, Copyright (1997) with permission from Elsevier.

kingly high in comparison with a Pt single crystal electrode Pt(110) as shown in Fig. 18. The activity for CO_2 reduction is in the sequence of Pd (110) > Pd(111) > Pd(100). It is also of interest that (111) surface gives higher activity than (100) in CO_2 reduction, as distinguished from other platinum group metal electrodes.

(*iii*) *Other Platinum Group Metals*

Ruthenium has been studied for methane or methanol formation in CO_2 reduction.[116] The partial current densities reported are below 0.1 mA cm^{-2}. Rh was employed as an electrode under 30 atm in 0.1 M $KHCO_3$, giving CO and $HCOO^-$ as the major products in constant current electrolysis at 163 mA cm^{-2}.[99] Re deposited on Au was employed for CO_2 reduction in methanol based electrolyte, giving CO as the product.[150]

Single crystal electrodes of Ir and Rh have also been studied by anode stripping.[106, 105] Ir and Rh reduce CO_2 to adsorbed CO in the reaction with adsorbed hydrogen in a similar way with Pt electrodes. In 0.1 M $HClO_4$ electrolyte, (110) orientation shows higher activity than (100) and (111) both with Ir and Rh metals.

3. Ni and other CO Formation Metals

Nickel electrode does not continuously give products in CO_2 reduction in aqueous electrolytes under 1 atm at ambient temperature except CH_4 and $HCOO^-$ with low faradaic efficiencies as shown in Table 3.[23,29,31] Ni electrode reduces CO_2 to adsorbed CO on the electrode surface both in linear and bridged configuration as evidenced by infrared spectroscopic measurements.[160,203] The adsorbed CO prevents further reduction of CO_2 as well as HER in a way similar to Pt electrode. The adsorbed CO is gradually reduced at low rate by cathodic polarization. The rate of reduction of the adsorbed CO was measured by cyclic voltammetry and infrared spectrometry, in good agreement with those obtained by macroscopic electrolysis.[103,160] If covered with submonolayer of Cd, Ni electrode can continuously reduce CO_2 to CO;[69] this fact suggests that a small degree of modification or contamination of Ni electrode may lead to CO formation in CO_2 reduction.

Azuma et al. argued that low temperature electrolysis is effective for reduction of CO_2 on Ni electrode, since a Ni electrode can continuously reduce CO_2 to CO at −2.2 V vs. SCE at 2°C with faradaic efficiency of 21%.[35] Hori and Murata reexamined the observation, and reported such an effective reduction of CO_2 is not reproduced from a Ni electrode at 1°C.[103]

Under elevated pressure, a Ni electrode can reduce CO_2 mainly to CO and $HCOO^-$ with slight amount of CH_4 in aqueous media according to Hara et al.,[99,104] as discussed in Section III.4(*i*). Saeki et al. communicated that a Ni electrode gives high yields of CH_4, C_2H_4 and C_2H_6 under 41 atm in CO_2-methanol mixture; the sum of the faradaic efficiencies of hydrocarbons amounted to 23% at the current density 200 mA cm^{-2}. They argued that higher ratio C/H derived from elevated pressure on the electrode surface may lead to the production of hydrocarbons.[129]

Iron electrode does not reduce CO_2 in aqueous electrolytes under 1 atm at ambient temperature; HER is prevalent in CO_2 saturated solution.[23, 29, 31] Under elevated pressure such as 30 atm, CO_2

is reduced giving $HCOO^-$ as the major product in constant current electrolysis above 120 mA cm^{-2}, accompanied with less amount of CO. Small amount of hydrocarbons are also produced.[102] In connection with this fact, CO can be adsorbed on a Fe electrode in CO saturated solution as evidenced by infrared spectroscoppy.[160] CO is electrochemically reduced to CH_4, C_2H_4 and C_2H_6.[100]

The elements Zr, Cr, W, Co, C, and n-Si do not reduce CO_2 in aqueous media at 1 atm, but these electrodes can provide activity in CO_2 reduction under elevated pressure according to Hara et al.[99] The major products are CO and $HCOO^-$.

The elements Mo and Ru were reported to reduce CO_2 to methanol or CH_4 by Frese et al. with low partial current density below 0.1 mA cm^{-2},[116,204] but the observations have not been reproduced by other workers.[31,99]

4. Non-metallic Electrode Materials for CO_2 Reduction

Ruthenium dioxide is stable under cathodic conditions. Some workers employed RuO_2 as an electrocatalyst for CO_2 reduction in aqueous electrolyte solutions, and have communicated that CO_2 can be reduced with low overpotential. But the partial current density of CO_2 reduction is mostly low to date. Bandi used an electrode RuO_2 + TiO_2 coated on Ti substrate, and reported that CH_3OH is produced in CO_2 reduction at –900 mV vs. Hg/Hg_2SO_4 in 0.05 M H_2SO_4 with the faradaic efficiency of 24% of the current density 0.52 mA cm^{-2}.[119] Later Popic et al. reported that electrodes of RuOx related oxide deposited on Ti substrate reduce CO_2 to CH_3OH at –0.8 V vs. SCE. The yield of CH_3OH increased with the lapse of the electrolysis time.[120] Spataru et al. applied an electrode RuO_2 deposited on boron doped diamond to CO_2 reduction. The current yields of the CO_2 reduction are 40% and 7.7% for $HCOO^-$ and CH_3OH respectively, at the potential –0.6 V vs. SCE with the current density 1 to 4 mA cm^{-2}.[121] Qu et al. used carbon nanotube as a substrate of an electrode coated with RuO_2/TiO_2, and showed that CH_3OH is formed at –0.8 V vs. SCE with faradaic efficiency 60% of the current density 6.5 mA cm^{-2}.[122]

Some nonmetallic electrocatalysts used for gas diffusion electrode are referred to in Section IX.

Koleli et al. studied conducting polymer films as electrode materials for CO_2 reduction.[153] They showed that polyaniline supported by a Pt sheet reduces CO_2 to HCOOH, HCHO and CH_3COOH in methanol based electrolyte at −0.4 V vs. SCE under ambient to elevated pressures up to 20 kbar. The current density is 1 to 2 mA cm^{-2} with the total faradaic efficiency of the products close to 100%. They obtained similar results in polypyrrol electrode as well.[152]

Schrebler et al. used Re or Re-Cu dispersed polypyrrole electrodes supported by Au substrates for CO_2 reduction in methanol based electrolyte. The current density was 1 mA cm^{-2} or so at −1.35 V vs. SCE. The products were CO and CH_4 with good faradaic balances.[151]

VIII. MECHANISTIC STUDIES OF ELECTROCHEMICAL REDUCTION OF CO_2 AT Cu ELECTRODE

1. Formation of CO as an Intermediate Species

Copper metal electrode can reduce CO_2 to CH_4, C_2H_4 and alcohols in aqueous electrolytes. The product distribution ranges widely, depending on electrolyte, potential and temperature. The unique properties of Cu electrode have been extensively studied.

Figure 21 shows a result of controlled potential reduction of CO_2 at a Cu electrode in 0.1-M $KHCO_3$. The faradaic yields of CO and $HCOO^-$ rise at –0.9 V, reaching maximum at –1.20 to –1.25 V, and drop with the increase of the cathodic potential. C_2H_4 begins to increase at –1.1 V; CH_4 starts at –1.2 V, rising steeply with the potential. These features suggest that CO and $HCOO^-$ may be precursors to hydrocarbons and alcohols.

$HCOO^-$ is not reduced at all at a Cu electrode, whereas CO is electrochemically reduced to hydrocarbons and alcohols with a similar product distribution with CO_2 reduction (Table 11).[19,205] Infrared spectroscopic measurements showed that CO is adsorbed at a Cu electrode at –0.9 to –1.0 V vs. SHE in CO_2 saturated electrolyte solution.[206] Ito and his coworkers confirmed the results that linearly adsorbed CO is present on a Cu polycrystal electrode at

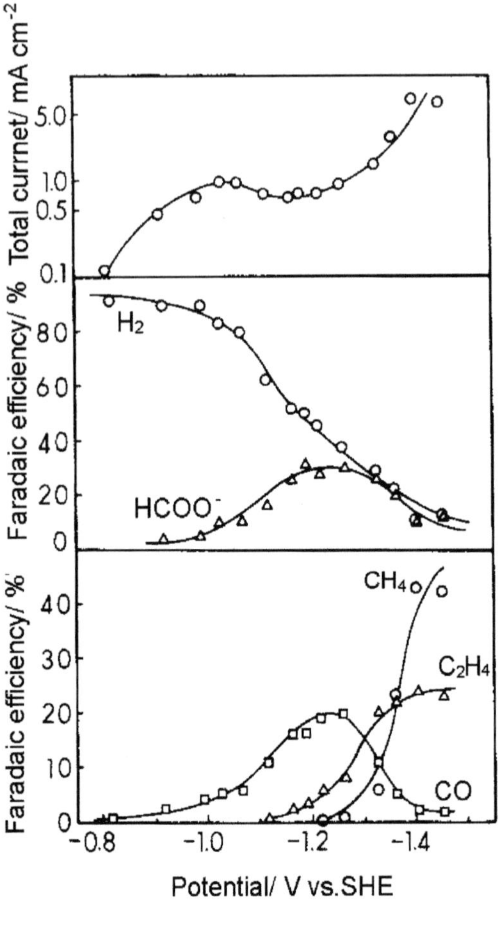

Figure 21. Variation of the faradaic efficinecies of various products in electrochemical reduction of CO_2 at a Cu electrode in controlled potential electrolysis, 0.1-mol kg^{-1} $KHCO_3$, 1 atm CO_2, 18 °C. The potential axis negative to the right. Reference 19–Reproduced by permission of The Royal Society of Chemistry.

–0.6 V vs. SHE by surface enhanced Raman spectroscopy, and on a Cu deposited Pt electrode by infrared absorption spectroscopy.[207] These facts established that CO is intermediately formed at a copper electrode during CO_2 reduction and further reduced to hydrocarbons and alcohols.

Why only Cu can yield hydrocarbons and alcohols from CO_2 with high current density among various metal electrodes? The reason has not yet been fully elucidated, tentatively explained as below.

Au, Cu, Ni and Pt electrodes yield CO from CO_2. Thus the electrocatalytic activities of these metals are compared for electroreduction of CO in reference to the heat of adsorption of CO (Table 12).[208] On the basis of the heats of adsorption,[161] CO will scarcely interact with Au surface. Thus CO will not be reduced at Au. Ni and Pt have high heats of adsorption, and adsorb CO strongly. CO, highly stabilized on the surface, can not be further reduced. The heat of adsorption of Cu is intermediate among these metals, leading to CO adsorption with moderate strengh; CO is effectively reduced to hydrocarbons at Cu. Watanabe et al. studied

Table 11
Faradaic Efficiencies of Various Products from Electrochemical Reduction of CO at a Cu Electrode in Aqueous Solutions. Reference 19– Reproduced by permission of The Royal Society of Chemistry

Electrolyte	pH	Potential V vs. SHE	Faradaic efficiency/ %						
			CH_4	C_2H_4	EtOH	n-PrOH	HCHO	H_2	Total
Current density: 2.5 mA cm^{-2}									
Phosphate[a]	6.1	−1.18	9.3	2.3	0.0	0.0	0.2	81.1	92.9
$KHCO_3$[b]	9.6	−1.40	16.3	21.2	10.9	1.5	0.1	45.5	95.5
Current density: 5.0 mA cm^{-2}									
Phosphate[a]	6.0	−1.21	16.8	1.7	0.0	0.0	0.02	75.4	93.9
$KHCO_3$[b]	9.6	−1.45	16.2	5.5	2.7	0.3	0.03	65.4	90.1
KOH[c]	12.9	−1.47	1.0	14.1	5.8	1.1	0.05	70.7	92.8

[a]Phosphate buffer solution containing KH_2PO_4 (0.17 mol dm^{-3}) and K_2HPO_4 (0.03 mol dm^{-3})
[b]$KHCO_3$ (0.1 mol dm^{-3}).
[c]KOH (0.1 mol dm^{-3})

Table 12
Electrochemical Reduction of CO in 0.1-M $KHCO_3$ at Various Metal Electrodes.[a] Reprinted with permission from Ref. 208, Copyright (1987) Chemical Society of Japan

Electrode	Potential V vs.SHE	Faradaic efficiency/ %				CO heat of adsorption kcal mol^{-1} [c]
		CH_4	C_2H_4	Other HCs and alcohols[b]	H_2	
Au	−1.49	0.0	0.0	0.0	101.6	9.2
Cu	−1.40	16.3	21.2	12.5	45.5	17.7
Ni	−1.46	2.6	0.3	0.7	94.2	40.8
Pt	−1.29	0.1	0.0	0.0	96.8	46.6

[a]Current density 2.5 mA cm^{-2}
[b]C_2H_6, C_2H_5OH, n-C_3H_7OH, HCHO
[c]Data obtained from Ref. 161

a molecular orbital calculation of CO adsorption on Cu and Ag. They showed that higher adsorption strength of CO on Cu than Ag leads to more effective formation of hydrocarbons.[209, 210]

Carbon monoxide is weakly adsorbed on Cu electrode, in equilibrium with gaseous CO dissolved in the electrolyte solution close to the electrode. When the electrolyte is stirred, CO desorbs easily from the surface.[19] Sakata et al. reported that the yield of CO from a Cu electrode is greatly enhanced when the electrolyte solution is stirred, as is compatible with the weak adsorption of CO on the electrode surface.[211]

2. CO_2 Reduction at Cu Electrode Affected by the Potential and the CO_2 Pressure

The product distribution of CO_2 reduction at a Cu electrode changes with the potential. Such a potential dependence of the product distribution is derived from the transfer coefficients. The rate of formation of a product P is presented by the partial current i_P in a Tafel type equation,

$$i_P = A_P \exp[-\alpha_P F(E - E^0)/RT] \tag{41}$$

where A_P is a constant, α_P the transfer coefficient for the product P, and the other symbols have the usual meanings. The value of α_P depends on the reaction mechanism for formation of the product P. Unless α_P of two or more products takes an equal value, the product distribution changes naturally with the potential.

Ito and his coworkers studied the product distribution of CO_2 reduction at a Cu electrode with constant potential electrolysis in 0.1 M $KHCO_3$, and showed that the product distribution is greatly affected with the potential.[30] $HCOO^-$ and CO prevail at the potential less negative than −1.50 V vs. Ag/AgCl in saturated KCl; ethylene becomes a main product between −1.55 and −1.65 V vs. Ag/AgCl, and CH_4 is the most favorable at −1.70 V. HER is prevalent below −1.70 V, since the cathodic current increases and the transport of CO_2 to the electrode can not meet the rate of the CO_2 reduction. CH_3CHO, C_2H_5OH, C_2H_5CHO, and n-C_3H_7OH show trends similar to C_2H_4. The potential dependence of the product distribution described by Ito et al. agree well with those given in Fig. 21, taking into account the potential of the reference electrode Ag/AgCl in saturated KCl 0.199 V vs. SHE.

Kyriacou and Anagnostopoulos reported that the faradaic efficiencies of CH_4 and C_2H_4 are proportional to the partial pressure of CO_2 between 0.1 and 1 atm.[78]

3. Electrolyte Solution, Anionic Species

The product distribution of Cu electrode is greatly affected by anionic and cationic species of the electrolyte solution. Table 13 shows the results of electrochemical reduction of CO_2 at a constant current density 5-mA cm^{-2} in various 0.1-M electrolyte solutions. $KHCO_3$, KCl, $KClO_4$, and K_2SO_4 solutions favor the CO_2 reduction. K_2HPO_4 solutions highly promote HER rather than CO_2 reduction at less negative potential; 0.5-M K_2HPO_4 gives much higher H_2 yields than 0.1-M K_2HPO_4.

Such an enhancement of HER is attributed to lower pH value at the electrode/ electrolyte interface, as mentioned in Section II.1(*iii*). The pH would rise locally at the interface due to OH^- generation in cathodic reactions in aqueous media. Nevertheless, the buffer action of HPO_4^{2-} neutralizes the OH^-, keeping the pH at a

Table 13
Faradaic Efficiencies of Products from the Electroreduction of CO_2 at a Cu Electrode at 5 mA cm^{-2} in Various Solutions at 19°C. Reference 19–Reproduced by permission of The Royal Society of Chemistry

Solution	Conc. M	pH[a]	Potential V vs.SHE	Faradaic efficiency (%)							
				CH_4	C_2H_4	EtOH	PrOH	CO	$HCOO^-$	H_2	Total
$KHCO_3$	0.1	6.8	−1.41	29.4	30.1	6.9	3.0	2.0	9.7	10.9	92.0
KCl	0.1	5.9	−1.44	11.5	47.8	21.9	3.6	2.5	6.6	5.9	99.8
KCl	0.5		−1.39	14.5	38.2	[b]	[b]	3.0	17.9	12.5	
$KClO_4$	0.1	5.9	−1.40	10.2	48.1	15.5	4.2	2.4	8.9	6.7	96.0
K_2SO_4	0.1	5.8	−1.40	12.3	46.0	18.2	4.0	2.1	8.1	8.7	99.4
K_2HPO_4	0.1	6.5	−1.23	17.0	1.8	0.7	tr	1.3	5.3	72.4	98.5
K_2HPO_4	0.5	7.0	−1.17	6.6	1.0	0.6	0.0	1.0	4.2	83.3	96.7

[a]pH values were measured for bulk solution after electrolysis.
[b]Not analyzed

lower value. KCl, $KClO_4$, and K_2SO_4 solutions do not have buffer ability, and thus the pH at the electrode/electrolyte interface rises. HER also goes up with the concentration of $KHCO_3$ owing to its buffer action as indicated in Fig. 22. No significant change in the product distribution is observed with the variation of KCl concentration as shown in Table 13.

In addition to the competition between CO_2 reduction and HER, product selectivity is also observed in accordance with anionic species. KCl, $KClO_4$, and K_2SO_4 solutions provide higher yields of C_2H_4 and alcohols than CH_4 in comparison with 0.1-M $KHCO_3$. A locally high pH region, formed at the electrode in these electrolytes, contributes to reduction of the intermediate CO to C_2H_4 and alcohols rather than to CH_4 as shown in Table 11.

4. Effects of Cationic Species in Electrolyte Solution

The cationic species also affects the CO_2 reduction at metal electrodes. Paik et al. studied electroreduction of CO_2 to $HCOO^-$ at a Hg pool electrode in Li^+, Na^+ and $(C_2H_5)_4N^+$ hydrogen-carbonate solutions.[22] The electrode potential at a constant current increases in the positive direction with the sequence of $Li^+ < Na^+ < (C_2H_5)_4N^+$. The results were discussed in connection with the vari-

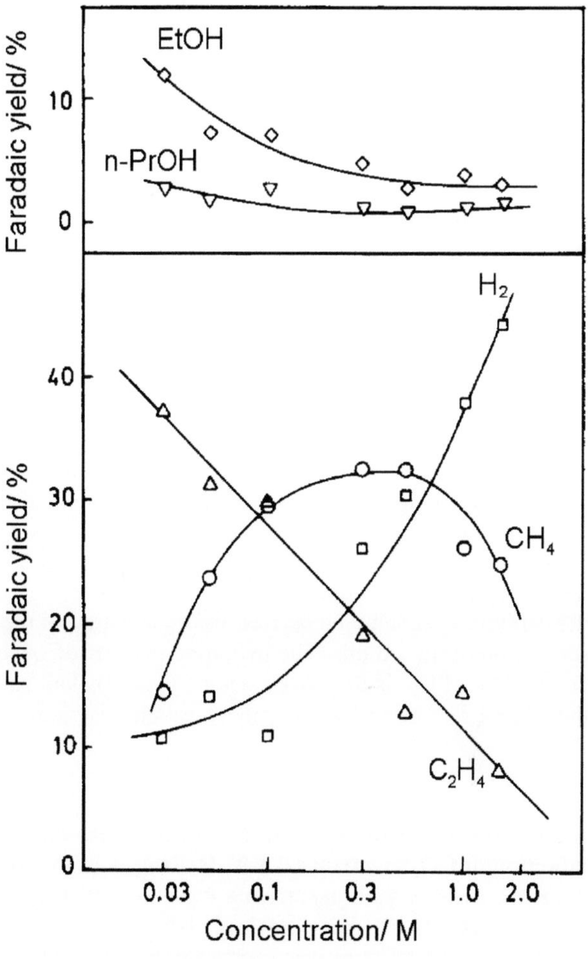

Figure 22. Faradaic yields of the products in the CO_2 reduction at a Cu electrode at 19°C in $KHCO_3$ aqueous solutions of various concentrations. Current density: 5 mA cm^{-2}. Reference 19–Reproduced by permission of The Royal Society of Chemistry.

ation of the potential at the outer Helmholtz plane (OHP), and qualitatively ascribed to the greater specific adsorption of the larger cations. Hori and Suzuki later confirmed Paik et al.'s observation at a Hg electrode.[15] Augustynski et al. reported that the magnitude of the cathodic current peak at a Pd electrode, measured in 0.5 M alkali hydrogen-carbonate solution, are in the order $Li^+ < Na^+ < K^+ < Cs^+$.[212]

The features of CO_2 reduction at a Cu electrode also depends remarkably on cationic species. Table 14 shows faradaic efficiencies of CO_2 reduction at a Cu electrode in 0.1 M various alkali hydrogen-carbonate solutions at a constant current density 5.0 mA cm^{-2}.[75] Hydrogen formation is prevalent in the Li^+ electrolyte.

The formation of C_2H_4 and alcohols increases relatively in the order $Li^+ < Na^+ < K^+ < Cs^+$. The CH_4 and C_2H_4 ratio in the faradaic efficiency C_1/C_2 obviously shows a trend $Li^+ > Na^+ > K^+ > Cs^+$.

The potential during the electrolysis is in the order $Li^+ < Na^+ < K^+ < Cs^+$. An identical feature appears in electrochemical reduction of CO, an intermediate species in CO_2 reduction.[75] Such a cationic effect in CO_2 reduction at a Cu electrode was confirmed by Kyriacou and Anagnostpoulos,[78] and by Ohta and his coworkers with methanol based electrolyte solutions.[213]

The cationic effect may be derived from the extent of the specific adsorption of the cations. The hydration number of Li^+ is the highest, and that of Cs^+ is the lowest among these 4 alkali cations. Cs^+, the largest and softest among these cations, is more easily

Table 14
Faradaic Efficiencies of the Products in the Electrochemical Reduction of CO_2 in various 0.1-M Hydrogen-carbonate Solutions.[a] Reprinted with permission from Ref. 75, Copyright (1991) Chemical Society of Japan

Electrolyte	Potential V vs. SHE	Faradaic efficiency (%)								C_1/C_2
		CH_4	C_2H_4	CO	EtOH	PrOH	$HCOO^-$	H_2	Total	
$LiHCO_3$	−1.45	32.2	5.2	tr	1.6	tr	4.7	60.5	104.2	6.19
$NaHCO_3$	−1.45	55.1	12.9	1.0	4.2	0.6	7.0	25.1	105.9	4.27
$KHCO_3$	−1.39	32.0	30.3	0.5	10.9	1.6	8.3	14.5	98.1	1.06
$CsHCO_3$	−1.38	16.3	30.5	2.4	7.2	4.4	15.8	24.4	101.0	0.53

[a]Current density: 5.0 mA cm^{-2}. Temperature: 18.5°C.

specifically adsorbed on the electrode than others. Thus the potential at the OHP will be more positive in Cs^+ solution than the other solutions. Hence the concentration of H^+ at OHP may naturally be lower in the order; $Li^+ > Na^+ > K^+ > Cs^+$. The effect of the potential at OHP on an electrode reaction was discussed by Frumkin.[214] Thus hydrogen species will be supplied to the electrode in the order $Li^+ > Na^+ > K^+ > Cs^+$, eventually leading to the C_1/C_2 ratio as mentioned above. HER occupies a higher faradaic efficiency in $LiHCO_3$.

5. Reaction Mechanism at Cu Electrode

As described in the previous Section, CO_2 is initially reduced to CO on Cu electrode, further reduced to hydrocarbons and alcohols. Various metal electrodes reduce CO_2 to CO, but only Cu electrode can effectively yield hydrocarbons and alcohols. Thus the unique property of Cu must be derived from interaction of Cu electrode and CO. Figure 23 presents voltammograms of a Cu electrode in phosphate buffer solution (pH 6.8) saturated with argon and CO. HER is severely suppressed by the presence of CO in the electrolyte solution. The surface of the Cu electrode is covered with CO with the coverage more than 90% as estimated from the current at −1.0 V with and without CO in Fig. 23.[215]

Figure 21 indicates that CH_4 formation starts at a more negative potential than C_2H_4, and rises more steeply with the increase of the negative potential. Table 11 shows that C_2H_4 formation is more favorable than CH_4 in high pH solutions. Such different features strongly suggest that CH_4 and C_2H_4 are produced via different reaction paths from the common starting substance, CO.

The partial current of C_2H_4 formation is related with the electrode potential, providing a linear Tafel relationship (Fig. 24) regardless of pH of the electrolyte.[216] The slope of the straight line gives a transfer coefficient of 0.35. A similar correlation between the partial current of C_2H_5OH and the potential gave a transfer coefficient of ca. 0.6. A value of the transfer coefficient between 0 and 1 indicates that the rate determining step is involved with the first electron transfer.[217]

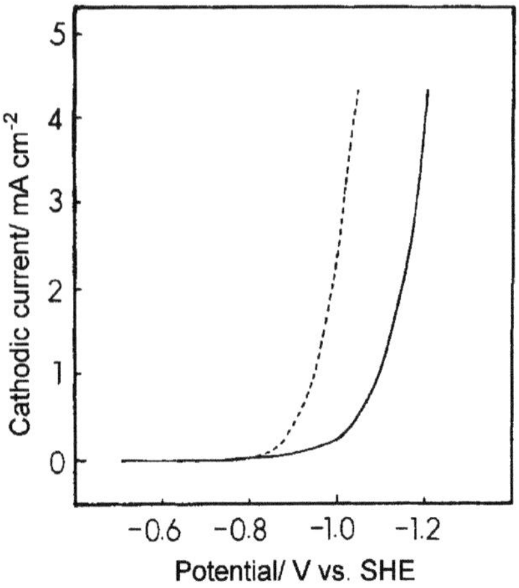

Figure 23. Voltammograms of a Cu electrode in phosphate buffer solution (0.1 M KH_2PO_4 + 0.1 M K_2HPO_4, pH 6.8) saturated with argon (-----) and CO (——).The potential axis negative to the right. Reference 19–Reproduced by permission of The Royal Society of Chemistry.

Electronation of CO in the gas phase readily gives rise to carbon atoms and O^- ions.[218] Kim et al. proposed "electrochemical splitting of CO" in CO reduction on a Cu electrode. They rationalized an elementary step of surface carbon atom formation from CO molecule on a Cu elecrode,[34]

$$CO(ads) + H_2O + e^- \rightarrow C(ads) + OH(ads) + OH^- \quad (42)$$

followed by

$$OH(ads) + e^- \rightarrow OH^- \quad (43)$$

C(ads) will readily be reduced to CH_2.

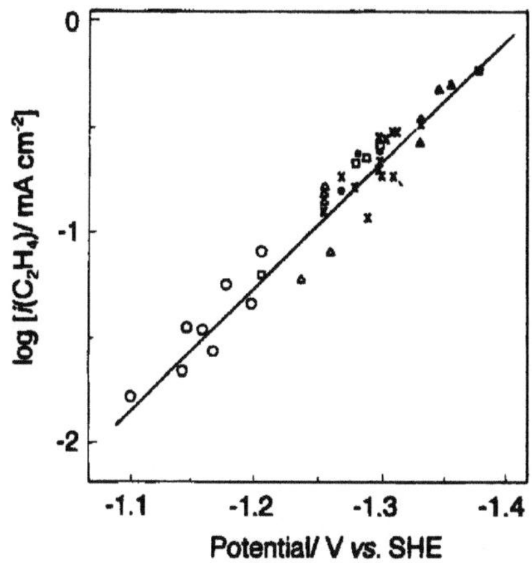

Figure 24. Partial current densities of C_2H_4 formation in electorochemical reduction of CO at a Cu electrode correlated with the electrode potential in various electrolyte solution: pH 6.0-6.3 (○), pH 7.1-7.7 (Δ), pH 8.0-8.6 (□), pH 8.7- 8.9 (x), pH 9.0-9.3 (●), pH 10.5-11.3 (▲), pH 12.2 (■). The potential axis negative to the right. Reprinted with permission from Ref. 216, Copyright (1997) American Chemical Society.

The partial current of CH_4 formation is widely scattered when plotted against the potential. With an assumption that the partial current is proportional to proton activity, a linear Tafel relationship is obtained as shown in Fig. 25.[216] The transfer coefficient is determined as 1.33. These facts indicate that the rate determining step of CH_4 formation is involved with the second electron transfer to a hypothetical intermediate species such as COH in an electrochemical equilibrium with a CO adsorbed on the electrode and a proton from the electrolyte,

$$CO(ads) + H^+ + e^- = COH(ads) \tag{44}$$

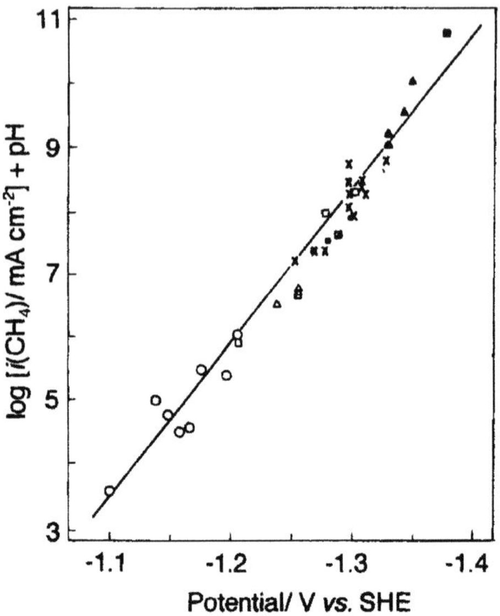

Figure 25. log $i(CH_4)$ + pH, obtained from CO reduction at a Cu electrode, correlated with the electrode potential in various electrolyte solution: pH 6.0-6.3 (○), pH 7.1-7.7 (Δ), pH 8.0-8.6 (□), pH 8.7- 8.9 (x), pH 9.0-9.3 (●), pH 10.5-11.3 (▲), pH 12.2 (■). The potential axis negative to the right. Reprinted with permission from Ref. 216, Copyright (1997) American Chemical Society.

Neither C_2H_6 nor CH_3OH is produced from a Cu electrode, and Cu is an inert metal for thermochemical Fischer-Tropsch reaction.[219] Thus few adsorbed hydrogen atoms will be present on Cu electrode surface. One can presume that hydrogen species is composed of two hydrogen atoms, such as H_2^+,[220] whereas the presence of H_2^+ has been disputed in connection with HER by many workers and has not received experimental confirmation.

The production of alcohols is accompanied with the corresponding aldehydes in both CO and CO_2 reduction. Since aldehydes are electrochemically reduced to the corresponding alcohols at Cu electrodes,[216] alcohols will be formed with intermediate formation of aldehydes. Aldehydes would be produced with insertion of CO to adsorbed CH_2 with reference to ethanol formation in the gas phase from CO + H_2 catalyzed by Rh/TiO_2.[221]

The discussion above leads to a molecular reaction pathway diagram as shown in Fig. 26.

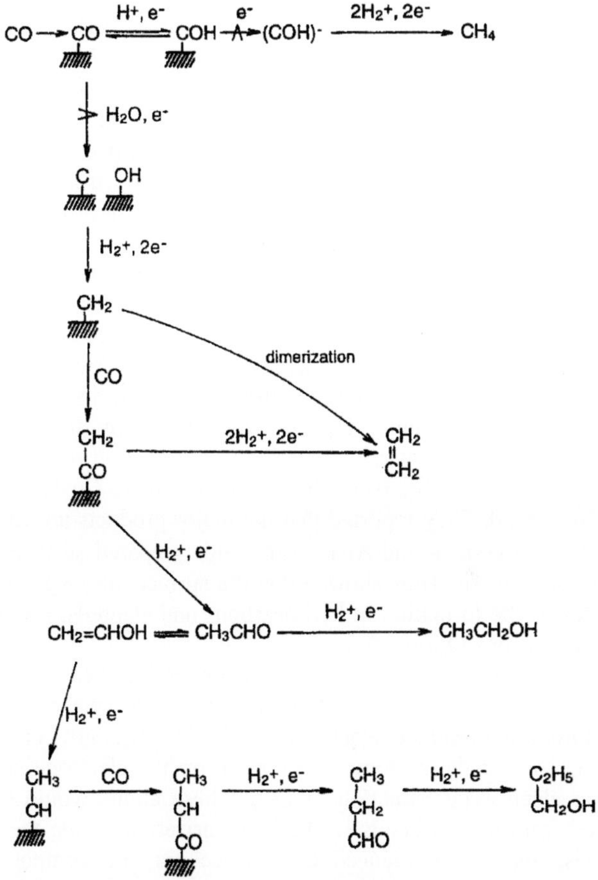

Figure 26. Molecular reaction pathway diagram of CO reduction. H_2^+ represents H_2^+ or a combined H(ads) + H^+. The → with a ∧ through it denotes a rate-determining step. The H_2O and OH^- resulting from the reactions are excluded for brevity. Reprinted with permission from Ref. 216, Copyright (1997) American Chemical Society.

6. Surface Treatment, Alloying and Modification of Cu Electrode

Surface treatments also affect CO_2 reduction. Acid treatment and thermal treatment are often applied in order to remove surface contamination as well as surface stress caused by mechanical polishing. The product selectivity of Cu electrode depends greatly on these surface treatments.[34,72,77] Frese reported formation of CH_3OH at intentionally preoxidized Cu electrodes. The maximum partial current of CH_3OH production reached 15 mA cm^{-2}.[76] However, Koga et al. did not detect CH_3OH from Cu electrodes oxidized in various manners.[72] Steady formation of CH_3OH from Cu electrodes at a high current density has not yet been confirmed by other workers to date.

Copper-based alloys have been used for CO_2 reduction. Cu-Ni and Cu-Fe alloys, formed by in-situ deposition during CO_2 reduction, gradually lose CH_4 and C_2H_4 yields simultaneously with increased H_2 evolution with the increase of Ni or Fe coverage on the Cu surface.[94] A Cu-Cd electrode, also prepared by in-situ deposition, produced CH_4 and C_2H_4. The yields gradually dropped with the increase of Cd coverage, whereas CO formation rose.[95] Watanabe et al. studied various Cu based alloys; Cu-Ni, Cu-Sn, Cu-Pb, Cu-Zn, Cu-Cd. They reported that the major products are CO and $HCOO^-$.[96] Kyriacou and Anagnostopoulos prepared surface alloy electrodes Cu-Au. They showed that the surface alloying severely suppresses the formation of hydrocarbons and alcohols, leading to the increase of CO formation.[97]

Modifications of Cu electrode have been reported with introduction of oxygen, sulfur, or chlorine atoms to the surface in order to improve the product selectivity.[76,77,81,90,92,222] It is not confirmed whether or not these atoms still remain on the electrode surface during electrolytic reduction of CO_2. These atoms would be removed from the surface by cathodic polarization during the electrolysis, and newly reduced Cu surface may be composed of crystallites with favorable atomic configurations for preferred selectivity.

Table 15
Product Distribution in CO_2 Reduction at a Series of Copper Single-Crystal Electrodes, Cu(S)-[n(100) × (111)] and Cu(S)-[n(100) × (110)]. Reprinted from Ref. 91, Copyright (2003) with Permission from Elsevier

Crystal orientation	n^a	Potential V vs. SHE	Faradaic efficiency (%)^e							
			CH_4	C_2H_4	CO	H_2	Aldh^b	Alch^c	HCOOH	HAc^d
		Cu(S)-[n(100) × (111)]								
(100)	∞	−1.40	30.4	40.4	0.9	6.8	4.4	12.0	3.0	1.0
(11 1 1)	6	−1.37	8.9	50.2	1.8	8.8	4.4	17.7	3.2	2.1
(711)	4	−1.34	5.0	50.0	1.1	15.6	6.4	14.2	4.6	0.9
(311)	2	−1.37	36.0	23.8	2.6	13.3	3.4	5.2	14.0	0.6
(111)	–	−1.55	46.3	8.3	6.4	16.3	2.7	3.3	11.5	1.5
		Cu(S)-[n(100) × (110)]								
(810)	8	−1.38	6.4	45.1	1.4	8.7	2.0	28.8	1.5	1.6
(610)	6	−1.37	7.6	44.7	0.9	9.0	2.7	29.3	1.4	1.6
(510)	5	−1.38	8.1	42.3	2.1	10.5	5.6	29.5	2.8	2.1
(210)	2	−1.52	64.0	13.4	2.2	7.0	1.5	7.3	5.5	0.7

Electrolyte solution: 0.1 M-$KHCO_3$, current density: 5 mA cm^{-2}
^{a}n in Cu(S)-[n(100) × (111)] and Cu(S)-[n(100) × (110)]
^{b}Aldehides: acetaldehyde, propionaldehyde
^{c}Alcohols: ethanol, allyl alcohol, n-propanol
^{d}HAc: acetic acid
^{e}The total values of the Faradaic efficiencies are between 95 and 103%

7. CO_2 Reduction at Cu Single-Crystal Electrodes

Several reactions proceed competitively at copper electrodes, and the electrocatalytic activity for the individual reactions depends on the atomic configuration of the electrode surface. Thus fundamental study using single crystal electrodes is relevant to reveal the unique electrocatalytic natures of copper metal. Frese reported that CH_4 formation in the electrochemical reduction of CO_2 is favored in the order of Cu(111), Cu(110) and Cu(100).[5] Later Hori et al. extensively studied the reaction with various series of Cu single crystal electrodes.[87, 88, 91, 223]

Atomic configurations of some Cu crystal orientations are unstable, and easily reconstructed during surface pretreatment, e.g. by the presence of oxygen and ultrasonic rinsing in the pretreatment procedure and so on. Cu(110) is particularly vulnerable among various Cu single crystal orientations, and the selectivity in CO_2 reduction on Cu(110) is hardly reproducible. Takahashi et al.

established a surface pretreatment procedure of Cu single crystals.[88] They obtained reproducible results of CO_2 reduction at Cu single crystal electrodes of various orientations.[91]

Table 15 demonstrates faradaic efficiencies of various products in CO_2 reduction at two series of Cu crystal orientations, Cu(S)[*n*(100) × (111)] and Cu(S)[*n*(100) × (110)]; the notation of single crystals is given in Section VII.1(*i*). C_2H_4 is produced more favorably than CH_4 on the (100) electrode, and CH_4 is predominantly produced on the (111) surface. Introduction of (111) steps to (100) basal planes enhances the formation of C_2H_4 and alcohols as exemplified by (11 1 1) and (711). The results of Cu(S)[*n*(100) × (110)] also show that insertion of (110) steps to (100) basal planes leads an identical effect as well. Figure 27 shows the selectivity ratio C_2H_4/ CH_4 as a function of the angle of the crystal orientation with reference to Cu(100). C_2H_4/ CH_4 takes a maximum at (711) and (810) in *n*(100)-(111) and *n*(100)-(110), respectively. The introduction of step atoms to (100) terrace, regardless of (111) or (110) step, promotes C_2H_4 formation.

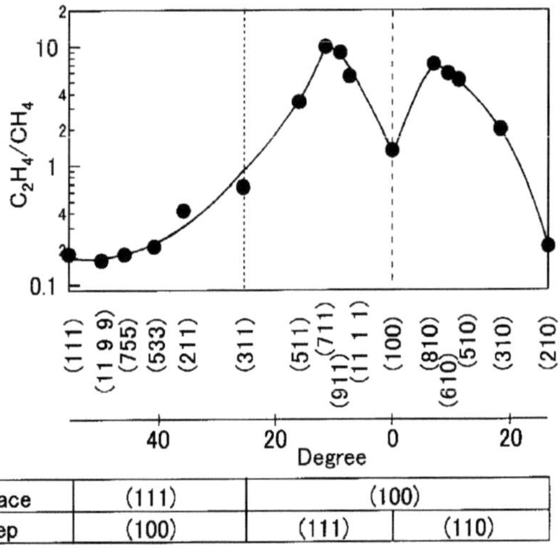

Figure 27. Variation of C_2H_4/CH_4 in terms of the faradaic efficiency and the electrode potential with the angle of the crystal orientation with reference of Cu(100). Reprinted from Ref. 91, Copyright (2003) with permission from Elsevier.

Table 16 tabulates the product distribution from Cu(S)[*n*(111) x (111)], Cu(S)[*n*(111) x (100)] and Cu(S)[*n*(110) x (100)] series. Cu(S)[*n*(111) x (100)], as exemplified by (755), gives lower fraction of C_2H_4, aldehydes and alcohols. Cu(110) yields high faradaic efficiency of CH_3COOH as well as high aldehydes and alcohols. The sum of aldehydes and alcohols is in the same trend with CH_3COOH as shown for Cu(S)[*n*(111) x (111)] and Cu(S)[*n*(110) x (100)] series in Table 16. Thus these three substances may be derived from a common intermediate species. The sum of CH_3COOH, acetoaldehyde and ethanol is reversely correlated with that of CH_4 as shown in Fig. 28.

Table 16
Product Distribution in CO_2 Reduction at a Series of Copper Single-Crystal Electrodes Cu(S)-[*n*(111) × (111)], Cu(S)-[*n*(111) × (100)] and Cu(S)-[*n*(110) × (100)]. Reprinted from Ref. 91, Copyright (2003) with permission from Elsevier

Crystal orientation	n^a	Potential V vs.SHE	Faradaic efficiency (%)[e]							
			CH_4	C_2H_4	CO	H_2	Aldh[b]	Alch[c]	HCOOH	HAc[d]
Cu(S)-[n(111) × (111)]										
(111)	∞	−1.55	46.3	8.3	6.4	16.3	2.7	3.3	11.5	1.5
(332)	6	−1.51	39.6	9.9	6.1	10.3	5.3	7.6	9.4	3.4
(331)	3	−1.55	13.8	16.6	7.7	5.7	7.6	16.0	9.1	7.5
(110)	2	−1.58	6.9	13.5	13.9	3.1	21.2	10.5	10.1	20.8
Cu(S)-[n(111) × (100)]										
(755)	6	−1.43	62.9	11.5	4.4	6.9	1.8	6.6	12.3	0.5
(533)	4	−1.42	62.9	13.0	3.0	4.7	1.3	2.8	9.7	0.5
(311)	2	−1.37	36.0	23.8	2.6	13.3	3.4	5.2	14.0	0.6
Cu(S)-[n(110) × (100)]										
(650)	6	−1.59	10.5	16.2	14.5	2.5	17.0	11.0	6.1	20.6
(210)	2	−1.52	64.0	13.4	2.2	7.0	1.5	7.3	5.5	0.7

Electrolyte solution: 0.1-M $KHCO_3$; current density: 5 mA cm^{-2}.
[a]*n* in Cu(S)-[*n*(111) × (111)], Cu(S)-[*n*(111) × (100)] and Cu(S)-[*n*(110) × (100)].
[b]Aldehydes.: acetaldehyde, propionaldehyde
[c]Alcohols.: ethanol, allyl alcohol, n-propanol.
[d]HAc: acetic acid.
[e]The total values of the Faradaic efficiencies are between 92 and 107%, except (331) of 84.0%.

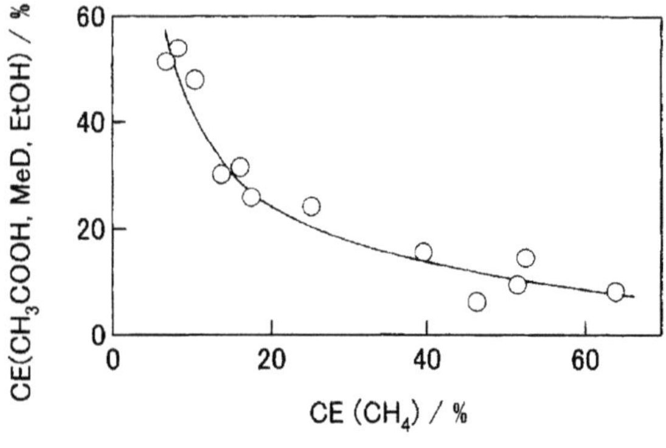

Figure 28. Correlation of the sum of the current effciency of CH_3COOH, CH_3CHO and C_2H_5OH with that of CH_4 for Cu(S)-[*n*(111) x (111)] and Cu(S)-[*n*(110)x(100)] electrodes. Reprinted from Ref. 91, Copyright (2003) with permission from Elsevier.

8. Adsorption of CO on Cu Electrode: Voltammetric and Spectroscopic Studies

Carbon monoxide is produced during CO_2 reduction, weakly and reversibly adsorbed on Cu electrode surface. Adsorbed CO is easily desorbed from Cu electrode surface by purging CO from the electrolyte solution, whereas HER at Cu electrode is significantly suppressed by the adsorption of CO.

CO adsorption on Cu electrode surface is interfered with by specifically adsorbed anions. CO can be adsorbed below a certain definite potential, determined by the adsorption strength of CO and the anion. When CO molecules displace the specifically adsorbed anions on Cu electrode, a voltammetric peak is observed; as exemplified for Cu(100) in CO saturated phosphate buffer solution in comparison with N_2 saturated solution (Fig. 29). Subtractively normalized interfacial Fourier transform infrared spectroscopy (SNIFTIRS) spectra in Fig. 30 demonstrates that CO is adsorbed at –0.8 V vs. SHE but not at –0.4 V, and adsorbed phosphate anion vice versa.[224] This process is equivalent to “charge displacement adsorption” of CO on Pt electrode revealed by Clavilier et al.[225] The profile of the voltammogram depends greatly on the crystal

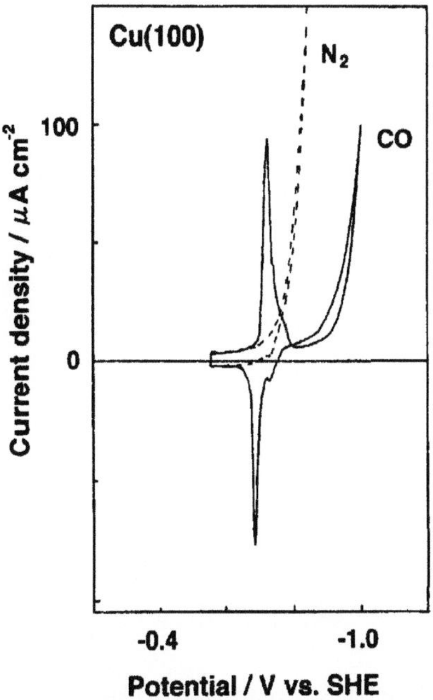

Figure 29. Voltammogram of a Cu(100) electrode in CO saturated phosphate buffer solution (pH 6.8, 0°C). Scan rate 50 mV s^{-1}. The potential axis negative to the right. Cathodic current upward. Reprinted from Ref. 224, Copyright (1998) with permission from Elsevier.

orientation of Cu electrode surface, and can be used as a finger print of individual crystal orientations (Fig. 31).[226]

Infrared absorption spectra would contribute to elucidation of the electrochemical reduction of CO_2 as well as CO. However, the potential range of the spectroscopic measurement is restricted; the measurement must be made below the potential of the charge displacement adsorption. And the measurement is actually impossible

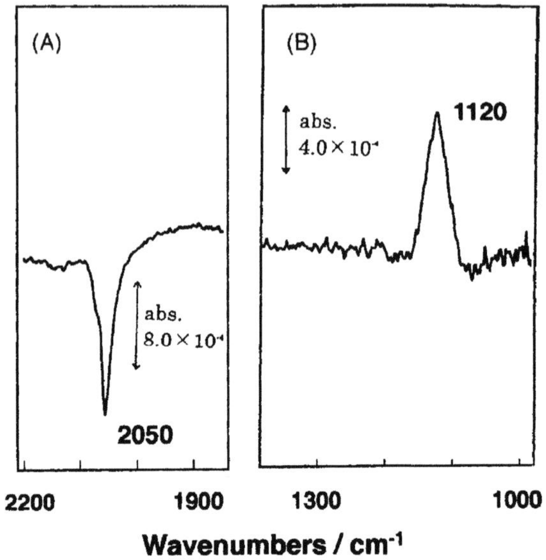

Figure 30. SNIFTIRS spectra from a Cu(100) electrode in CO saturated phophate solution (pH 6.8, 1.6 to 2.0°C). The spectra were acquired between −0.4 V vs. SHE (upward absorption band) and −0.8 V (downward absorption band). Spectrum (A) corresponds to adsorbed CO, and (B) to adsorbed phosphate anion. Reprinted from Ref. 224, Copyright (1998) with permission from Elsevier.

in the potential range where any gas evolving reaction proceeds, i.e., CO_2 reduction and HER.

Figure 32 presents spectra from Cu single crystal electrodes n(111)-(111) series.[226] It is remarkable that the infrared absorption intensity of adsorbed CO grows at the wavenumber 2057 to 2075 with the decrease of n value, or the increase of the step atom density. Figure 33 gives spectra from n(111)-(100) series. The spectra from n(111)-(100) series also give an identical feature with n(111)-(111) series. The absorption intensity at 2073 to 2077 increases with the step atom density. No infrared band is detected from (111) terraces including (111) surface, whereas voltammetric measurements evidently show that HER is heavily suppressed at the (111)

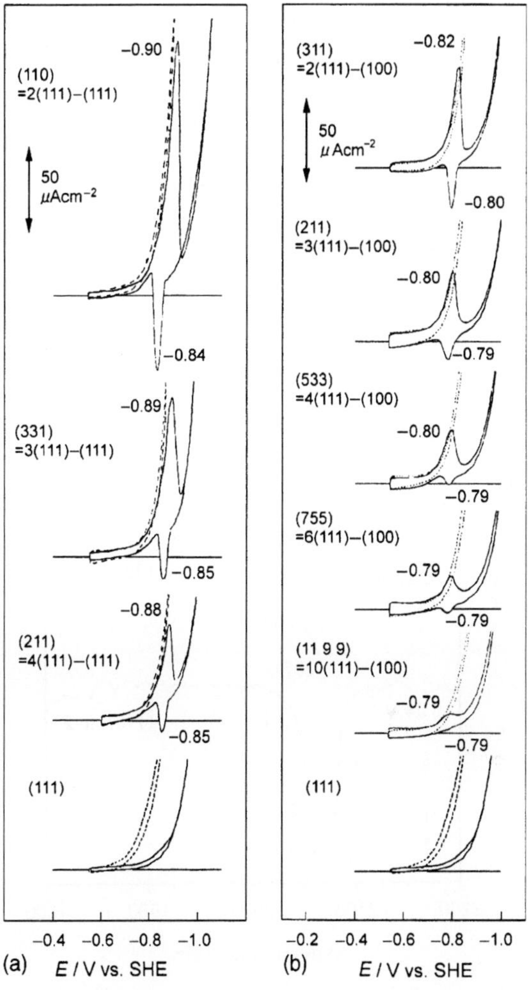

Figure 31. Voltammograms of Cu single crystal electrodes in a phosphate buffer solution (0.1 M KH_2PO_4 + 0.1 M K_2HPO_4) saturated with Ar (broken line) and with CO (solid line) at 0°C with the scan rate 50 mV s^{-1}; (a) n(111)-(111) series, and (b) n(111)-(100) series. The potentials of the redox peaks are given in the figures. The potential axis negative to the right. Cathodic current upward. Reprinted from Ref. 226, Copyright (2005) with permission from Elsevier.

electrode in CO saturated solution. This fact strongly suggests the presence of adsorbed CO on the surfaces (Fig. 31). Two rationalizations may be possible. Since the measurements are made with a potential difference technique, the infrared absorption at two potentials may possibly compensate with each other, and no infrared absorption band is detected. Otherwise, CO molecule may be adsorbed on the surface with the molecular axis parallel to the electrode surface. Such adsorbed CO molecules are invisible by the reflective infrared spectroscopy.

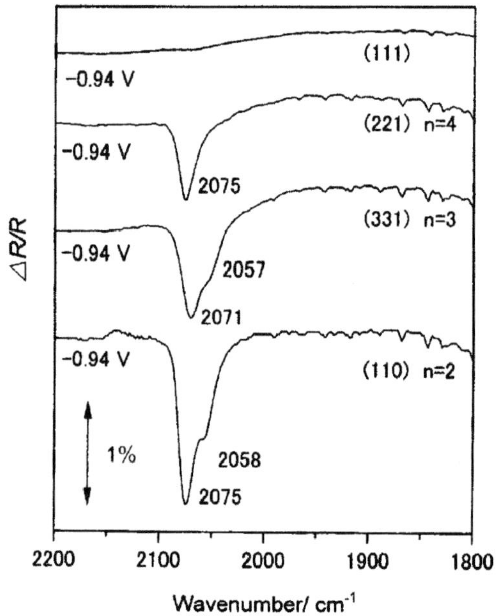

Figure 32. IR spectra of CO adsorbed on Cu single crystal electrodes *n*(111)-(111) with saturation coverage. The reference potential is –0.55 V vs. SHE. Reprinted from Ref. 226, Copyright (2005) with permission from Elsevier.

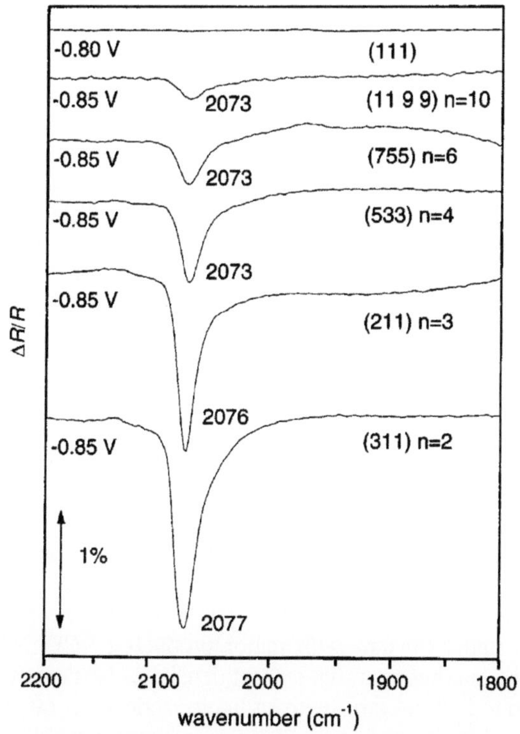

Figure 33. IR spectra of CO adsorbed on Cu single crystal electrodes *n*(111)-(100) electrodes with saturation coverage. The reference potential is –0.55 V vs. SHE. Reprinted from Ref. 226, Copyright (2005) with permission from Elsevier.

IX. ATTEMPTS TO ENHANCE THE TRANSPORT PROCESS IN CO_2 REDUCTION

The solubility of CO_2 in water is as low as 30 mM at 1 atm at the ambient temperature, restricting the transport of CO_2 to the electrode. Under moderate experimental conditions, the highest current density will be limited to 20 mA cm^{-2} or so for formation of CO or $HCOO^-$ for example. Such a low transport process must be improved by any means, if the CO_2 reduction is utilized for a practical process. Various solutions have been proposed, application of high pressure, nonaqueous electrolyte, gas diffusion electrode and solid polymer electrolyte and so on.

1. Elevated Pressure

The solubility of CO_2 increases under elevated pressure, leading to enhanced CO_2 transport rate. Thus CO_2 reduction will naturally proceed at higher current density. CO_2 reduction with simple bulk metal electrodes under elevated pressure was discussed in the previous Sections from the viewpoint of the electrocatalytic aspects. The present Section mentions enhancement effect of the transport process.

Ito and his coworkers applied elevated pressure up to 20 atm to CO_2 reduction in aqueous and nonaqueous electrolyte solutions with Zn, In, Sn and Pb electrodes.[39, 126, 227] They showed enhanced current density of 30 mA cm^{-2} at –1.7 V vs. SCE under 20 atm. The faradaic efficiency of HCOOH formation increased with the increase of CO_2 pressure.

Sakata and his coworkers studied CO_2 reduction under elevated pressure extensively.[99, 101] They could enhance the current density to 163 mA cm^{-2} with a Cu electrode at –1.64 V vs. Ag/AgCl under 30 atm with major products of CH_4, C_2H_4 and CO.[211] The maximum partial current density of HCOOH formation amounted to 560 mA cm^{-2} with an In electrode under 60 atm.[40]

The solubility of CO_2 is high in methanol. Fujishima and his coworkers employed CO_2 methanol mixtures under elevated pressure as the electrolyte solution. Tetraalkyl ammonium salts were used for the supporting electrolytes. They showed that CO_2 reduction can proceed with a Cu electrode at 200 to 500 mA cm^{-2} under 40 to 60 atm. The major products were CO and methyl formate.[156]

Later Mazin et al., and Li and Prentice also published studies on CO_2 reduction at Cu electrodes in methanol based electrolyte and ethanol based one under elevated pressure.[143,144]

2. Gas-Diffusion Electrode

Gas-diffusion electrode (GDE) is a porous composite electrode developed for fuel cell technology, usually composed of Teflon bonded catalyst particles and carbon black. GDEs have been applied to CO_2 reduction by many workers, successfully enhancing the rate of the process.

Mahmood et al. reported GDE with Pb particles as the catalyst which can reduce CO_2 to HCOOH. The highest current density was 100 mA cm^{-2} or more at −1.8 V vs. SCE.[228] In and Sn showed lower catalytic activity. Furuya et al. employed Ru-Pd alloy particles as the catalyst, giving HCOOH at the current density 80 mA cm^{-2} at −1.1 V vs. SHE.[229] They studied simultaneous electrolysis of CO_2 and nitrite or nitrate ions using GDEs with Cu catalyst particles and various metal phthalocyanines. They obtained urea as the reaction product other than CO and NH_3.[230-232]

Ni-, Co- and Sn- phthalocyanines were tested for CO_2 reduction as the catalyst particle. These catalysts gave CO (for Ni- and Co- phthalocyanines) or HCOOH (Sn-, Cu- and Zn- phthalocyanine) as the redution products at high current density such as ca. 100 mA cm^{-2}.[233-235]

GDEs with Cu particles as the catalyst produce C_2H_4 and C_2H_5OH from CO_2 at high current densities with high faradaic efficiencies. Sammells et al. communicated the current density 667 mA cm^{-2} in KOH electrolyte solution with the current yield of 53% for C_2H_4 formation at −4.75 V vs. Ag/AgCl without referring to the IR drop correction for the electrode potential.[236] Ikeda et al. showed that C_2H_4 and C_2H_5OH are the main products with the current density 300 mA cm^{-2} at −1.25 V vs. SHE.[237]

Schwarz et al. incorporated Perovskite-type electrocatalysts $A_{1.8}A'_{0.2}CuO_4$ (A = La, Pr, and Gd; A' = Sr and Th) in GDEs with the electrolyte solution 1 M KOH. They reported that CO_2 was reduced to CH_3OH, C_2H_5OH and *n*-C_3H_7OH with the total current yield of alcohols 40% of the current density 180 mA cm^{-2}.[238] They studied various perovskite type oxides, and argued that such oxides without Cu are not active for CO_2 reduction.

Sakata and his coworkers employed GDEs purchased from Tanaka Noble Metal Co. under elevated pressure. The GDE, loaded with Pt catalyst, can reduce CO_2 at elevated pressure higher than 5 atm in 0.5 M $KHCO_3$, whereas the GDE scarcely reduces CO_2 at 1 atm. GDE without Pt catalyst is not active under 20 atm, and only HER takes place. Under 20 atm, CO_2 is reduced mainly to CH_4 with the current density of 600 mA cm^{-2} with the faradaic efficiency of ca. 50% for total CO_2 reduction at −1.93 V vs. Ag/AgCl.[239, 199] They further studied Fe[240], Pd and Ag[241] for the catalyst particles for GDEs, and demonstrated that the current density of CO_2 reduction to CO amounts to 3.05 A cm^{-2} with an Ag loaded GDE.[241]

Tryk et al. studied GDEs composed of active carbon fiber and loaded with catalysts Ni, Fe, Pd metals, porphyrins, and phthalocyanines.[242,244] The GDEs gave partial current density of CO_2 reduction up to 80 mA cm^{-2} with production of CO under atmospheric pressure. They presumed that the nanopores present in active carbon fiber may provide quasi high pressure atmosphere due to nanoscale effect.[245] Thus Ni electrocatalyst, which is practically inert for CO_2 reduction under atmospheric pressure, may be activated in a similar manner as observed with Ni electrode under elevated pressure.[104]

3. Solid-Polymer Electrolytes

Ion-exchange membrane coated with porous catalyst metal (Solid Polymer Electrolyte, SPE) as well as GDE can provide gas phase electrolysis of CO_2. The first attempt was published by Ito et al., communicating thin porous Au layer electrode coated on a cation exchange membrane. The SPE could not enhance the cathodic current of CO_2 reduction.[246] DeWulf et al. applied an SPE with Cu as the catalyst layer on a cation exchange membrane (CEM) Nafion 115. The SPE reduced CO_2 to CH_4 and C_2H_4 for a while, but the current density for CO_2 reduction dropped below 1 mA cm^{-2} after 70 min electrolysis.[247]

Sammells et al. prepared a Cu coated SPE electrode using a CEM Nafion 117.[248, 249] They reduced CO_2 to C_2H_4 and C_2H_6 with the current density of 10 to 30 mA cm^{-2} with the terminal voltage 1.5 to 3.5 V. The faradaic efficiency for CO_2 reduction remained less than 10%.

Kunugi et al. studied Cu coated SPEs prepared from a CEM (Nafion 117) and an anion exchange membrane (AEM) Selemion AMV, attempting to reduce CO_2 from combustion flue gas.[250] The SPE from Nafion 117 formed C_2H_4 as the product with the maximum partial current 2 mA cm^{-2} at −1.8 V vs. SCE. Both types of SPE retained stable activity in CO_2 reduction for 5 hr.

Hori et al. prepared porous silver coated SPEs with an AEM (Selemion AMV). The silver electrode layer became more porous by applying ultrasonic treatment during the electroless deposition of silver layer on the membrane, and afforded higher current density of CO_2 reduction than without the ultrasonic treatment. Their silver coated SPE reduced CO_2 mainly to CO with the partial current density 60 mA cm^{-2} at –2.8 V vs. SHE.[251]

It should be pointed out that a CEM based SPE is deteriorated in a short period of CO_2 reduction, when the SPE is combined with electrolyte solution other than strong acid. If $KHCO_3$ solution is used for the electrolyte solution for example, K^+ will be accumulated at the electrode metal/ CEM interface in the cathode gas chamber during the electrolysis, since cations (K^+) are transported across CEM. OH^- is simultaneously generated at the interface by the cathodic CO_2 reduction. KOH, thus formed in this process, is neutralized by CO_2 to form $KHCO_3$ at the interface. The electrode metal/ CEM interface is easily destroyed by the accumulation of $KHCO_3$.[251]

A modification of SPE was reported, using a porous membrane made of polyvinylidene fluoride polymer deposited with thin porous Au film.[193] CO_2 was reduced to CO with the faradaic efficiency 75% at the partial current density 20 mA cm^{-2} as estimated from the difference of the two currents measured under argon and CO_2 atmospheres separately.

4. Three-Phase Electrodes

Three phase electrodes were devised aiming at enhancement of the transport process in CO_2 reduction. Ogura et al. reported Cu and Ag net cathodes with CO_2 bubbled from the bottom of the electrolysis cell. They showed that the cathodic current is enhanced by such a configuration.[58,252] Koleli et al. used fixed bed reactor packed with Pb or Sn granules. CO_2 was supplied from the bottom of the reactor with K_2CO_3 or $KHCO_3$ electrolyte solution.[41,253] Their reactor reduced CO_2 mainly to HCOOH with high faradaic efficiency. Akahori et al. examined another three phase electrode device, composed of In impregnated Pb coated carbon fiber or Pb coated glass fiber electrode.[42]

X. CONCLUSIONS

The electrochemical reduction of CO_2 is a hopeful technique as a chemical process. The features of CO_2 reduction at metal electrodes may be summarized next.

1. Electroactive species in CO_2 reduction in aqueous media is CO_2 molecule. HCO_3^- is electrochemically reduced after decomposition to CO_2 and OH^- as evidenced with Hg and Au electrodes. The rate of decomposition of HCO_3^- is low. If the rate of decomposition is enhanced, CO_2 reduction will be accelerated.
2. Initial process of CO_2 reduction leads to formation of CO or formate in aqueous electrolytes; the process involves H_2O, not H^+, as revealed with Hg and Au electrodes.
3. Metal electrodes are divided into 4 groups in accordance with the selectivity in CO_2 reduction in aqueous media. A major difference between CO formation and $HCOO^-$ formation may be derived from the stability of adsorbed intermediate $CO_2{}^{\cdot -}$ on the electrode metals. Hydrocarbon formation at Cu electrode proceeds with intermediate formation of adsorbed CO on the electrode surface.
4. Deactivation of Cu and other metal electrodes is derived primarily from electrodeposition on the electrode surface of impurity metals originally contained in chemical reagents used for the electrolyte, such as Fe and Zn. Contamination of the electrode surface leads to severe deactivation of the electrodes. Neither intermediate species nor product from CO_2 reduction causes the deactivation. Purification of the electrolyte solution is effective to prevent the deactivation process.
5. The low reaction rate due to the low solubility of CO_2 in aqueous media can be enhanced by elevation of the pressure and utilization of 3-dimensional electrodes, such as gas diffusion electrodes, solid polymer electrolytes, and packed bed electrodes.
6. The high overpotential needs to be reduced in order to improve the low energy utilization efficiency. The formation of anion radical $CO_2{}^{\cdot -}$ is presumed to be responsible for the high overpotential. The overpotential depends on electrode

metal and crystal orientation. Stabilization of $CO_2^{\cdot -}$ adsorption on electrodes may lead to reduction of the overpotential. In addition, production of hydrocarbons and alcohols is preferred to CO or formic acid. Cu is an attractive electrode metal for hydrocarbon and alcohol formation at present, whereas the product distribution is wide. Mechanistic studies have revealed the features of Cu electrode process. Novel electrode materials have been proposed by numerous workers, but most of these materials lack scientifically sound and basic information. Fundamental studies will contribute to enhancement of our understanding of the overpotential and product selectivity of the electrodes.

ACKNOWLEDGMENT

The author is grateful to Professor O. Koga of Chiba University and Dr. M. Gattrell of National Research Council of Canada for discussions during the preparation of the manuscript. Professor Koga also assisted with collecting recent publications for writing this article. Dr. M. Gattrell provided an opportunity for the author to review his article before the submission.

LIST OF ABBREVIATIONS

AEM	anion-exchange membrane
AN	acetonitrile
CEM	cation-exchange membrane
DMF	dimethyl formamide
DMSO	dimethyl sulfoxide
ETAAS	electrothermal atomic absorption spectroscopy
GDE	gas-difusion electrode
HER	hydrogen-evolution reaction
HPLC	high-performance liquid chromatography
PC	propylene carbonate
RHE	reversible-hydrogen electrode
SCE	saturated-calomel electrode
SHE	standard-hydrogen electrode
SNIFTIRS	subtractively-normalized interfacial Fourier transform spectroscopy

SPE	solid-polymer electrode
$TBABF_4$	tetrabutyl ammonium tetrafluoroborate
TEAP	tetraethylammonium perchlorate
UPD	underpotential deposition
UV	ultraviolet
XPS	X-ray photoelectron spectroscopy

REFERENCES

[1]P. G.Russell, N. Kovac, S. Srinivasan, and M. Steinberg, *J. Electrochem. Soc.* **124**(1977) 1329.

[2]Y. Hori, in *Proceedings of the Symposium on Environmental Aspects of Electrochemistry and Photoelectrochemistry*, Ed. by M. Tomkiewicz et al., The Electrochemical Society, 1993, Vol. 93-18 p. 1.

[3]Jean-Paul Randin, in *Encyclopedia of Electrochemistry of the Elements*, Ed. by A. J. Bard, Marcel Dekker, 1976, Vol. 7, p. 152.

[4]I. Taniguchi, in *Modern Aspects of Electrochemistry*, Plenum, New York, 1989, No. 20, p. 327.

[5]K. W. Frese, Jr., in *Electrochemical and Electrocatalytic Reactions of Carbon Dioxide*, Ed. by B. P. Sullivan, K. Krist and H. E. Guard, Elsevier, New York, 1993, p. 191.

[6]M. M. Halmann, *Chemical Fixation of Carbon Dioxide, Methods for Recycling CO_2 into Useful Products*, CRC Press, 1993.

[7]V. S. Bagotzky, N. V. Osetrova, *Russ. J. Electrochem.* (Transl. of Elektrokhimiya), **31**(1995) 409.

[8]M. Jitaru, D. A. Lowy, M. Toma, B. C. Toma, L. Oniciu, *J. Appl. Electrochem.* **27**(1997) 875.

[9]M. M. Halmann and M. Steinberg, *Greenhous Gas, Carbon Dioxide Mitigation, Science and Technology*, Lewis Publishers, 1999.

[10]Y. Hori, in *Handbook of Fuel Cells, Fandamentals, Technology and Applications*, Ed. by W. Vielstich, A. Lamm and H. A. Gasteiger, Wiley, 2003, Vol. 2, p.720.

[11]R. P. S. Chaplin and A. A. Wragg, *J. Appl. Electrochem.*, **33**(2003) 1107.

[12]M. A. Scibioh, B. Viswanathan, *Proceedings of the Indian National Science Academy, Part A: Physical Sciences*, **70**(2004) 407.

[13]M. Gattrell, *J. Electroanal. Chem.* **594**(2006) 1.

[14]M. Pourbaix, *Atlas D'Equilibres Electrochimiques*, Gauthier-Villars, 1963.

[15]Y. Hori and S. Suzuki, *Bull. Chem. Soc. Jpn.* **55**(1982) 660.

[16]*Lange's Handbook of Chemistry*, 13th ed., Ed. by J. A. Dean, McGraw-Hill (1985), p.6-2, p.9-4, p.9-107.

[17]S. T. Han and L. J. Bernardin, *T. A. P. P. I.* (Technical Asociation of the Pulp and Paper Industry, New York) **41**(1958) 540.

[18]B. R. W. Pinsent, L. Pearson and F. J. W. Roughton, *Trans. Faraday Soc.* **52**(1956) 1512.

[19]Y. Hori, A. Murata and R. Takahashi, *J. Chem. Soc. Faraday Trans. I* **85**(1989) 2309.

[20]Y. Hori, Y. Yoshinami and H. Kawarada, *Abstract of Annual Meeting of ISE*, 1990.
[21]N. Gupta, M. Gattrell and B. MacDougall, *J. Appl. Electrochem.* **36**(2006) 161.
[22]W. Paik, T. N. Andersen, and H. Eyring, *Electrochim. Act.* **14**(1969) 1217.
[23]Y. Hori, H. Wakebe T. Tsukamoto and O. Koga, *Electrochim. Act.* **39**(1994) 1833.
[24]A. T. Kuhn, in *Electrochemistry, The Past Thirty and the Next Thirty Years*, Edited by H. Bloom and F. Gutmann, Plenum Press, New York, 1977, p.139.
[25]J. O'M. Bockris, S. U. M. Khan, *Surface Electrochemistry, A Molecular Level Approach*, Plenum Press, New York, 1993, p. 9.
[26]J. O'M. Bockris and A. K. N. Reddy, *Modern Electrochemistry*, Vol. 2, Plenum Press, New York, 1970.
[27]K. J. Vetter, *Electrochemical Kinetics, Theoretical and Experimental Aspects*, Academic Press, New York, 1967, p.556.
[28]K. Ito, T. Murata, and S. Ikeda, *Bull. Nagoya Inst. Techn.* **27**(1975) 209.
[29]Y.Hori, K. Kikuchi and S. Suzuki, *Chem. Lett.* (1985) 1695.
[30]H. Noda, S. Ikeda, Y. Oda, and K. Ito, *Chem. Lett.* (1989) 289.
[31]H. Noda, S. Ikeda, Y. Oda, K. Imai, M. Maeda, and K. Ito, *Bull. Chem. Soc. Jpn.* **63**(1990) 2459.
[32]R. L.Cook, R. C. MacDuff, and A. F. Sammells, *J. Electrochem. Soc.* **134**(1987) 1873.
[33]R. L. Cook, R. C. MacDuff, and A. F. Sammells, *J. Electrochem. Soc.* **134**(1987) 2375.
[34]J. J. Kim, D. P. Summers, and K. W. Frese, Jr., *J. Electroanal. Chem.* **245**(1988) 223.
[35]M. Azuma, K. Hashimoto, M. Hiramoto, M. Watanabe, and T. Sakata, *J. Electroanal. Chem.* **260**(1989) 441.
[36]M. Azuma, K. Hashimoto, M. Hiramoto, M. Watanabe, T. Sakata, J.Electrochem. Soc. 137(1990) 1772.
[37]A.Bewick, and G.P.Greener, *Tetrahedron Letters* (1970) No. 5 391.
[38]A. W.B. Aylmer-Kelly, A. Bewick, P. R. Cantrill and A. M. Tuxford, *Faraday Disc. Chem. Soc.* **No. 56**(1973) 96.
[39]K. Ito, S. Ikeda, T. Iida, and H. Niwa, *Denki Kagaku*, **49**(1981) 106.
[40]M. Todoroki, K. Hara, A. Kudo, and T. Sakata, *J. Electroanal. Chem.* **394**(1995) 199.
[41]F. Koleli, T. Atilan, N. Palamut, A. M. Gizir,R. Aydin, and C. H. Hamann, *J. Appl. Electrochem.* **33**(2003) 447.
[42]Y. Akahori, N.Iwanaga, Y. Kato, O. Hamamoto, and M. Ishii, *Electrochemistry* (Tokyo, Japan), **72**(2004) 266.
[43]T. E.Teeter, and P. Van Rysselberghe, *J. Chem. Phys.* **22**(1954) 1759.
[44]J. Jordan, and P. T.Smith, *Proc. Chem. Soc.* (1960) 246.
[45]A.Bewick, and G.P.Greener, *Tetrahedron Letters* (1969) No.53 4623.
[46]F. Wolf, and J. Rollin, *Z.Chem.***17**(1977) 337.
[47]Y. Hori and S. Suzuki, *J. Electrochem. Soc.* **130**(1983) 2387.
[48]B. R Eggins, E. M. Brown, E A. McNeil1, and J. Grimshaw, *Tetrahedron Letters* **29**(1988) 945.
[49]A. F. Cherashev and A. P. Khrushch, *Russ. J. Electrochem.* (Transl. of Elektrokhimiya) **33**(1997) 181.
[50]Y. Hori, A. Murata, R. Takahashi and S. Suzuki, *J. Chem. Soc. Chem. Commun.* (1987) 728.

[51]M. Fujihira, T. Noguchi, *Chem. Lett.* (1992) 2043.
[52]P. Kedzierzawski and J. Augustynski, *J. Electrochem. Soc.* **141**(1994) L58.
[53]H. Noda, S. Ikeda, A. Yamamoto, H. Einaga, and K. Ito, *Bull. Chem. Soc. Jpn.,* **68**(1995) 1889.
[54]R. Shiratsuchi, S. Ishimaru, and G. Nogami, *Denki Kagaku* **66**(1998) 668.
[55]T. Ohmori, A. Nakayama, H. Mametsuka, and E. Suzuki, *J. Electroanal. Chem.* **514**(2001) 51.
[56]R. Kostecki, and J. Augustynski, *Ber. Bunsen-Ges. Phys. Chem.* **98**(1994) 1510.
[57]N. Hoshi, M. Kato, and Y. Hori, *J. Electroanal. Chem.* **440**(1997) 283.
[58]H. Yano, F. Shirai, and K. Ogura, *J. Electroanal. Chem.* **533**(2002) 113.
[59]A. Hattori, H. Noda, S. Ikeda, M. Maeda, and K. Ito, *Denki Kagak*, **6**(1991) 528.
[60]S. Ikeda, A. Hattori, K. Ito, and H. Noda, *Electrochemistry* (Tokyo) **67**(1999) 27.
[61]W. M. Ayers, and M. Farley, *ACS. Sym. Ser.* **363**(1988) 147.
[62]K. Ohkawa, K. Hashimoto, and A. Fujishima, *J. Electroanal. Chem.* **345**(1993) 445.
[63]M. Azuma, K. Hashimoto, M. Watanabe, and T. Sakata, *J. Electroanal. Chem.* **294**(1990) 299.
[64]K. Ohkawa, Y. Noguchi, S. Nakayama, K. Hashimoto, and A. Fujishima, *J. Electroanal. Chem.* **369**(1994) 247.
[65]B. I. Podlovchenko, E. A. Kolyadko, and S. Lu, *J. Electroanal. Chem.* **373**(1994) 185.
[66]H. Yoshitake, T. Kikkawa, and K. Ota, *J. Electroanal. Chem.* **390**(1995) 91.
[67]N. Hoshi, M. Noma, T. Suzuki, and Y. Hori, *J. Electroanal. Chem.* **421**(1997) 15.
[68]C. Iwakura, S. Takezawa, and H. Inoue, *J. Electroanal. Chem.* **459**(1998) 167.
[69]Y. Hori and A. Murata, *Chem. Lett.* (1991) 181.
[70]Y. Hori, K. Kikuchi, A. Murata, and S. Suzuki, *Chem. Lett.* (1986) 897.
[71]R. L. Cook, R. C. MacDuff, and A. F. Sammells, *J. Electroanal. Chem.* **135**(1988) 1320.
[72]O. Koga, K. Nakama, A. Murata and Y. Hori, *Denki Kagaku* (Electrochemistry), **57**(1989) 1137.
[73]D. W. DeWulf, T. Jin, and A. J. Bard, *J. Electrochem. Soc.* **136**(1989) 1686.
[74]S. Wasmus, E. Cattaneo, and W. Vielstich, *Electrochem. Acta* **35**(1990) 771.
[75]A. Murata and Y. Hori, *Bull. Chem. Soc. Jpn.* **64**(1991) 123.
[76]K. W.Frese, Jr., *J. Electrochem. Soc.***138**(1991) 3338.
[77]G. Kyriacou and A.Anagnostopoulos, *J. Electroanal. Chem.* **322**(1992) 233.
[78]G. Z. Kyriacou and A. K. Anagnostopoulos, *J. Appl. Electrochem.* **23**(1993) 483.
[79]R. Shiratsuchi, Y. Aikoh, and G. Nogami, *J. Electrochem. Soc.* **140**(1993) 3479.
[80]B. Jermann and J. Augustynski, *Electrochim. Act.* **39**(1994) 1891.
[81]K. Hara, A. Tsuneto, A. Kudo, and T. Sakata, *J. Electroanal. Chem.* **434**(1997) 239.
[82]P. Friebe, P. Bogdanoff, N. Alonso-Vante, and H. Tributsch, *J. Catal.* **168**(1997) 374.
[83]Y. Terunuma, A. Saitoh, and Y. Momose, *J. Electroanal. Chem.* **434**(1997) 69.
[84]K. Ohta, K. Suda, S. Kaneco, and T. Mizuno, *J. Electrochem. Soc.* **147**(2000) 233.
[85]J. Lee and Y. Tak, *Electrochim. Act.* **46**(2001) 3015.
[86]S. Kaneco, N. Hiei, S. Itoh, T. Suzuki, and K. Ohta, *ITE Letters on Batteries, New Technologies & Medicine* **2**(2001) 83.
[87]Y. Hori, I. Takahashi, O. Koga, and N. Hoshi, *J. Phys Chem. B* **106**, 2002, 15.
[88]I. Takahashi, O. Koga, N. Hoshi, and Y. Hori, *J. Electroanal. Chem.* **533**(2002) 135.

[89]H. Yano, F. Shirai, M. Nakayama, and K. Ogura, *J. Electroanal. Chem.* **519**(2002) 93.
[90]Y. Momose, K. Sato, and O. Ohno, *Surface and Interface Analysis* **34**(2002) 615.
[91]Y. Hori, I. Takahashi, O. Koga, and N. Hoshi, *J. Mol. Cat.* **199**(2003) 39.
[92]K. Ogura, H. Yano, and F. Shirai, *J. Electrochem. Soc.* **150**(2003) D163.
[93]P. Dube and G. M. Brisard, *J. Electroanal. Chem.* **582**(2005) 230.
[94]Y. Hori, A. Murata, S. Ito, Y. Yoshinami, and O. Koga, *Chem. Lett.* (1989) 1567.
[95]Y. Hori and A. Murata, *Chem. Lett.* (1990) 1231.
[96]M. Watanabe, M. Shibata, A. Kato, M. Azuma, and T. Sakata, *J. Electrochem. Soc.* **138**(1991) 3382.
[97]G. Kyriacou and A. Anagnostopoulos, *J. Electroanal. Chem.* **328**(1992) 233.
[98]S. Ishimaru, R. Shiratsuchi, and G. Nogami, J. Electrochem. Soc. **147**(2000) 1864.
[99]K. Hara, A. Kudo, and T. Sakata, J. Electroanal. Chem. **391**(1995) 141.
[100]A. Murata and Y. Hori, *Denki Kagaku,* **59**(1991) 499.
[101]S. Nakagawa, A. Kudo, M. Azuma, and T. Sakata, *J. Electroanal. Chem.* **308**(1991) 339.
[102] K. Hara, A. Kudo, and T. Sakata, *J. Electroanal. Chem.* **386**(1995) 257.
[103]Y. Hori and A. Murata, *Electrochim. Acta,* **35**(1990) 1777.
[104]A. Kudo, S. Nakagawa, A. Tsuneto, and T. Sakata, *J. Electrochem. Soc.* **140**(1993) 1541.
[105]N. Hoshi, H. Ito, T. Suzuki, and Y. Hori, *J. Electroanal. Chem.* **395**(1995) 309.
[106]N. Hoshi, T. Uchida, T. Mizumura, and Y. Hori, *J. Electroanal. Chem.* **381**(1995) 261.
[107]J. Giner, *Electrochim. Act.* **8**(1963) 857.
[108]M. W. Breiter, *Electrochim. Act.* **12**(1967) 1213.
[109]S. Taguchi and A. Aramata, *Electrochim. Act.* **39**(1994) 2533.
[110]A. Rodes, E. Pastor, and T. Iwasita, *J. Electroanal. Chem.* **369**(1994) 183.
[111]N. Hoshi, T. Mizumura, Y. Hori, *Electrochim. Act.* **40**(1995) 883.
[112]N. Hoshi, T. Suzuki, and Y. Hori, *Electrochim. Act.* **41**(1996)1647.
[113]N. Hoshi, T. Suzuki, and Y. Hori, *J. Phys. Chem. B* **101**(1997) 8520.
[114]N. Hoshi and Y. Hori, *Electrochim. Act.* **45**(2000) 4263.
[115]G. M. Brisard, A. P. M. Camargo, F. C. Nart, and T. Iwasita, *Electrochem. Commun.* **3**(2001) 603.
[116]K.W. Frese, Jr. and S. Leach, *J. Electrochem. Soc.* **132**(1985) 259.
[117]S. Komatsu, T. Yanagihara,Y. Hiraga, M. Tanaka, and A. Kunugi, *Denki Kagaku* **63**(1995) 217.
[118]K. Hara, A. Kudo, and T. Sakata, *J. Electroanal. Chem.* **421**(1997) 1.
[119]A. Bandi, *J. Electrochem. Soc.* **137**(1990) 2157.
[120]J. P. Popic, M. L. Avramov-Ivic, and N. B.Vukovic, *J. Electroanal. Chem.* **421**(1997) 105.
[121]N. Spataru, K. Tokuhiro, C. Terashima, T. N. Rao, and A. Fujishima, *J. Appl. Electrochem.* **33**(2003)1205.
[122]J. Qu, X. Zhang, Y. Wang and C. Xie, *Electrochim. Act.* **50** (2005) 3576.
[123]Von U. Kaiser and E. Heitz, Ber. Bunsenges. *Phys. Chem.* **77**(1973) 818.
[124]J. C. Gressin, D. Michelet, L. Nadjo, and J. M. Savéant, *Nouv. J. Chim.* (1979) 545.
[125]C. Amatore and J.-M. Savéant, *J. Am. Chem. Soc.* **103**(1981) 5021.
[126]K. Ito, S. Ikeda, T. Iida, and A. Nomura, *Denki Kagaku,* **50**(1982), 463.
[127]S. Ikeda, T. Takagi, and K. Ito, *Bull. Chem. Soc. Jpn.* **60**(1987) 2517.

[128]B. R. Eggins, C. Ennis, R. McConnell, and M. Spence, *J. Appl. Electrochem.* **27** (1997) 706.
[129]T. Saeki, K. Hashimoto, N. Kimura, K. Omata, and A. Fujishima, *J. Electroanal. Chem.* **404**(1996) 299.
[130]S. Kaneco, R. Iwao, K. Iiba, S. Ito, K. Ohta, and T. Mizuno, *Energy* (Oxford) **23** (1998) 1107.
[131]L. V.Haynes and D. T.Sawyer, *Anal. Chem.* **39**(1967) 332.
[132]J. O'M. Bockris and J. C. Wass, *J. Electrochem. Soc.* **136**(1989) 2521.
[133]D. A. Tyssee, J. H. Wagenknecht, M. M. Baizer, and J. L. Chruma, *Tetrahedron Letters*, (1972) No.47 4809.
[134]A. Gennaro, A. A. Isse, M.-G. Severin, E. Vianello, I. Bhugun, and J. -M. Savéant, *J. Chem. Soc. Faraday Trans.* **92**(1996), 3963.
[135]Y. Tomita, S. Teruya, O. Koga and Y. Hori, *J. Electrochem. Soc.* **147**(2000) 4164.
[136]Y. Hori and T. Kakehata, unpublished data.
[137]T. Mizuno, M. Kawamoto, S. Kaneco, and K. Ohta, *Electrochim. Act.* **43**(1998) 899.
[138]A. Naitoh, K. Ohta, T. Mizuno, H. Yoshida, M. Sakai, and H. Noda, *Electrochim. Act.* **38**(1993) 2177.
[139]T. Saeki, K. Hashimoto, Y. Noguchi, K. Omata, and A. Fujishima, *J. Electrochem. Soc.* **141**(1994) L130.
[140]T. Saeki, K. Hashimoto, N. Kimura, and K. Omata, A. Fujishima, *Chem. Lett.* (1995) 361.
[141]T. Saeki, K. Hashimoto, A. Fujishima, N. Kimura, and K. Omata, *J. Phys. Chem.* **99** (1995) 8440.
[142]S. Kaneco, K. Iiba, M. Yabuuchi, N. Nishio, H. Ohnishi, H. Katsumata, T. Suzuki, and K. Ohta, *Industrial & Engineering Chemistry Research* **41**(2002) 5165.
[143]V. M. Mazin, E. I. Mysov, and V. A. Grinberg, *Russ. J. Electrochem.* **33**(1997) 779.
[144]J. Li, G. Prentice, *J. Electrochem. Soc.* **144**(1997), 4284.
[145]K. Ohta, K. Kawamoto, T. Mizuno, and D. A. Lowy, *J. Appl. Electrochem.* **28** (1998) 717.
[146]S. Kaneco, K. Iiba, K. Ohta, T. Mizuno, and A. Saji, *Electrochim. Act.* **44**(1998) 573.
[147]S. Kaneco, R. Iwao, H. Katsumata, T. Suzuki, and K. Ohta, *ITE Letters on Batteries, New Technologies & Medicine* **2**(2001) 640.
[148]S. Kaneco, R. Iwao, K. Iiba, S. Ito, K. Ohta, and T. Mizuno, *Environ. Eng. Sci.* **16** (1999) 131.
[149]S. Kaneco, K. Iiba, K. Ohta, T. Mizuno, and A. Saji, *J. Electroanal. Chem.* **441** (1998) 215.
[150]R. Schrebler, P. Cury, F. Herrera, H. Gomez, and R. Cordova, *J. Electroanal. Chem.* **516**(2001) 23.
[151]R. Schrebler, P. Cury, C. Suarez, E. Munoz, H. Gomez, and R. Cordova, *J. Electroanal. Chem.* **533**(2002)167.
[152]R. Aydin and F. Koleli, *Synthetic Metals* **144**(2004) 75.
[153]R. Aydin and F. Koleli, *J. Electroanal. Chem.* **535**(2002) 107.
[154]F. Koleli, T. Ropke, and C. H. Hamann, *Synthetic Metals* **140**(2004) 65.
[155]I. Taniguchi, B. Aurian-Blajeni, and J. O'M. Bockris, *J. Electroanal. Chem.* **161**(1984) 385.

[156]T. Saeki, K. Hashimoto, N. Kimura, K. Omata, A. Fujishima, J. Electroanal. Chem. 390(1995) 77.
[157]A. Gennaro, A. A. Isse, J. -M. Savéant, M. -G. Severin, E. Vianello, J. Am. Chem. Soc. 118(1996) 7190.
[158]*Encyclopedia of Electrochemistry*, Ed. by A. J. Bard, H. Lund, Vol. XI Organic section, Marcel Dekker, New York and Basel, 1978, p. 208.
[159]B. Beden, A. Bewick, M. Razaq, and J. Weber, *J. Electroanal. Chem.* **139**(1982) 203.
[160]O. Koga, T. Matsuo, H. Yamazaki, and Y. Hori, *Bull. Chem. Soc. Jpn.* **71**(1998) 315.
[161]G. A. Somorjai, *Catal. Rev.-Sci. Eng.* **18**(1978) 173.
[162]Perry's Chemical Engineers' Handbook, 6th ed. McGraw-Hill, 1984.
[163]T. Mizuno, A. Naitoh, and K. Ohta, *J. Electroanal. Chem.* **391**(1995) 199.
[164]P. Van Rysselberghe, G. J. Alkire, and J. M. McGee, *J. Amer. Chem. Soc.* **68**(1946) 2050.
[165]K. S. Udupa, G. S. Subramanian, and H. V. K. Udupa, *Electrochim. Act.* **16** (1971) 1593.
[166]C. J. Stalder, S. Chao, and M. S. Wrighton, *J. Am. Chem. Soc.* **106**(1984) 3673.
[167]M. Spichiger-Ulmann and J. Augustynski, *J. Chem. Soc. Faraday Trans. 1* **81**(1985) 713.
[168]B. R. Eggins, E. M. Bennett, and E. A. McMullan, *J. Electroanal. Chem.* **408**(1996) 165.
[169]Y. Hori and S. Suzuki, *J. Res. Inst. Catalysis, Hokkaido University* **30**(1982) 81.
[170]Y. Hori, H. Konishi, T. Futamura, A. Murata, O. Koga, H. Sakurai, and K. Oguma, *Electrochim. Acta* **50**(2005) 5354.
[171]Y. Hori and T. Futamura, unpublished data.
[172]D. T. Sawyer, A. Sobkowiak, and J. L. Roberts, Jr., *Electrochemistry for Chemists*, 2nd ed. John Wiley & Sons, 1995.
[173]J. Pacansky, U. Wahlgren, and P. S. Bagus, *J. Chem. Phys.* **62**(1975) 2740.
[174]E. Lamy, L. Nadjo, and J. -M. Savéant, *J. Electroanal. Chem.* **78** (1977) 403.
[175]H. A. Schwarz and R. W. Dodson, *J. Phys. Chem.* **93**(1989) 409.
[176]P. S. Surdhar, S. P. Mezyk, and D. A. Armstrong, *J. Phys. Chem.* **93**(1989) 3360.
[177]S. D. Babenko, V. A. Benderskii, A. G. Krivenko, and V. A. Kurmaz, *J. Electroanal. Chem.* **159**(1983) 163.
[178]S.Kapusta and N.Hackerman, *J. Electrochem. Soc.* **130**(1983) 607.
[179]D. J. Schiffrin, *Faraday Disc. Chem. Soc.* **No. 56** (1973) 75.
[180]G. V. Buxton and R. M. Sellers, *J. Chem. Soc. Far. Trans. I* **69**(1973) 555.
[181]J. Ryu, T. N. Andersen, and H. Eyring, *J. Phys. Chem.* **76**(1972) 3278.
[182]Yu. B.Vassiliev, V. S. Bagotzky, N. V. Osetrova, O. A. Khazova, and N. A. Mayorova, *J. Electroanal. Chem.* **189**(1985) 271.
[183]S. Sakaki, *J. Am. Chem. Soc.* **114**(1992) 2055.
[184]M. Beley, J.-P. Collin, R. Ruppert, and J.-P. Sauvage, *J. Am. Chem. Soc.***108** (1986) 7461.
[185]F. Solymosi, *J. Molecular Catalysis*, **65**(1991) 337.
[186]J. Krause, D. Borgmann, and G. Wedler, *Surf. Sci.* **347** (1996) 1.
[187]H. -J. Freund and M. W. Roberts, *Surf. Sci. Rep.* **25**(1996) 225.
[188]B. Aurian-Blajeni, M. A. Habib, I. Taniguchi, and J. O'M. Bockris, *J. Electroanal. Chem.* **157**(1983), 399.
[189]A. J. McQuillan, P. J. Hendra, and M. Fleischmann, *J. Electroanal. Chem.*, **65** (1975) 933.

[190]R. T. Sanderson, *Inorganic Chemistry*, Reinhold, New York (1967).
[191]M. Hammouche, D. Lexa, M. Momenteau, and J. -M. Savéant, *J. Am.Chem. Soc.* **113**(1991) 8455.
[192]Y. Hori, T. Kaneda, and Y. Tomita, unpublished data.
[193]G. B. Stevens, T. Reda, and B. Raguse, *J.Electroanal. Chem.* **526**(2002) 125.
[194]B. N. Nikolic, H. Huang, D. Gervasio, A. Lin, C. Fierro, R. R. Adzic, and E.Yeager, *J. Electroanal. Chem.* **295** (1990) 415 .
[195]A. Aramata, M. Enyo, O. Koga, and Y. Hori, *Chem. Lett.* (1991) 749.
[196]T. Iwasita, F. C. Nart, B. Lopez, and W. Vielstich, *Electrochim. Act.* **37**(1992) 2361.
[197]A. Rodes, E. Pastor, and T. Iwasita, *J. Electroanal. Chem.* **377** (1994) 215.
[198]M. C. Arevalo, C. Gomis-Bas, F. Hahn, B. Beden, A. Arevalo, and A .J. Arvia, *Electrochim. Act.* **39**(1994) 793.
[199]K. Hara and T. Sakata, *J. Electrochem. Soc.* **144**(1997) 539.
[200]I. Toyoshima and G. A. Somorjai, *Catal. Rev. -Sci. Eng.* **19**(1979) 105.
[201]H. Yoshitake, K. Takahashi, and K. Ota, *J. Chem. Soc. Far.* **90**(1994) 155.
[202]S. Taguchi, A. Aramata, and M. Enyo, *J. Electroanal. Chem.* **372**(1994) 161.
[203]Y. Hori, O. Koga, A. Aramata and M. Enyo, *Bull. Chem. Soc. Jpn.*, **65**(1992) 3008.
[204]D. P. Summers, S. Leach, and K. W. Frese, Jr. *J. Electroanal. Chem.* **205**(1986) 219.
[205]Y. Hori, A. Murata, R. Takahashi, and S. Suzuki, *J. Amer. Chem. Soc.* **109**(1987) 5022.
[206]Y. Hori, O. Koga, H. Yamazaki, and T. Matsuo, *Electrochim. Acta* **40**(1995) 2617.
[207]I. Oda, H. Ogasawara, and M. Ito, *Langmuir* **12**(1996) 1094.
[208]Y. Hori, A. Murata, R. Takahashi, and S. Suzuki, *Chem. Lett.* (1987) 1665.
[209]K. Watanabe, U. Nagashima, and H. Hosoya, *Chem. Phys. Lett.* **209**(1993), 109.
[210]K. Watanabe, U. Nagashima, and H. Hosoya, *Appl. Surf. Sci.* **75**(1994) 121.
[211]K. Hara, A. Tsuneto, A. Kudo, and T. Sakata, *J. Electrochem. Soc.* **141**(1994) 2097.
[212]M. Spichiger-Ulmann and J. Augustynski, Nouv. *J. Chim.* **10**(1986) 487.
[213]S. Kaneco, H. Katsumata, T. Suzuki, and K. Ohta, *Electrochim. Act.* **51**(2006) 3316.
[214]K. J. Vetter, *Electrochemical Kinetics, Theoretical and Experimental Aspects*, Academic Press, New York, p.556 (1967). Originally, A. N. Frumkin, *Z. Phys. Chem. Abt. A* **164**(1933) 121.
[215]Y. Hori, A. Murata, and Y. Yoshinami, *J. Chem. Soc. Faraday Trans.* **87**(1991) 125.
[216]Y. Hori, R. Takahashi, Y. Yoshinami, and A. Murata, *J. Phys. Chem. B* **101** (1997) 7075.
[217]J. O'M. Bockris and S. U. M. Khan, *Quantum Electrochemistry*, Plenum, New York, 1979, p.49.
[218]P. J. Chantry, *Physical Review* **172**(1968) 125.
[219]R. C. Brady and R. Pettit, *J. Am. Chem. Soc.* **102**(1980) 6181.
[220]H. Kita and T. Kurusu, *J. Res. Inst. Catal., Hokkaido Univ.* **21**(1973) 200.
[221]A. Takeuchi and J. Katzer, *J. Phys. Chem.* **86**(1982) 2438.
[222]H. Yano, T. Tanaka, M. Nakayama, and K. Ogura, *J. Electroanal. Chem.* **565**(2004) 287.
[223]Y. Hori, H. Wakebe, T. Tsukamoto, and O. Koga, *Surf. Sci.* **335**(1995) 258.

[224]Y. Hori, O. Koga, Y. Watanabe, and T. Matsuo, *Electrochim. Act.* **44**(1998) 1389.
[225]J. Clavilier, R. Alabat, R. Gomez, J. M. Orts, J. M. Feliu, and A. Aldaz, *J. Electroanal. Chem.* **330**(1992) 489.
[226]O. Koga, S. Teruya, K. Matsuda, M. Minami, N. Hoshi, and Y. Hori, Electrochim. Act. 50(2005) 2475.
[227]K. Ito, S. Ikeda and M. Okabe, Denki Kagaku, 48(1980) 247.
[228]M. N. Mahmood, D. Masheder, and C. J. Harty, J. Applied Electrochem. 17(1987) 1159.
[229]N. Furuya, T. Yamazaki, and M. Shibata, J. Electroanal. Chem. 431(1997) 39.
[230]M. Shibata, K. Yoshida, and N. Furuya, J. Electrochem. Soc. 145(1998) 595.
[231]M. Shibata and N. Furuya, *J. Electroanal. Chem.* **507**(2001) 177.
[232]M. Shibata and N. Furuya, *Electrochim. Act.* **48**(2003) 3953.
[233]M. N. Mahmood, D. Masheder, and C. J. Harty, *J. Applied Electrochem.* **17**(1987) 1223.
[234]N. Furuya, K. Matsui, and S. Motoo, *Denki Kagaku* **56**(1988) 288.
[235]E. R. Savinova, S. A. Yashnik, E. N.Savinov, and V. N. Parmon, *React. Kinet. Catal. Lett.* **46**(1992) 249.
[236]R. L.Cook, R. C. MacDuff, and A. F.Sammells, *J. Electrochem. Soc.* **137**(1990) 607.
[237]S. Ikeda, T. Ito, K. Azuma, K. Ito and H. Noda, *Denki Kagaku* **63**(1995) 303.
[238]M. Schwartz, R. L. Cook, V. M. Kehoe, R. C. MacDuff, J. Patel, and *A. F. Sammells*, J. Electrochem. Soc. **140**(1993) 614.
[239]K. Hara, A. Kudo, T. Sakata, and M. Watanebe, *J. Electrochem. Soc.* **142**(1995) L57.
[240]K. Hara and T. Sakata, *Anal. Sci. Technol.* **8**(1995) 683.
[241]K. Hara and T. Sakata, *Bull. Chem. Soc. Jpn.* **70**(1997) 571.
[242]T. Yamamoto, K. Hirota, D. A. Tryk, K. Hashimoto, A. Fujishima, and M. Okawa, *Chem. Lett.* (1998) 825.
[243]D. A. Tryk, T. Yamamoto, M. Kokubun, K. Hirota, K. Hashimoto, and M. Okawa, A. Fujishima, *Appl. Organomet. Chem.* **15**(2001) 113.
[244]T. V. Magdesieva, T. Yamamoto, D. A. Tryk, and A. Fujishima, *J. Electrochem. Soc.* **149**(2002) D89.
[245]K. Kaneko, R. F. Cracknell, and D. Nicholson, *Langmuir* **10**(1994) 4606.
[246]M. Maeda, Y. Kitaguchi, S. Ikeda, and K. Ito, *J. Electroanal. Chem.* **238**(1987) 247.
[247]D. W. DeWulf and A. J. Bard, *Catalysis Letters* **1**(1988) 73.
[248]R. L. Cook, R. C. MacDuff, and A. F. Sammells, *J. Electrochem. Soc.* **135**(1988) 1470.
[249]R. L. Cook, R. C. MacDuff, and A. F. Sammells, *J. Electrochem. Soc.* **137**(1990) 187.
[250]S. Komatsu, M. Tanaka, A. Okumura, and A. Kunugi, *Electrochim. Act.* **40**(1995) 745.
[251]Y. Hori, H. Ito, K. Okano, K. Nagasu, and S. Sato, *Electrochim. Act.* **48**(2003) 2651.
[252]K. Ogura, H. Yano, and T. Tanaka, *Catalysis Today* **98**(2004) 515.
[253]F. Koleli and D. Balun, *Applied Catalysis A* **274**(2004) 237.

4

Induced Codeposition of Alloys of Tungsten, Molybdenum and Rhenium with Transition Metals

Noam Eliaz* and Eliezer Gileadi**

*Biomaterials and Corrosion Laboratory, School of Mechanical Engineering,
**School of Chemistry, Faculty of Exact Sciences,
Tel-Aviv University, Ramat Aviv, Tel-Aviv 69978, Israel

I. INTRODUCTION

In this section, the process of electrodeposition is reviewed briefly, and its place in the general context of electrode reactions and charge transfer across the metal/solution interface is set (Section 1.1). In Section 1.2, special emphasis is given to deposition of alloys, and particularly to anomalous deposition of alloys (Sections 1.2.3 and 1.2.4). Next, the phenomenon of induced codeposition is defined, and possible mechanisms are discussed briefly (Section 1.2.5). Several electroless (Section 1.2.6) and electrodeposition processes, in which induced codeposition plays a role, are mentioned. A more extensive discussion of electrodeposition of W-, Mo- and Re-based alloys is included in Section 2. Typical

Modern Aspects of Electrochemistry, Number 42, edited by C. Vayenas et al., Springer, New York, 2008.

bath compositions and operating conditions are listed in Appendices A through C, respectively.

Many books have been dedicated over the years to the topic of electrodeposition (see, for example, Refs. 1-6). These books deal with a variety of sub-topics such as surface preparation of the substrate prior to deposition, thermodynamics and kinetics of electrodeposition, the reactions that take place on an atomistic level, the mechanisms of growth, the effect of bath chemistry and operating conditions, the deposition of specific metals and alloys, the structure and properties of deposits, etc.

Electrodeposition has been practiced in industry for more than 150 years. For many years it was considered as an empirical, low-level, technology. Although many useful plating baths have been developed, and additives were identified for specific purposes—macro and micro leveling, brightening, stress relieving, inhibition of hydrogen embrittlement (HE) and so on, most of the advancements in the field were achieved by ingenious trial-and-error methods, rather than by attempting to gain a profound understanding of the scientific aspects of the field. A turning point in this approach can be associated with the announcement by IBM in 1997 of replacing vapor-deposited aluminum by electrodeposited copper wiring in ultra-large-scale integration (ULSI) silicon chips. Following this announcement, electrodeposition gained much interest in the microelectronic industry as a potentially attractive manufacturing process compared to other technologies. At the same time, the awareness of possible advancements through research and development increased, both in academia and in industry, and it stopped being perceived as mostly an empirical technology.[7]

Electrodeposition offers several important advantages compared to most other plating technologies. These include:

- relatively low cost
- fairly simple and available equipment
- the laws governing scaling up and scaling down of electrochemical processes are well understood[7]
- porous, geometrically complex or non line-of-sight surfaces can be coated
- proper design of the cell and the counter electrode can ensure that metal is deposited only where it is needed[7]

- the throwing power is higher than in physical vapor deposition[7]
- production of high-aspect-ratio structures with good precision[7]
- relatively low processing temperature, allowing the formation of highly crystalline deposits, with possibly lower residual stresses
- the thickness, composition and microstructure of the deposit can be controlled precisely
- dense materials with high purity, low defect density and narrow distribution of grain size can be produced.

Representative applications of electrodeposition include, among others, gold-plated brass jewelry; copper plating for fabrication of interconnects in electronic packaging; hard chromium plating of aircraft landing gears made of alloy steels; zinc-nickel alloy plating on steel components; silver-plated mirrors; tin-lead coatings for soldering on printed-circuit boards; plating of nickel, nickel-iron and copper in fabrication of micro-electro-mechanical systems (MEMS) by LIGA, etc.[6,7]

Current distribution on the cathode is a major variable in electrodeposition. It is determined by several parameters, such as the geometry of both the anode and the cathode, as well as their relative position in the bath, the kinetics of charge transfer and the conditions of mass transport.[7] It determines the thickness of the coating and its uniformity, as well as the local chemical composition in alloy deposition. In the past two or three decades, specific interest in pulse plating and periodic reverse pulse plating (where the current is either interrupted or reversed periodically, respectively)[8] has increased significantly. These techniques may have a significant influence on the composition and structure of electrodeposited alloys[7,9-11] as well as on the surface morphology and the micro-throwing power.

A clarification of nomenclature and sign convention, which may often be confusing, is called for in this context. It can be stated categorically that the cathode is always the electrode at which a reduction process (e.g., hydrogen evolution or metal deposition) takes place. Similarly, the anode is always the electrode at which oxidation (e.g., oxygen evolution or metal dissolution) takes place. But, which is the positive and which is the negative electrode? The answer to that depends on the type of reaction taking place in the cell. If the cell is externally driven (namely, electric power is consumed to force a current through it, leading to chemical change), the cathode is the negative terminal and the anode is the positive terminal in a so-called *electrolytic cell.* This applies to electrodeposition, water electrolysis, electrosynthesis, battery charging, etc. It is reflected in Pourbaix (potential versus pH) diagrams and in potential versus current density curves from potentiodynamic polarization experiments by the fact that, as the potential becomes more positive (or less negative), oxidation reactions become more dominant. If, on the other hand, the reaction in the cell is self-driven (i.e., proceeds spontaneously), the anode is the negative terminal and the cathode is the positive terminal in a so-called *galvanic cell.* This applies to battery discharge and fuel cells, where chemical energy is converted to electric energy. It is reflected in the electromotive force (EMF) series (standard potentials) where, as the standard reduction potential becomes more positive (or less negative), the material is said to be more noble. Open-circuit potential (OCP) measurements yield often, but not exclusively, the reversible potential of the system. The reversible potential represents thermodynamic equilibrium, which is not time dependent. The OCP can be time dependent, as a result of changes of the electrode surface, e.g., the formation of a passive oxide layer. A shift of the OCP with time in the positive direction represents the formation of a more noble (better corrosion resistant) surface, and vice versa.

Consider, for example, the lead-acid battery in a car. In the fully charged state, the negative electrode is metallic lead (Pb^0), while the positive electrode is lead dioxide (PbO_2), in which lead is in the Pb^{4+} state. During discharge, the active material in the negative electrode is oxidized to $PbSO_4$, so by definition it is the anode, and that in the positive electrode is reduced to $PbSO_4$, so by definition it is the cathode. When the battery is charged, the reactions at both electrodes are reversed: in the negative electrode $PbSO_4$ is reduced to Pb^0, so it is the cathode, and in the positive electrode $PbSO_4$ is oxidized back to PbO_2, thus it becomes the anode. The polarity of the electrodes in the cell does not change, but the reactions are reversed, so the anode has become the cathode and vice versa (see Ref. 12 for a detailed discussion of this point).

1. Metal Deposition as a Class of its Own

Electrodeposition may be defined in a broad manner as *the process of depositing a substance upon an electrode by electrolysis.*[8] The essence of electrode kinetics is charge transfer across the interface. A profound understanding of the structure of the electrical double layer, as discussed by Helmholz,[13,14] Gouy-Chapman[15-17] and Stern[18] is needed, of course, in the discussion of the mechanism of charge transfer. There are other important factors, such as catalysis and adsorption, mass-transport limitations and so on, all of which may influence the rate and mechanism of charge transfer to some extent, but it is the very act of charge transfer that matters. It is the vehicle that allows the conversion of chemical to electrical energy, as in a fuel cell or a battery during discharge. It is also the venue by which electrical energy is used to produce desired chemicals, as in the chor-alkali industry, the production of aluminum, electroforming, electroplating and all other electrolytic industrial processes.

Electrode reactions can be classified in two groups:

(a) redox reactions, in which both reactants and products are in solution, and
(b) processes in which at least one of them is part of the electrode itself, or is rigidly attached to it.

(*i*) *Redox Reactions*

A good example of a redox reaction is the reduction of ferri-cyanide to ferro-cyanide, given by

$$\left[\mathrm{Fe(CN)_6}\right]^{3-}_{\mathrm{soln}} + \mathrm{e}^{-}_{\mathrm{crys}} \rightarrow \left[\mathrm{Fe(CN)_6}\right]^{4-}_{\mathrm{soln}} \tag{1}$$

This is a typical *outer-sphere charge-transfer reaction*, characterized by the fact that the close environment of the central cation is not changed as a result of charge transfer. Furthermore, it is noted that both reactant and product are on the solution side of the interface, specifically at the Outer Helmholtz Plane (OHP), believed to be at a distance of about 0.5-0.6 nm from the surface of the metal. Charge is transferred across the interface by an electron, and there is no reason to assume that the ionic species have crossed the interface at any point during the charge transfer process. Electron transfer is assumed to occur by tunneling, which is inherently a quantum mechanical process. Nevertheless, there is also a classical aspect of the process shown in Eq. (1), since the electron is treated as a particle that crosses the interface – it is either on the *ferri* or the *ferro* species. No intermediate state corresponding to partial charge transfer is considered.

The example given above is not the only kind of redox reaction where both reactant and product are in solution. Consider the hydrogen evolution reaction, which can be written as

$$2\,H_3O^+_{soln} + 2\,e^-_{crys} \rightarrow H_2 + 2\,H_2O \tag{2}$$

This is a typical *inner sphere* redox reaction. The proton is heavily solvated (and probably exists in solution as $[H_9O_4]^+_{soln}$, while the interaction of molecular hydrogen with water is minimal. The mechanism of hydrogen evolution has been studied in great detail on different metals and under widely different conditions, and will not be discussed here. It is interesting to point out one major characteristic of this type of reaction, compared to outer-sphere redox reactions. The rate of hydrogen evolution is sensitive to the type of metal electrode used. The exchange current density, i_0, can be as low as 10^{-13} A cm^{-2} on Pb, and as high as 10^{-3} A cm^{-2} on Pt, in 1-M $HClO_4$. Note that a high value of i_0 (that is, higher than, say, 1 μA cm^{-2}) indicates a fast reaction and a non-polarizable electrode. Thus, the electrode clearly acts as a heterogeneous catalyst, since an adsorbed hydrogen atom is formed as an intermediate in this reaction. In comparison, the ferri/ferro redox couple has an exchange current density similar to that of hydrogen evolution on Pt, but it is not sensitive to the type of the electrode used.[†] This is not surprising, considering that both reactant and product are located at a distance of at least 0.5 nm from the electrode surface, far beyond the range of covalent bonding.

Hydrogen evolution is clearly an intermediate case between outer-sphere charge transfer and metal deposition. On the one

[†] Some apparent dependence has occasionally been reported, but this is all but eliminated when proper correction is made for the diffuse double-layer effect.

hand, both the reactant (a solvated proton) and the product (molecular hydrogen) are on the solution side of the interface. On the other hand, an adsorbed hydrogen atom is believed to be formed as an intermediate, indicating that both charge and mass have crossed the interface. This case has been discussed recently by one of the present authors.[19]

(ii) Metal Deposition and Dissolution

Metal deposition, for the simple case of a monovalent metal ion, is commonly written as:

$$\left[Ag(H_2O)_n\right]^+_{soln} + e^-_{crys} \rightarrow Ag^0_{crys} + n \cdot H_2O \tag{3}$$

The number n of water molecules solvating the ion is not important in this context, as long as it is realized that the energy of solvation is very high—about 5 eV for a monovalent ion, and around 20 eV for a divalent ion (482 and 1,927 kJ mol^{-1}, respectively). It is implicit in Eq. (3) that charge is transferred by an electron crossing the interface, just as in the case of the redox reaction given by Eq. (1). But there is a very great difference between Eq. (1) and Eq. (3). In the former, the ligands around the iron ion are held with a total energy that is much higher than the hydration energy of a silver ion, but they are not removed as a result of electron transfer. Admittedly, there is a solvent rearrangement energy involved (which plays a major role in determining the Gibbs energy of activation, according to the theory of charge transfer developed by Marcus,[20-22] Dogonadze et al.[23-25] and Levich,[26] but this is just a small fraction of the total energy that would be needed to break up the ferri-cyanide complex to its components (i.e., of the solvation energy of the iron ion).

In the reaction represented by Eq. (3), all the solvent molecules must be removed, to allow formation of the product—a neutral silver atom, and its incorporation in the metal lattice. But charge transfer, if it were to occur by electron transfer across the interface, would have to be represented by two steps, as shown in Eqs. (4) and (5):

$$\left[Ag(H_2O)_n\right]^+_{soln} + e^-_{crys} \rightarrow \left[Ag(H_2O)_n\right]^0_{soln} \quad (4)$$

followed by

$$\left[Ag(H_2O)_n\right]^0_{soln} \rightarrow Ag^0_{crys} + n \cdot H_2O \quad (5)$$

The electron-transfer step represented by Eq. (4) would be very fast, of the order of 1 fs (10^{-15} s). This is too short for atoms to move. On the other hand, it should take about 10^5 fs for the atom-transfer reaction shown by Eq. (5) to occur, i.e. for the water molecules around the neutral atom to relax to their equilibrium position in bulk water, and for the silver atom to reach the surface and be incorporated in it.

It has been stated by several noted authors in electrochemistry that, in the case of metal deposition, charge is carried across the interface by ions rather than by electrons.[27-30] Unfortunately, the above authors did not implement the consequence of this difference in the analysis of the mechanism of metal deposition and dissolution. In one instance,[29] the author went as far as to state that

> "...although charge is transferred across the interface by the metal ions, the mechanism will be treated as though it were electron transfer, for convenience."

This approach cannot be sustained, since electron and ion transfer represent two physically different phenomena, and there is no justification to assume that they would follow the same rules.

The mechanism of charge transfer during metal deposition and dissolution was treated recently by Gileadi,[19,31-34] and will not be discussed here in detail. It was shown that the Gibbs energy of formation of an isolated neutral silver atom in solution is about 2.55 eV higher than that of formation of the same atom in bulk silver.* Hence, the reversible potential for formation of an isolated atom in solution would be –2.55 V vs. Ag^+/Ag, and it could not occur anywhere near the reversible potential for silver deposition.

---

*In general, the difference between these two quantities is approximately equal to the energy of sublimation of the metal considered, which is typically in the range of 1–4 eV.

Moreover, according to the Marcus theory of charge transfer, electron transfer can only occur when the initial and final states are brought temporarily to the same energy level. This follows from the *time-resolved nature of the kinetics of charge transfer*. If electron transfer causes a change in the overall Gibbs energy of the system, it would violate the law of conservation of energy, since rearrangement of the atoms that could provide or dissipate the difference in Gibbs energy occurs on a much longer time scale. Consequently, it was concluded by Gileadi et al.[19,31-34] that in metal deposition and dissolution reactions, charge is carried across the interface by ions rather than by electrons.

Accepting that charge is carried across the interface by ions has a major impact on the way the mechanism of metal deposition and dissolution reactions should be interpreted. First, it can no longer be assumed that the symmetry factor β (that reflects the fraction of total electrochemical Gibbs energy added to the system, which is used to change the electrochemical Gibbs energy of activation of the charge-transfer reaction) is roughly equal to 0.5. Such a value implies that the transition state is formed midway between the reactants and products. To be sure, the experimental evidence for the use of this value in metal deposition was never satisfactory, and the theory of Marcus[20-22] and of Dogonadze et al.[23-25] and Levich,[26] which can predict values close to 0.5 under certain limiting conditions, was developed for outer-sphere charge transfer processes, not for metal deposition. But these theories were at least developed to apply for electron transfer. Moreover, when the discharge of a divalent ion is concerned, it has been the common practice to assume that electrons can only be transferred across the interface one at a time, and mechanisms were postulated to fit the data to this hypothesis (still assuming that $\beta \cong 0.5$). Accepting that charge is carried across the interface by the divalent ion (in the deposition of nickel, for example) renders the question of whether simultaneous two-electron transfer can or cannot occur redundant during metal deposition.* In such a process, the *effective charge* on the ion decreases gradually as the ion approaches the metal sur-

*However, the question of simultaneous two-electron transfer in outer-sphere charge-transfer processes, such as $Tl^{3+} + 2e^- \rightarrow Tl^+$, vs. consecutive transfer of two electrons, one at a time, is still open, as discussed elsewhere.[35]

face, and there is no singular point at which a full electronic charge would have been transferred.

(a) Morphology of the surface

Another distinct way in which metal deposition differs from redox reactions in solution is that the morphology of the surface may change during deposition. The surface roughness usually increases during deposition, particularly if the current distribution is not uniform, and whenever the current applied is close to the mass-transport limited current density in the same system. This would decrease the true current density, which is calculated per unit of *real surface area* taking into account the changes in the roughness factor, although the total current applied is maintained constant, thus distorting the shape of the Tafel plot.

(b) The nature of the substrate

In most cases the nature of the substrate upon which a metal is deposited is of little consequence, since it is rapidly covered by a layer of the metal being deposited. The continuous renewal of the surface can be used beneficially in the context of maintaining a clean surface. It was shown by Gileadi[36] that for a dropping mercury electrode with a drop time of 1 s, an impurity level of 10 μM can be tolerated without having a significant effect on the results, since the maximum mass-transport limited flux of the impurity can lead to no more than 1% of surface being covered by the impurity, per second. A similar calculation can be made for metal deposition on a solid substrate. Thus, for a divalent ion, a monolayer corresponds to approximately 0.5 mC cm^{-2}. When plating at a rate of, say, 10 mA cm^{-2}, about 20-atomic layers are formed per second. Although it cannot be assumed that the metal is deposited in an orderly fashion, layer by layer, it would be safe to assume that the whole surface would be renewed at least once a second, and probably more often than that. Allowing an impurity level of 10 μM would limit the coverage by impurity to just a few percents of a monolayer per second. Indeed, one of the major stumbling blocks in the study of redox reactions on solid electrodes is the difficulty in maintaining the surface free of impurities during a series of measurements that may take several minutes. During metal deposition this difficulty is by and large eliminated, since the surface is being renewed continuously, and deposition occurs always on a virgin surface.

(*c*) *Underpotential deposition (UPD)*

A unique feature observed in metal deposition on a foreign substrate is underpotential deposition. It is found that a metal can be deposited on a foreign substrate at potentials positive with respect to the reversible potential for deposition of the metal in the same solution. Considering, for example, the deposition of silver on platinum, which can be written as

$$\left[Ag(H_2O)_n\right]^+_{soln} + e^-_{crys} + Pt \rightarrow Ag_{ads}/Pt + n \cdot H_2O \qquad (6)$$

it is implied here that the discharged silver atom is adsorbed on the surface of the platinum substrate and is chemically bound to it. The reaction represented by Eq. (6) can occur at potentials positive with respect to the reversible potential for silver (i.e., at an underpotential) only if the Ag-Pt bond is stronger than an Ag-Ag bond. This is not always the case, but when it is, it will naturally be the first step in metal deposition, and may have an important role in the binding between the plated coating and the surface of the substrate.

Underpotential deposition, as defined by Eq. (6), should be inherently terminated when a full monolayer has been formed, since the second layer is no longer deposited on the substrate. This is usually the case, but in certain instances (notably, for UPD of Ag on Pt) as much as two atomic layers can be deposited before the reversible potential is reached and bulk deposition takes over.[37] This observation can be rationalized, considering that the properties of Ag atoms in the first layer on top of a Pt surface could be quite different from those in the bulk of silver. In other words, the surface of Pt below a single atomic layer of Ag could influence the chemical properties of this layer and its energy of bonding to the second layer. The effect is, however, short ranged, and it is not expected to extend further.

Underpotential deposition has been studied extensively,[38-42] mostly for single-crystal substrates. The behavior of different crystal faces has been examined,[43-48] two-dimensional phases were observed, and correlation between the UPD potential shift Δ_{UPD} (defined as the potential difference between the reversible potential and the potential observed at a partial surface coverage $\theta = 0.5$) was related to the difference in the work function of the two metals.[38] While these are issues of great importance for the fundamental understanding of UPD formation, applied issues such as the dependence of adhesion of a coating on the formation of a UPD layer have not been discussed.

(*d*) *A complexing agent*

In a large majority of practical plating baths, a complexing agent is used to improve the quality of the product, in particular to obtain smooth and bright deposits. On the other hand, when deposition at a high rate is needed, as for electroforming, the metal is deposited from a simple solution containing no complexing agent. Until about the middle of the 20th century, the most commonly used complexing agent was cyanide. Many metals such as Au, Ag, Cu, Ni, Co, Cd and Zn were plated from alkaline baths containing KCN. This practice was abandoned for environmental considerations, in spite of the fact that it was most satisfactory from the purely engineering point of view. Cyanide has been replaced by other complexing agents, mostly organic poly-acids, having two or more carboxylic groups, such as citrate, oxalate, etc. The main purpose of formation of the complex is to slow down the kinetics of the electrodeposition reaction. It should be noted that in most industrial electrolytic processes, such as the production of metals, the chlor-alkali industry and in organic electrosynthesis, as well as in batteries and fuel cells, fast kinetics is an advantage – decreasing the overpotential and thereby reducing energy consumption. The opposite is true in metal deposition. Decreasing the exchange current density is equivalent to increasing the Faradaic resistance, i.e. the resistance associated with the charge transfer process. This generally leads to improved uniformity of the deposited layer and enhanced smoothness and brightness of the deposit.

(*e*) *The types of overpotential and the relevant scale*

The processes taking place in an electrochemical cell during metal deposition can be modeled by a combination of three resistors in series, each associated with an overpotential given by the product of the applied current density and the relevant resistance:

- *Ohmic solution resistance, R_{soln}.* It is determined by the specific resistivity of the solution and by the configuration of the cathode and the anode with respect to each other, as well as their respective shapes. The corresponding contribution to the observed overpotential is called the iR_{soln} potential drop or η_{iR}, the resistance overpotential. When this is the largest resistance in the system, the process is said to occur under conditions of *primary current distribution*. The relevant scale to be considered is of the order of a few millimeters up to several centimeters. Primary current distribution usually leads to non-uniform current distribution on the cathode, resulting in non-uniform thickness of the coating.
- *Faradaic resistance, R_F.* This resistance is directly related to the Faradaic reaction taking place. It is also called the charge-transfer resistance, R_{ct}, or the (non-Ohmic) activation resistance R_{ac}, since it is associated with the finite rate of the electrode reaction per se. The corresponding overpotential is usually referred to as the activation, or charge-transfer overpotential, η_{ac} or η_{ct}, respectively. Since charge transfer occurs over a distance of about 0.6 nm, the relevant scale is in the range of a few nanometers. When the Faradaic resistance is the largest resistance in the system, the reaction is said to be *activation controlled* and *secondary current distribution* is maintained. It should be obvious that secondary current distribution leads to uniform thickness of the deposit, on the macroscopic scale. For example, in a barrel plating process for plating screws, primary current distribution will lead to excessive coating on the top of the grooves and no coating at the bottom. In contrast, secondary current distribution will lead to much more uniform coating thickness throughout the groove.

The transition from primary to secondary current distribution can be characterized by a dimensionless parameter called the *Wagner number* (W_a), defined as

$$W_a \equiv \frac{\kappa(\partial\eta/\partial i)_{c,T}}{l} = R_F/R_{soln} \tag{7}$$

where κ is the specific conductance of the solution (S cm^{-1}) and l is a characteristic length, usually related to the dimensions of the electrodes or their distance apart. Values of $W_a > 10$ correspond to secondary current distribution, while $W_a < 0.1$ indicates primary current distribution.

- *Concentration resistance, R_{conc}.* The third factor determining the nature of the deposit is mass transport. The corresponding resistance is referred to as the *concentration resistance*, R_{conc}, which results from the depletion of the electro-active species at the cathode surface, caused by mass-transport limitation. The mechanism of mass transport of the electro-active species (either charged or uncharged) could be diffusion, convection or migration, or some combination of these mechanisms. For the simple one-dimensional case (corresponding to semi-infinite linear diffusion) at steady state, the rate of mass transport, expressed as the current density, can be written as

$$i = nFD\frac{(c_{bulk} - c_{surf})}{\delta} \tag{8}$$

where D is the diffusion coefficient (cm^2 s^{-1}), c_{bulk} and c_{surf} (mol cm^{-3}) represent the concentrations of the reactant in the bulk of the solution and at the surface of the cathode, respectively, and δ (cm) is a characteristic length, called *the Nernst diffusion layer thickness*, which is determined by the conditions of the experiment. For example, in a rotating disc experiment, δ is in the range of 5–50 μm, for rotation rates of 1×10^4 to 1×10^2 rpm, respectively.*

*The exact value depends on the diffusion coefficient of the electro-active species and on the viscosity of the solution, as given by the well-known Levich equation. The above range was calculated for $D = 1\times10^{-5}$ cm^2 s^{-1} and $\nu = 1\times10^{-2}$ cm^2 s^{-1}, which is a reasonable approximation to dilute aqueous solutions at room temperature.

In industrial operations, stirring is typically implemented by moving the electrodes, stirring the solution, or pumping it through the bath (the latter provides an opportunity to filter the solution and remove particulate material that could damage the quality of the metal coating, but that issue is outside the scope of this chapter). The typical values of δ, under industrial plating conditions, may be 50-150 μm, placing it between the values relevant for primary and secondary current distributions. If the experiment is set up so that diffusion is the sole mode of mass transport, then δ is given by

$$\delta = \sqrt{\pi D t} \tag{9}$$

yielding values of δ of 80–180 μm after 10 seconds, for a typical range of values of the diffusion coefficient in dilute aqueous solutions.

The limiting current density, i_L, which represents the highest rate at which the metal can be deposited under given experimental conditions, can be derived from Eq. (8) by setting $c_{surf} = 0$

$$i_L = nFD(c_{bulk} / \delta) \tag{10}$$

In practice, deposition of metals is conducted at current densities well below the limiting current, in the range of $i \leq 0.3\ i_L$. At higher current densities, the deposits tend to be rough, powdery or friable.

As in the cases discussed above, one can define a resistance characterizing the mass-transport limitation and a corresponding concentration overpotential η_{conc}. The value of this parameter, compared to the Faradaic resistance, determines the brightness and smoothness of the deposit, but has little influence on the uniformity of the thickness, since the characteristic length associated with it (ca. 0.01 cm) is two or three orders of magnitude smaller than that associated with primary current distribution.

It is hardly necessary to point out that non-uniform current distribution can lead to poor performance in any industrial process.* For alloy deposition this may be even more critical, considering that the composition of the alloy is often a function of current density. Thus, a non-uniform current distribution might lead to non-uniformity of the alloy composition, in addition to variation of the thickness of the coating.

(f) *The roughness factor*

The roughness factor is an important parameter in the context of the study of electrode kinetics. It is defined as the ratio of *the real to geometric* surface area. But what is the so-called *real surface area*? That depends on the relative values of the three overpotentials, or more precisely, the values of the three resistances associated with them. Thus, under conditions of primary current distribution (or in the transition region between primary and secondary current distribution, where $0.1 < W_a < 10$), maintaining a uniform thickness may be the most important issue. This is determined by the uniformity of the distance between the anode and the cathode, on a scale of about 0.01 cm (i.e. about 1% of the total distance between the electrodes). Whether the cathode is rough or smooth on the atomic scale of 1 nm will have absolutely no effect on the thickness. At the mass-transport limited current, where the characteristic length is 5–150 μm, the roughness on the atomic scale is also irrelevant. On the other hand, when charge transfer or adsorption are concerned, the roughness on the atomic scale is the most important parameter. Specifically, it is important to note that the roughness factor, as determined by the maximum amount of atomic hydrogen or oxygen adsorbed on the surface, or by measurement of the double-layer capacitance, is quite irrelevant when the mass-transport limited current density is considered.

---

* Except in cases where the non-uniformity is intentionally built into the process, to increase the thickness of the plating where it is needed, and vice versa.

This raises some important possibilities, which have not escaped the attention of the electroplating community. For example, while metal deposition is conducted in fairly concentrated solutions of the metal being plated, and at current densities well below the mass-transport limit, additives acting as inhibitors for metal deposition are often introduced at concentrations that are several orders of magnitude lower, to ensure that their supply to the surface will be mass-transport limited. In this way, the tendency for increased rate of metal deposition on certain features on the surface, such as protrusions, will be moderated by the faster diffusion of the inhibitor to the very same areas. Furthermore, if deposition occurs in the region of mixed control, which is usually the case, it must be remembered that the relevant roughness factor is quite different for the charge-transfer and the mass-transport processes, and this may well be a function of current density, since the Faradaic resistance is inherently potential dependent.

(*g*) *Hydrogen evolution*

In most electroplating baths of practical interest, hydrogen evolution occurs as a side reaction. Under cathodic overpotentials, the following reactions can lead to discharge of protons and adsorption of atomic hydrogen on the cathode:

$$H_3O^+ + e^-_{crys} \rightarrow H_{ads} + H_2O \tag{11}$$

$$H_2O + e^-_{crys} \rightarrow H_{ads} + OH^- \tag{12}$$

While Eq. (11) is relevant for relatively low pH values (pH ≤ 3),[*] Eq. (12) applies at higher values of the pH. The adsorbed hydrogen atoms, formed as intermediates, can recombine and evolve as gas bubbles (molecular hydrogen) in solution, following one of two possible mechanisms:

$$H_{ads} + H_{ads} \rightarrow H_2 \tag{13}$$

---

[*]This pH value was chosen because at higher pH, the reduction of the protonium ion may become mass-transport limited and reaction 12 would occur instead.

$$H_{ads} + H_3O^+ + e^-_{crys} \rightarrow H_2 + H_2O \quad (14a)$$

$$H_{ads} + H_2O + e^-_{crys} \rightarrow H_2 + OH^- \quad (14b)$$

Equation (13) is called atom-atom recombination, which is not directly dependent on potential, but is influenced by it through the dependence of the fractional surface coverage by atomic hydrogen on potential. Equations (14a) and (14b) represent electrochemical charge transfer steps at low and high values of pH, respectively. Finally, a (typically small) portion of the *ad*sorbed hydrogen may be *ab*sorbed in the metal:

$$H_{ads} \rightarrow H_{abs} \quad (15)$$

This absorbed hydrogen can then be transported inside the metal electrode via a diffusion mechanism, which is enhanced by the high subsurface concentration of hydrogen that can be predicted by Sieverts' Law* for high fugacity. The extent of hydrogen absorption depends on bath composition, temperature and pH, the applied current density, surface barriers (e.g., oxides), and surface poisoning agents such as compounds of As, P, Sb, Te, Se and S, which are known to reduce the partial surface coverage on the one hand, and to interfere with surface recombination of hydrogen atoms (cf., Eq. 13) on the other hand, thus enhancing the alternative reaction pathway of absorption of hydrogen in the bulk of the metal (cf., Eq. 15).

Once inside the metal electrode, diffusing hydrogen atoms can recombine around defects such as micro-voids, inclusions, interfaces and grain boundaries, forming molecular hydrogen. High-

---

* Sieverts' Law: $c = K_S\sqrt{P}$, where *c* is the subsurface concentration (solubility) of the dissolved atom in the solid metal, *P* is the partial pressure of the diatomic gas (sometimes replaced by the fugacity, *f*), and K_S is the solubility constant (temperature dependent), which is the chemical equilibrium constant between the molecular species in the gas phase and the atomic species within the metal lattice. This empirical relation was first demonstrated by Sieverts in 1929 for the solubility of hydrogen in iron. Departures from this law occur at high gas pressures and/or high concentrations of dissolved atoms.

pressure bubbles may be formed inside the cathode,[49-52] ultimately leading to crack initiation and/or propagation, even in the absence of significant applied loads.[53,54] Other mechanisms, which do not involve bubble formation, can lead to hydrogen embrittlement of electrodeposited parts.[54-56] These include the decohesion (i.e., reduction in the lattice cohesion forces in regions of high hydrogen concentration)[57,58] or adsorption (i.e. reduction in surface energy due to hydrogen adsorption)[59] mechanisms, the hydride formation mechanism (i.e. formation and cleavage of brittle hydrides),[60-62] and mechanisms of hydrogen interaction with dislocations[63-66] (e.g., establishment of drag forces,[67] local hardening at existing crack tips due to hydrogen,[68] hydrogen-enhanced localized plasticity,[62,69] sweep of hydrogen atoms by dislocations[70] that enables local accumulation of hydrogen – exceeding a critical concentration, etc.). The term *hydrogen embrittlement* refers to reduced ductility, non-ductile fracture mode, and reduced tensile strength caused by exposure to hydrogen. Hydrogen-related stresses may also induce shear stresses at the substrate/coating interface. These high stresses might overcome the adhesion strength and lead to delamination of the coating. Cracks in the deposit also provide channels for penetration of humidity and oxygen that, upon reaching the interface between the coating and the substrate metal, can initiate corrosion and/or delamination of the coating. Hydrogen evolution also leads to local increase in pH, increasing the concentration of hydroxide ions that can be incorporated in the electrodeposit, thus changing its properties.

(*h*) *The current efficiency*

The current efficiency, also called the Faradaic efficiency (FE), is the fraction of the total current used to deposit the metal (or metals, during alloy deposition). The danger of hydrogen-induced cracking (HIC), also referred to as hydrogen-assisted cracking (HAC), increases in electrodeposition systems with lower cathode current efficiencies (such as hard chromium). In highly efficient systems, such as Cu and Ag baths, codeposition of hydrogen occurs only when i_L is exceeded, or when the added complexing agents shift the potential of metal deposition to sufficiently negative values. The higher the hydrogen overpotential on a given metal, the lower the amount of hydrogen absorbed in it.

The effects of internal hydrogen are most significant in high-strength steels (yield strength greater than 1.17 GPa, or 170 kpsi). In general, the higher the strength level of the steel, the greater its susceptibility to hydrogen embrittlement. Although hydrogen diffusion in martensitic and ferritic structures is relatively rapid, it is not always a simple task to remove this hydrogen once it has entered the steel. Baking is required following electrodeposition for all steel parts hardened to at least 40 HRC. This process is performed at a temperature in the range of 177-205°C for at least 3 h, and should be applied not later than 4 h after the completion of the plating process.[71] Unfortunately, this post-treatment is not always sufficient. Although hydrogen diffusivity increases exponentially with increasing temperature, the required time to reduce the hydrogen concentration to a desired level increases with the square of the thickness of the part being coated. For thick sections this may mean hundreds of hours of baking. Even then, there is no guarantee that permanent damage or irreversible hydrogen embrittlement has not already occurred.[56] ASTM has suggested a standard test method for mechanical testing of plated specimens in order to identify hydrogen embrittlement due to plating processes and service environments.[72]

It should be noted that the FE is often found to depend on current density. In this context, a decrease in FE with increasing current density could lead to improved uniformity of plating. While the current density is higher on a protrusion, the current efficiency is lower, hence the rate of deposition of the metal may remain constant or even decrease. On a recessed area the opposite behavior is expected, of course. Thus, the variation of the FE could act as a negative feedback, enhancing the uniformity of the thickness of the coating. An increase in current efficiency with increasing current density would have an opposite effect of creating a positive feedback, leading to rapid roughening of the surface. Low FE also leads to increased consumption of electricity, but this is usually not a major issue in electroplating.

2. Specific Issues in Electrodeposition of Alloys

(i) *History*

The first alloys to be electrodeposited were brass (Cu-Zn alloys) and alloys of the noble metals. Faust[48] reviewed the principles of alloy deposition. In order to codeposit two metals, they must be in a bath in which the individual reversible potentials are reasonably close to each other. This is the case when the standard potentials E^0 of the two metal are close, as for deposition of tin-lead alloys where the values of E^0 are –0.126 V and –0.136 V vs. SHE for Pb and Sn, respectively. Changing the concentration of one of the metals in solution can bring the reversible potentials closer to each other. However, since the Nernst equation allows for only (59/n) mV per decade change in concentration, where n is the number of electrons needed to deposit a metal atom, this has a limited range of applicability. When the two metals forming the alloy have widely different values of E^0, their reversible potentials can be shifted closer to each other by adding a complexing agent that forms complexes with different stability constants, since the reversible potential in the presence of a ligand that forms a suitable complex is given by

$$E_{\text{rev}} = E^0 + \frac{2.3\,RT}{nF}\log K + \frac{2.3\,RT}{nF}\log c_{Me^{z+}} \qquad (16)$$

where R is the ideal gas constant (8.314 J $K^{-1}mol^{-1}$), T is the absolute temperature, F is Faraday's constant, $c_{Me^{z+}}$ is the bulk concentration of the metal ion being deposited (neglecting activity coefficients), and K is the stability constant of the complex, which is different for different metals. Equation (16) can be rewritten as

$$E_{\text{rev}} = E^{0'} + \frac{2.3\,RT}{nF}\log c_{Me^{z+}} \qquad (17)$$

where

$$E^{0'} \equiv E^{0} + \frac{2.3RT}{nF}\log K \tag{18}$$

is the effective standard potential in the presence of the complexing agent.*

(ii) Special Considerations Related to Alloy Deposition

Alloy deposition is similar to metal deposition in the sense that the surface is being renewed continuously during formation of the deposited layer. Hence, the nature of the substrate is of little importance, except in special situations, where a single-crystal substrate is employed and epitaxial growth of the deposit takes place. Epitaxy, or oriented overgrowth, is a special case of heterogeneous nucleation, where the deposit grows with a crystal structure that conforms to that of the substrate, at least up to a certain thickness. Depending on the binding energy and the crystallographic misfit between the deposit and the substrate, the growth may take place layer-by-layer according to the Frank-van der Merwe model,[73] by three-dimensional islands formation according to the Volmer-Weber model,[74] or by a combination of both according to the Stranski-Krastanov model.[75]

Considerations of the mechanism of charge transfer discussed for metal deposition apply also to alloys, but there are some differences. First, it must be realized that alloy deposition is a complex process, in which at least two parallel reactions take place simultaneously (i.e., the deposition of the two metals constituting the alloy), and in many cases hydrogen evolution constitutes a third parallel reaction.

When a complexing agent is employed, which is usually the case, attention should be paid to the solution chemistry in the mul-

---

* Activity coefficients can, as a rule, be neglected for moderately dilute solutions (say, up to 1–2 M), since they appear in the Nernst equation in logarithmic form. Thus, for example, if the activity coefficient is only 0.8, instead of an assumed value of 1.0, the resulting error in the value of E_{rev} is only $(5.7/n)$ mV. It should be borne in mind, however, that in highly concentrated electrolytes, such as 30% KOH or 85% H_3PO_4, used in some batteries and fuel cells, such an approximation is no longer valid and may introduce significant errors.

ti-component plating baths and to possible formation of complexes containing both metals, in addition to the usual complexes of each metal with the ligand. For two metals Me_1, Me_2 and a monovalent negatively charged ligand L^-, there could typically be several species in solution, such as

$$Me_1^{z+};\ Me_2^{z+};\ [Me_1L_n]^{(z-n)};\ [Me_2L_n]^{(z-n)};\ [Me_1Me_2L_m]^{(2z-m)} \tag{19}$$

and each of the above species could be electro-active, allowing deposition of one of the metals or both. Thus, the resulting alloy may be formed by deposition of each metal separately, or from the complex containing both metals. Evidently, there could be several parallel reactions taking place simultaneously. Measurement of the current-potential relationship in such complex systems would not be meaningful. Even if the experimental data can be *forced* to provide a linear Tafel region, from which an apparent Tafel slope is obtained, this will have little relevance to the mechanism of formation of the alloy. Although the above observation is well known and could be considered rather obvious, it has been overlooked in many publications.

The above does not imply that it is impossible to study the mechanism of alloy deposition; it only shows that conclusions cannot be drawn from the usual interpretation of the directly observed current-potential relationship employed in the analysis of electrode kinetics. The partial currents for deposition of each of the alloying elements should be determined as a function of potential and other experimental parameters via determination of the atomic composition of the alloy and the FE. The FE during alloy deposition can be different from that of single-metal deposition of one or both metals involved in the process. Hence, the FE can be expected to depend on the composition of the alloy, and the thickness distribution may differ from that expected according to the current distribution.[42]

The Faradaic efficiency can be calculated using the equation:

$$FE = \frac{w}{It_d}\sum \frac{x_i n_i F}{M_i} \times 100 \tag{20}$$

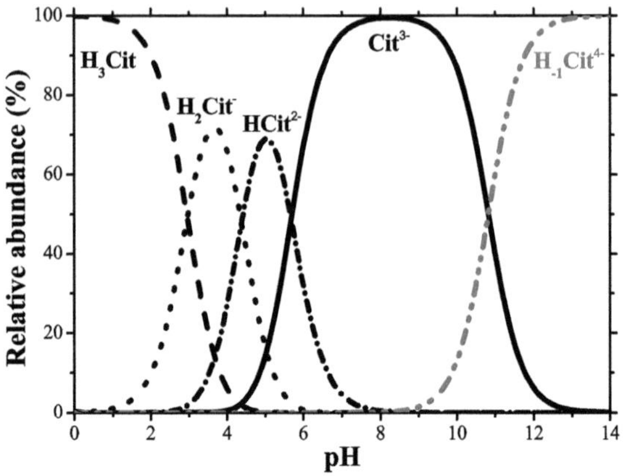

(a)

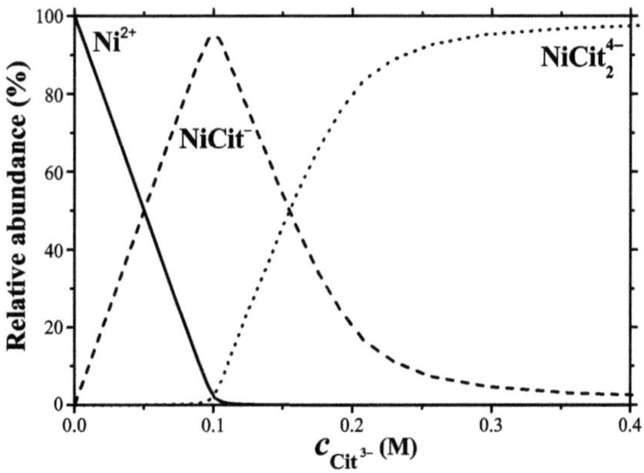

(b)

Figure 1. (a) Stepwise deprotonation of citric acid as a function of pH. The notation $H_{-1}Cit^{4-}$ refers to citrate ion in which all three acidic protons, as well as the proton on the alcoholic group, have been removed. (b) Concentration distribution of Ni^{2+}-Cit^{3-} complexes as a function of the overall citrate concentration (0.1 M $NiSO_4$, pH = 8.0).

where w is the measured weight of the deposit (g), I is the total current passed (A), t_d is the deposition time (s), x_i is the weight fraction of the element in the alloy deposit, n_iF is the number of coulombs per mole for the reduction of the element, and M_i is the atomic mass (g mol^{-1}). Detailed consideration of the species that exist at equilibrium in the plating bath, which can be calculated if the relevant stability constants of the complexes are known, would then provide further insight regarding the fundamental factors determining alloy composition, possible anomalies, changes in morphology, and so on.

In Fig. 1 we show two examples. Citric acid, which is a commonly used ligand, has three carboxylic groups and one alcoholic group. It can exist in solution as the neutral molecule or as ions carrying a negative charge of 1-4, depending on pH. The distribution of these species as a function of pH is shown in Fig. 1a. The corresponding pK_i values are given in Table 1. For pH values in the range of 7-10 the predominant species is the triply-charged anion Cit^{3-} ($C_6H_5O_7^{3-}$). The distribution of its complexes with Ni^{2+} is shown in Fig. 1b, as a function of the overall concentration of citrate at pH = 8. The corresponding $\log(\beta_n) \equiv \sum_{i=1}^{n} pK_i$ values, where β_n are the equilibrium constants for the reactions Me + $n \cdot$L $\rightarrow$ [MeL_n] that form the protonated complexes, are provided in Table 1. Two complexes are shown. Free Ni^{2+} and $[NiCit]^-$ are seen to be the predominant species, as long as the concentration ratio is Cit/Ni $\leq$ 1. The concentration of free Ni^{2+} falls almost to zero when the above ratio reaches unity. As the concentration of citrate is increased further, a different complex, $[Ni(Cit)_2]^{4-}$ becomes predominant. When the ratio Cit/Ni $\geq$ 4, all the Ni^{2+} ions are in this highly charged complex, which is very stable, and deposition of Ni from it is strongly impeded. Thus, although citrate is a very useful ligand for plating baths containing nickel, a large stoichiometric excess may be detrimental to their performance.

In this context, it is appropriate to draw attention to an error often committed in electroplating, and particularly in alloy plating. In plating transition metals and their alloys, a citrate bath is often used and the pH is in the range of 6–9. The purpose of using the citrate (or other organic poly-acids) is to form a complex and prevent deposition of hydroxides of the metals. An unspecified amount of ammonium hydroxide (NH_4OH) is often added to increase the FE and fine-tune the pH to the chosen value. The fact that NH_3 forms well-known complexes with most transition metals, as shown in Fig. 2a for the case of Ni^{2+} (see pK_a and log(β_n) values in Table 1), is ignored, in spite of the fact that it could change significantly the distribution of the metal ions in the different complexes that can be formed in the system. A detailed discussion of this issue is given below, in the section dealing with induced codeposition of Ni-W alloys. Finally, Fig. 3,[6,76,77] summarizes most of the metal pairs that have been codeposited electrochemically, either commercially or in laboratory studies.

Table 1
Equilibrium Constants for Acid Dissociation and Complex Formation. An Increased Index Number Reflects a Higher Deprotonation/Complexation State

Species/Complex	pK_a	$\log(\beta_n)$	Relevant figure	Ref.
$[(H)_n(Cit)]^{-(3-n)}$ n = −1 corresponds to $[H_{-1}Cit]^{4-}$ (c.f. caption to Fig. 1)	$pK_1 = 2.96$ $pK_2 = 4.38$ $pK_3 = 5.68$ $pK_4 = 10.82$	$\log(\beta_1) = 2.96$ $\log(\beta_2) = 7.34$ $\log(\beta_3) = 13.02$ $\log(\beta_4) = 23.84$	1a	193
$Ni(Cit)_n^{-(3n-2)}$	$pK_1 = 5.50$ $pK_2 = 2.30$	$\log(\beta_1) = 5.50$ $\log(\beta_2) = 7.80$	1b	195
$[Ni(NH_3)_n]^{2+}$	$pK_1 = 2.80$ $pK_2 = 2.24$ $pK_3 = 1.73$ $pK_4 = 1.19$ $pK_5 = 0.75$ $pK_6 = 0.03$	$\log(\beta_1) = 2.80$ $\log(\beta_2) = 5.04$ $\log(\beta_3) = 6.77$ $\log(\beta_4) = 7.96$ $\log(\beta_5) = 8.71$ $\log(\beta_6) = 8.74$	2a	194
NH_4^+ / NH_3	$pK_1 = 9.25$	$\log(\beta_1) = 9.25$	2b	193
$[(WO_4)(Cit)(H)_m]^{-(5-m)}$	$pK_1 = 4.64$ $pK_2 = 6.82$ $pK_3 = 10.21$	$\log(\beta_1) = 4.64$ $\log(\beta_2) = 11.46$ $\log(\beta_3) = 21.67$	8	137
$[(MoO_4)(Cit)(H)_m]^{-(5-m)}$	$pK_1 = 4.58$ $pK_2 = 6.83$ $pK_3 = 8.25$	$\log(\beta_1) = 4.58$ $\log(\beta_2) = 11.41$ $\log(\beta_3) = 19.66$	13a	196

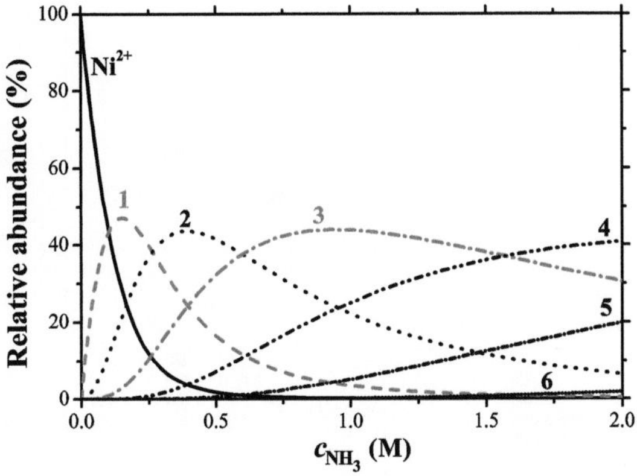

(a)

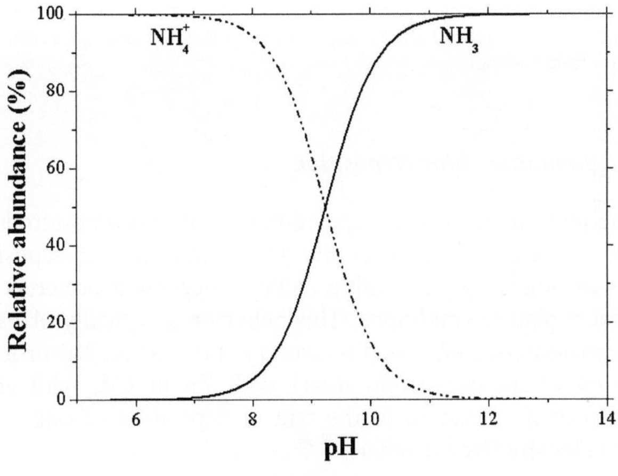

(b)

Figure 2. (a) Concentration distribution of $[Ni(NH_3)_n]^{2+}$ complexes as a function of the overall ammonia concentration (0.1 M $NiSO_4$, pH = 8.0, pK_a = 9.25). The numbers adjacent to the curves represent the values of n in the above formula. (b) The relative abundance of NH_3 and NH_4^+ as a function of pH (pK_a = 9.25).

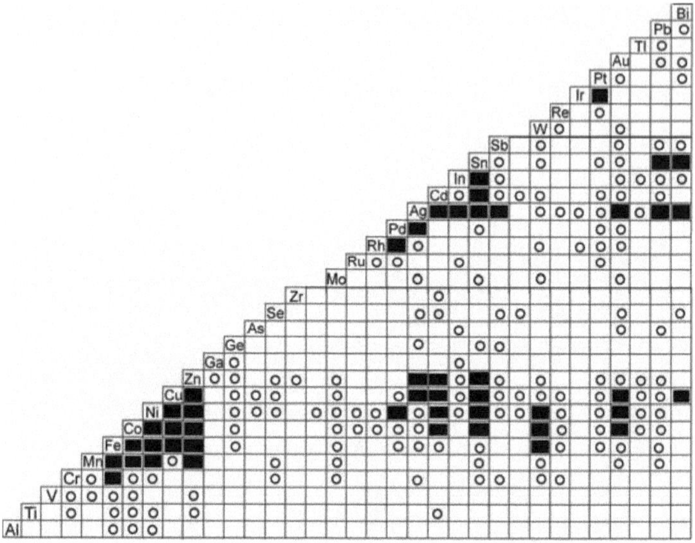

Figure 3. Metal pairs that have been codeposited electrochemically: O = Demonstrated in laboratory studies and/or small-scale applications, ■ = Technically interesting, widely employed alloys. Reprinted from Ref. 6, Copyright (2004) with permission from Elsevier.

(*iii*) *Anomalous Alloy Deposition*

The term anomalous codeposition (ACD) was first introduced by Abner Brenner,[78] to describe an electrochemical deposition process in which the less noble metal is deposited preferentially under most plating conditions. This behavior is typically observed in codeposition of iron-group metals (i.e. Fe, Co and Ni) or in codeposition of an iron-group metal with Zn or Cd, with either inhibition or acceleration of the rate of deposition of one of the alloying elements by the other.[7,78-84]

As a first approximation, one might expect that the composition of an electroplated alloy would be related to the current observed for each of the elements, when measured alone in the same solution at the same potential. Assume, for simplicity, that both metals are deposited at high negative overpotentials, within the linear Tafel region (where $\eta/b \geq 1$). Then, one could write the partial cathodic current densities in terms of the activation overpotentials as

$$i_{c,1} = -i_{0,1}\exp\left(-\frac{\alpha_{c,1}F}{RT}\eta_{c,1}\right) = -i_{0,1} \times 10^{(-\eta_{c,1}/b_1)} \tag{21}$$

$$i_{c,2} = -i_{0,2}\exp\left(-\frac{\alpha_{c,2}F}{RT}\eta_{c,2}\right) = -i_{0,2} \times 10^{(-\eta_{c,2}/b_2)} \tag{22}$$

where b is the Tafel slope (in units of V decade^{-1}), α_c is the cathodic transfer coefficient, and the subscripts 1 and 2 refer to the two elements. It can be noticed that the transfer coefficient is simply the inverse Tafel slope in dimensionless form:

$$\alpha \equiv \frac{2.3RT}{F}\frac{1}{b} \tag{23}$$

Typical values of b_c are in the range 30 to 300 mV decade^{-1}, corresponding to α_c values of 2 and 0.2, respectively, but values close to 0.1 V decade^{-1} are most commonly observed in metal deposition. When the exact value for a specific system is unknown, the approximation $b_a = |b_c| = 0.12$ V decade^{-1} has often been used, although there is no theoretical basis to support this choice, and it would be more accurate to obtain the value of b_a from α_a, employing the simple relationship

$$\alpha_c + \alpha_a = n \tag{24}$$

The exchange current densities and the Tafel slopes for two metals are in general different, although they may happen to be close to each other for a particular case. The overpotential is not the same for the two metals, of course, although deposition takes place at the same potential, measured with respect to a given reference electrode. In other words, at the deposition potential, E_{dep}, one has

$$\eta_1 = E_{dep} - E_{rev,1} \quad \text{and} \quad \eta_2 = E_{dep} - E_{rev,2} \tag{25}$$

Assuming, for simplicity, that the two Tafel slopes are equal, the atom ratio of the two elements in the alloy should be given by

$$\frac{i_{c,1}}{i_{c,2}} = \frac{i_{0,1}}{i_{0,2}} \exp(-\frac{E_{rev,1} - E_{rev,2}}{b_c}) \tag{26}$$

independent of the deposition potential. If this is observed experimentally to be the case, then alloy deposition should be considered *normal* or *ordinary*. As it turns out, such behavior is the exception rather than the rule!

In Fig. 4, the partial current densities for two metals are shown. Having the case of Ni-Fe alloy in mind, typical values were chosen for i_0 and b_c, as specified in the caption of this figure. A line for hydrogen evolution is also shown, from which the Faradaic efficiency could be calculated as

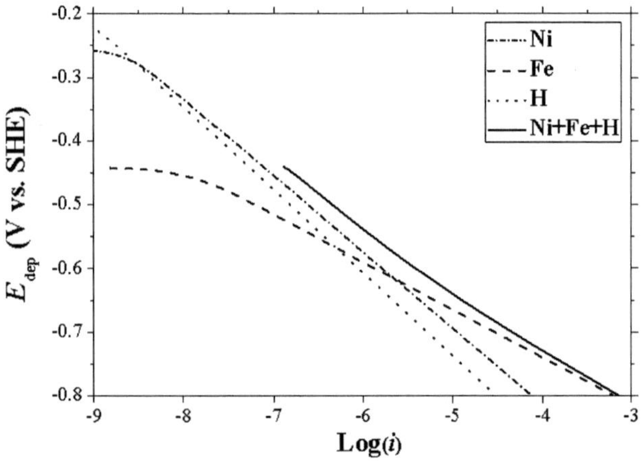

Figure 4. The effect of applied (deposition) potential on the partial cathodic current densities for codeposition of iron and nickel, as well as hydrogen evolution. Parameters used: $i_{0,Ni} = 2\times10^{-9}$ A cm^{-2}, $\alpha_{c,Ni} = 0.49$, $i_{0,Fe} = 1\times10^{-8}$ A cm^{-2}, $\alpha_{c,Fe} = 0.79$, $i_{0,H} = 5\times10^{-10}$ A cm^{-2}, $\alpha_{c,H} = 0.45$.

$$\mathrm{FE} = \frac{i_{\mathrm{Ni}} + i_{\mathrm{Fe}}}{i_{\mathrm{Ni}} + i_{\mathrm{Fe}} + i_{\mathrm{H}}} \tag{27}$$

The only quantity in this equation that can be measured directly is the total current density. The partial current densities for each of the three reactions are obtained by weighing the deposit and analyzing its composition. Each of the current densities is calculated from the Butler-Volmer equation

$$i_{\mathrm{j}} = i_{0,\mathrm{j}}\left[\exp(\alpha_a F\eta/RT) - \exp(-\alpha_c F\eta/RT)\right] \tag{28}$$

using the relation given in Eq. (24) to calculate the values of α_{a} corresponding to each assumed value of α_{c}. The total current density that would be measured (also shown in Fig. 4) is the sum of three partial current densities

$$i = i_{\mathrm{Ni}} + i_{\mathrm{Fe}} + i_{\mathrm{H}} \tag{29}$$

The point of presenting Fig. 4 is the following. First, it shows that alloy deposition is complex, including (usually) at least three reactions occurring in parallel. It cannot be over-emphasized that, in the usual interpretation of kinetic parameters such as the Tafel slope, reaction order, effect of pH etc., it is tacitly assumed that only one reaction is taking place. Thus, such analysis is inherently inapplicable to alloy deposition. When three parallel reactions take place simultaneously, each having its own exchange current density, overpotential and Tafel slope, there is no justification to assume that the Tafel plot would be linear. Moreover, its value (or, perhaps, one should say "its apparent value") cannot easily be associated with the mechanism of deposition of one of the alloying elements or the other, or of the alloy as such.

A well-known example of anomalous alloy deposition is the plating of Permalloy™.* Considering that the values of E^0 for

---

* A trademark of the Western Electric Co. for Ni-Fe alloys containing 20–60 at.% Fe, mostly used as a 80 at.% Ni – 20 at.% Fe alloy, which has high magnetic permeability and high electrical resistance.

nickel and iron are –0.25 V and –0.44 V vs. SHE, respectively, one might expect that nickel would be deposited more readily, and it would be difficult to reach a concentration of 20 at.% iron in the alloy, although the exchange current density of iron is five times larger than that of nickel (1×10^{-5} vs. 2×10^{-6} mA cm^{-2} in 2.0 N solutions of their respective sulfates at room temperature[85]). In fact, the opposite is found experimentally. If the plating bath contains equal concentrations of the two metals, the concentration of nickel in the alloy will be below the desired level. The proper alloy is deposited from a bath containing a much higher concentration of Ni^{2+} than Fe^{2+} ions.[86-88]

It would seem that mixing the two ions creates some interaction between them, which slows down the rate of deposition of the component having a more positive standard potential. It was shown by Landolt et al.[89,90] that adding 0.025 M $FeSO_4$ to a solution of 0.2 M $NiSO_4$ could reduce the partial current density for deposition of nickel by as much as a factor of ten. On the other hand, the same authors found that addition of 0.025 M $NiSO_4$ to a solution of 0.25 M $FeSO_4$ could increase the partial current density for deposition of iron by a factor of two or more. This is evidently a case of anomalous codeposition of two metals.

(*iv*) *Possible Causes of Anomalous Alloy Deposition*

Although this chapter is about induced codeposition, and is not meant to deal specifically with anomalous alloy deposition, a few general comments would seem to be appropriate.

(*a*) *The effect of partial mass-transport limitation*

It should be recalled that the concentration of the electroactive species at the surface of an electrode is always lower than its bulk concentration, following the simple relationship

$$\frac{c_{\text{surf}}}{c_{\text{bulk}}} = 1 - \frac{i}{i_L} \tag{30}$$

Thus, even if the concentrations of two metal ions in the plating bath are equal, their concentrations at the surface, which determine the rate of deposition, may be quite different, if their partial current densities are different. Plating of interest in industry is usually conducted at the highest current density compatible with production of high-quality coating ($i/i_L \approx 0.3$). However, this might not be a simple guideline to follow because:

- a limiting current may not be observed experimentally, as hydrogen evolution becomes predominant at negative potentials;
- the limiting current that is relevant is that measured for each of the alloying elements separately, and this could be quite different for the two metals. The reason for this is that the desired alloy does not necessarily contain equal amounts of the two metals, hence the partial current densities are different;
- considering that alloy deposition is often anomalous, the composition of the alloy may not scale linearly with the concentrations of the metal ions in solution.

The consequence of all of this is that, during alloy plating, the deposition of the two metals may be under different degrees of mass-transport limitation, while the hydrogen evolution reaction usually occurs under conditions of activation control. Should this be the case, the alloy composition would depend on the total current density applied and on the conditions of mass transport. As it often happens, this may be a mixed blessing. On the one hand, it requires well-controlled and reproducible conditions of mass transport in order to obtain deposited layers of uniform composition. On the other hand, it could be used as a tool for producing alloy coatings of graded or alternating composition in the same solution, which may improve bonding and enhance resistance to thermal shock, by changing the applied current density.

(*b*) *The effect of Faradaic efficiency*

Hydrogen evolution occurring in parallel with metal deposition is a sore point in electroplating. The situation is even worse in the case of alloy deposition. If the plating bath is strongly buffered, which is often the case, the rate of hydrogen evolution is largely activation-controlled over the whole operating range of the bath.

On the other hand, the rate for metal deposition could be partially controlled by mass transport. For deposition of a single metal, the immediate consequence is that the FE could be a function of the mass transport regime (stirring, movement of the cathode in an actual plating bath, and rotation rate in a laboratory experiment). Moreover, since the rate laws for the kinetics of hydrogen evolution and metal deposition are in general different, it follows that the FE would also be a function of both the applied current density (or potential) and the temperature. In the case of alloy deposition, if the FE is significantly lower that 100%, the situation could be more complicated, since the composition of the alloy could be a function of the FE. But the kinetics of hydrogen evolution is itself a function of alloy composition and surface morphology, since the two alloying elements will, in general, have different catalytic activities with respect to this reaction. If the alloy is a solid solution, one may expect that the exchange current density for hydrogen evolution on the surface of the alloy would be a weighted average of its values on the two elements. However, if the alloy consists of segregated phases, or if intermetallic compounds are formed during deposition, there is no telling how the rate of hydrogen evolution (and, hence, the FE) may change. The point to remember is that the overall FE observed will affect the partial currents of deposition of the different alloying elements, which will influence the composition of the alloy, thus possibly having a pronounced effect on the overall FE, and so on.

(*c*) *Formation of adsorbed intermediates*

The formation of a monovalent species in deposition and dissolution of divalent ions of the iron-group transition metals is commonly assumed in the literature. Since the monovalent ions (such as Fe^{+}) are unstable in solution, they are assumed to be adsorbed on the surface, either as the ion itself or as a hydroxide, such as $FeOH_{ads}$. This could stabilize the monovalent form of the element. Moreover, since a monovalent hydroxide is not known to exist in solution, one does not know its solubility product, and the possibility of the existence of this adsorbed species, even in solutions of low pH, cannot be ignored a priori. Unfortunately, the nature of the adsorbed intermediate, or even the evidence for its existence on the surface, is at best circumstantial.

In the work of Landolt et al.[89,90] inhibition of the partial current density of nickel by the addition of iron to the plating bath occurred when the concentration of Ni^{2+} ions was eight times larger than that of Fe^{2+} ions. This would seem to support the notion that inhibition is due to formation of some adsorbed species containing iron, since there is not enough Fe^{2+} in solution to interact with all the Ni^{2+}, reducing its rate of deposition.* It is more difficult to explain the mechanism of enhancement of deposition of iron by nickel. Although the above authors[89,90] did provide a simulation that could explain the behavior observed experimentally, at least partially, further detailed studies of this phenomenon would be needed to verify the mechanism proposed.

(*d*) *The solution chemistry*

Alloy deposition could involve rather complex solution chemistry, as noted above, which has not been investigated at sufficient depth in the analysis of the anomalies observed.

When a suitable complexing agent is used, different complexes could be formed. For example, complexes of the form $[Ni(NH_3)_n]^{2+}$ with n = 1-6 can be formed. In this case, the stability constants for each of the species are known, hence the relative concentrations of all the Ni-NH_3 complexes can be calculated as a function of the concentration of Ni^{2+} and of NH_3 (cf., Fig. 2a). Deposition of nickel can take place from each of these complexes, but the relative rate may depend on the number of ligands in the complex. When citric acid is added and the pH is adjusted to 8, the predominant species in solution is Cit^{3-}. This can form two different complexes with Ni^{2+}, either $[Ni(Cit)]^-$ or $[Ni(Cit)_2]^{4-}$ (cf., Fig. 1b). Nickel can readily be deposited from the former, but not from the latter. Moreover, when the molar ratio $Cit^{3-}/Ni^{2+} > 4$, most of the nickel is sequestered in the second complex above, inhibiting almost completely the deposition of nickel.

In alloy deposition, the possibility of formation of mixed-metal complexes containing ionic species of both metals and a

---

* It could well be that even a lower concentration of iron in solution would have a strong inhibiting effect, but this was unfortunately not tested in the above papers.[89,90]

suitable ligand should also be considered. For the induced codeposition of W-Ni alloys, the mixed-metal complex $[(Ni)(HWO_4)(Cit)]^{2-}$ was assumed to be the electro-active species,[91-93] as will be discussed below.

(*v*) *Induced Codeposition*

Certain elements, such as W, Mo, Ge and P cannot be deposited alone from their aqueous solutions. Nevertheless, they may readily be codeposited with iron-group elements. The term *induced codeposition* was coined by Brenner in 1963[78] to describe a situation where

> "a metal that cannot be deposited alone from its aqueous solution is codeposited in the presence of another metal, forming an alloy."

It was applied first to describe the electroless deposition of Ni-P alloys,[94] and later for electroplating of alloys of W and Mo with the iron-group metals.

It is of great interest from the scientific point of view, as well as for the development of plating baths, to understand the mechanism of induced codeposition. This may undoubtedly be considered to be *anomalous* in the sense that the composition of the alloy cannot be predicted from the electrochemical behaviors of the individual alloying elements. It cannot be measured by the criterion given in Eq. (26) above for anomalous codeposition of alloys, since one of the alloying elements cannot be deposited by itself. Nevertheless, some similarities do exist. Podlaha and Landolt[95-97] studied the induced codeposition of Mo-Ni alloys and concluded that the precursor for deposition of the alloy was an adsorbed complex containing both metals. In a later study of anomalous codeposition of iron-group transition metals,[89,90] Landolt et al. found an inverse influence of the two metals in solution on their respective rate of deposition. For example, adding Fe^{2+} to a nickel-plating bath inhibited the rate of deposition of nickel, while adding Ni^{2+} to an iron-plating bath accelerated the rate of deposition of iron. The catalytic effect of Ni^{2+} on the rate of deposition of iron was explained assuming a model similar to that proposed for the effect of Ni^{2+} on deposition of molybdenum.

In recent studies of induced codeposition of Ni-W alloys, Gileadi et al.[91,92] reported that increasing the concentration of Ni^{2+} in the bath led to a distinct increase of the partial current density for deposition of tungsten. However, unlike the case of anomalous codeposition of Fe-Ni alloys, it was also observed that increasing the concentration of WO_4^{2-} in solution led to an increase of the partial current density for deposition of nickel. Thus, induced codeposition of Ni-W alloys seems to be a true synergistic effect, where increasing the concentration of either metal ion in solution leads to an increase of the partial current density of the other. This behavior was explained by postulating the formation of a soluble complex containing both Ni^{2+} and WO_4^{2-}, which is the precursor for the deposition of the alloy.

(*vi*) *Electroless Deposition of Alloys*

Electroless deposition, or autocatalytic plating, may be defined as "deposition of a metal coating by a controlled chemical reduction, catalyzed by the metal or alloy being deposited." Electroless deposition has been known for a long time. One of its early uses was the deposition of a mirror-like layer of silver on the internal surfaces of Dewar flasks for improved thermal isolation, and as the back coating of mirrors. Later, it was used for deposition of different metals and alloys, and even for induced codeposition of alloys.

An electroless-plating bath consists of a soluble salt of the metal or metals being deposited and a suitable reducing agent. As in any plating bath, suitable additives are also used to improve the product, but these will not be discussed here. It is important to remember that electroless deposition is inherently an electrodeposition process, conducted under (nearly) potentiostatic conditions. The source of electrons for the reduction process is, of course, different—in electrodeposition it is the power supply, via the metallic electrode; in electroless deposition it is the reducing agent. In both cases the potential is maintained by and large constant, even if the current is the externally controlled parameter, as long as the solution is not exhausted from its electro-active components. In electroless deposition the potential across the interface is controlled through a corrosion-type mechanism, by the balance between the rate of oxidation of the reducing agent and the rate of reduction of the metal ions, either of which could be the rate-limiting factor, depending on the composition of the bath. Hence, the rate of metal deposition can be determined, in principle, by micro-polarization measurements, following the common method to determine the corrosion current.

The composition of the bath should be chosen such that the rate of metal deposition will be well below the mass-transport-limited rate. Proper selection of the type and concentration of the reducing agent can achieve this. The main advantage of electroless deposition, compared to electroplating, is that metals can be deposited on non-conducting surfaces. This is widely used as a first step for electrodeposition in many engineering applications, as well as for ornamental purposes. The other advantage is that the primary current distribution, caused by the shape of the part being plated and the distance of various parts from the anode, has no effect in electroless deposition, because there is obviously no anode. In this sense electroless deposition can be considered as being conducted under conditions equivalent to electrodeposition under secondary current distribution. On the other hand, partial mass-transport limitation is not eliminated, hence ternary current distribution, which determines the micro-throwing power and, in some cases, the surface morphology of the deposit, cannot be ignored.

Electroless deposition is usually a slow process, and it is therefore limited to formation of relatively thin layers. Since the development of surface roughness in metal deposition has a built-in positive feedback effect, causing the roughness to increase with increasing thickness of the deposit, this is less of a problem in the case of electroless deposition.

The main disadvantage of electroless deposition is that the plating bath is inherently unstable. Hence, a delicate balance must be struck between the desire to have a stable bath and, yet, allow reasonably high rates of deposition. The surface of the substrate being plated must be activated, to ensure that deposition will only occur where needed. The rate of electroless deposition cannot be controlled as readily as the rate of electroplating, although some control can be achieved by varying the concentration of the reducing agent in the bath and by controlling the rate of mass transport and the temperature.

In view of the inherent similarity between electroless deposition and electroplating, it is not surprising that anomalous codeposition and induced codeposition can be performed by both methods. A most important early case was the development of an electroless plating bath for deposition of amorphous Ni-P alloys by Brenner.[78,94] The reducing agent in this bath was sodium hypophosphite (NaH_2PO_2). The same reducing agent was used recently by Shacham-Diamand et al.[98-102] in electroless deposition of Co-W-P alloys as barrier layers for ULSI devices.

II. CASE STUDIES

1. Tungsten Alloys Containing Ni, Co and Fe

(*i*) *Properties of Tungsten Alloys*

Selected properties of tungsten (W) are listed in Table 2, in comparison to the respective properties of molybdenum (Mo) and rhenium (Re), which will be discussed in the following sections. The attribute ranges were taken from the Cambridge Engineering Selector[103] material database, and reflect different thermal conditions and suppliers of the commercially pure metals. Of all metals in the periodic table, tungsten possesses the highest melting point, the lowest linear thermal expansion coefficient, the highest tensile strength, the fourth Young's modulus of elasticity, and the sixth thermal conductivity. It maintains most of its strength and hardness at fairly high temperatures, and is also highly corrosion resistant, being stable in any single mineral acid at room temperature. At the same time, it is one of the densest metals, lacks ductility, and is oxidized in air only at temperatures above 1,000°C. It is fairly expensive and has limited availability.

Attempts to electrodeposit pure tungsten date approximately 140 years back. Nowadays, however, it is commonly accepted that this metal cannot be deposited alone from its aqueous solutions, but can be codeposited as an alloy, exhibiting a unique combination of properties. For example, Ni-W alloys have good mechanical properties (e.g., high tensile strength and premium

Table 2
Selected Properties of Tungsten, Molybdenum and Rhenium

Property	W	Mo	Re
Atomic number	74	42	75
Atomic mass, M (g mol^{-1})	183.8	95.9	186.2
Oxidation states	2,3,4,5,6	2,3,4,5,6	−1,2,4,6,7
Crystal structure	bcc	bcc	hcp
Atomic radius, r_{metal} (Å)	1.41	1.39	1.37
Density, ρ (g cm^{-3})	19.25–19.35	10.1–10.3	21.00–21.02
Melting temperature, T_m (°C)	3,410–3,420	2,607–2,622	3,157–3,181
Linear thermal expansion coefficient, α (°C^{-1})	4.2-4.6×10^{-6}	4.8-5.5×10^{-6}	6.00-7.25×10^{-6}
Thermal conductivity, κ (W m^{-1}K^{-1})	170-175	129-147	45-48
Specific resistivity, ρ (Ω·m)	5.4-6×10^{-8}	5.2-6×10^{-8}	18.7-20.0×10^{-8}
Tensile strength, σ_u (MPa)	1,670-3,900	380-2,100	1,000-2,500
Yield strength, σ_y (MPa)	1,350-3,500	170-2,000	280-2,350
Young's modulus of elasticity, E (GPa)	340-405	315-343	461-471
Strain at fracture, ε_f (%)	1-25	1-45	1-30
Poisson's ratio, ν	0.27-0.29	0.29-0.295	0.255-0.265
Hardness (MPa)	4,500-8,500	1,500-6,500	2,600-7,500
Fracture toughness, K_{Ic} (MPa $\sqrt{m}$)	120-150	20-40	120-150

hardness, as well as superior abrasion resistance), good resistance to strong oxidizing acids, and high melting temperature.[78,91-93,104,111] It was reported,[109] for example, that the corrosion rate of an amorphous Ni-W deposit in hydrochloric acid at 30°C is only 1/40 that of type 304 stainless steel, commonly used in industry. In general, the passivation current density (i_{pass}) drops remarkably with the addition of tungsten to nickel. While the hardness of the *as-plated* alloy is typically in the range 650-750 VHN, heat treatment at temperatures ranging from 190°C to 600°C for 12 to 24 hours can raise the hardness to 1,200–1,400 VHN due to precipitation hardening.[110,111]

(*ii*) *Applications of Tungsten Alloys*

Tungsten and its alloys, including those with iron-group metals, have been used in filaments of incandescent lamps, electrical contacts, resistors, heating elements, thermocouples, cutting tools, X-ray targets, balance weights, anti-vibration tooling, bearings, radiation shields, nozzles of rocket engines, heat sinks, mold inserts, magnetic heads and relays, crucibles, extrusion dies, high-strength wires and springs.[103] Tungsten is also used as an alloying element in high-speed tool steels and corrosion-resistant alloys.[78] Recently suggested applications include barrier layers or capping layers in copper metallization for ULSI devices or MEMS, electrodes catalyzing hydrogen evolution from alkaline solutions, and substitutes for hard chromium plating with a good combination of wear and corrosion resistance (e.g., in the aerospace industry).[91-93,104-116]

(*iii*) *Electrodeposition of Tungsten Alloys*

The first deposition of a tungsten alloy (W-Fe) may be attributed to Fink and Jones,[117] although these authors mistakenly claimed to have deposited pure tungsten. Soon afterwards it was realized that, although tungsten could not be electrodeposited from an aqueous solution of sodium tungstate (Na_2WO_4) or any other soluble compound containing this element, induced codeposition could take place if the plating bath contained iron-group metals, namely nickel, cobalt or iron. Golt'tz and Kharlamov[118] developed already in 1936 practical plating baths for alloys, by replacing the highly alkaline carbonate solutions with ammoniacal solutions. From these ammoniacal baths, they deposited W-Ni alloys 0.2-mm thick at FE of up to 30%. However, porous and weak deposits resulted from the excessively high current densities with respect to the low metal content of the bath. Next, several investigators found that the addition of organic poly-hydroxy acids (e.g., citric, tartaric, malic, gluconic, hydroxy acetic, or saccharic) into ammoniacal baths improved the FE and the solubility of the metal ions in the bath. Consequently, smooth, hard and thick deposits of the alloys could be formed at lower current densities and at FE approaching 100%.[78]

Here, we shall focus only on moderately alkaline solutions for electrodeposition of tungsten-based alloys. The reason is that acidic solutions have been reported to be of limited practical value, because of the poor deposits that were obtained from them. Nevertheless, for comparison to electrodeposition in moderately alkaline solutions, several characteristics of electrodeposition from acidic baths are summarized. Iron was found to induce codeposition of tungsten more effectively than nickel or cobalt. An increase of pH in the range of 2 to 5 had a negligible effect on the composition of the alloy, but at the same time increased significantly the FE. Variations in the current density and temperature had little effect on the composition of the alloys, except for the Fe-W alloys, for which the tungsten content decreased considerably as the current density was decreased. Increasing the temperature resulted in the production of deposits having a more metallic appearance, and in an increase in the FE.[78]

The binary alloys of tungsten with iron, cobalt and nickel are readily deposited from moderately alkaline baths. Typical bath compositions and operating conditions are listed in Appendix A. A complexing agent is typically added and forms a soluble complex ion with the iron-group metal. The W-Fe alloys are the most readily obtained with high W-contents, but the quality and thickness that could be achieved were the least satisfactory. The W-Ni alloys were reported to produce the lowest tungsten content, which was usually found to be in the range of 5–25 at.% (13–50 wt.%).[78] In order to further improve the tribological properties and thermal stability of the coating, it is sometimes desirable to increase its tungsten content. Unfortunately, this has been found difficult, even when the WO_4^{2-} ion in solution is in large excess compared to the Ni^{2+} ion. One possible way to increase the tungsten concentration in the alloy is to apply periodic reverse pulse plating, which may also increase the throwing power and the deposition rate, as well as improve the properties of the deposit (e.g., reduce residual stresses and porosity, refine the grain size, improve wear and corrosion resistance, etc.)[104] Another route was suggested by Gileadi and co-workers, who removed the NH_3 from the plating bath, while using citrate as a complexing agent, thus increasing the tungsten content of the alloy to nearly 50 at.% .[91,93,104-108] The best plating conditions for obtaining sound and thick deposits at good FE are high concentrations of the iron-group metal in the bath, elevated bath temperatures and moderate current densities. However, some of these conditions may lower the concentration of tungsten in the resulting alloy.

Brenner reviewed several variables that affect the tungsten content in the deposit.[78] The most important parameter is the ratio of concentrations of tungstate to nickel ions. In general, as this ratio is increased, the tungsten content in the deposit increases, until it approaches a limit of approximately 50–60 wt.% for Fe-W and Co-W alloys, and 30 wt.% for Ni-W alloys. Another variable, the type of complexing agent, had a moderate effect on the tungsten content in the deposit. The composition of the alloy deposits obtained from ammoniacal baths, on the other hand, was not significantly affected by variations of pH between 8.0 and 10.0. This claim of Brenner, however, is contradictory to the recent findings of Gileadi et al.[92,93] Typically, the tungsten content in Co-W and Ni-W alloys increases slightly and the FE decreases with increased current density. For Fe-W alloys, the W-content in the deposit was found to be less affected by current density. The bath temperature has an important effect on the FE and soundness of deposits. In addition, the tungsten content of the deposit usually increases at elevated temperatures.

The effects of several variables on the FE, tungsten concentration in the alloy, the deposit structure, thickness and hardness on stationary working electrodes were studied recently by Eliaz et al., for ammonia-free, citrate-containing Ni-W plating baths.[104,108] The FE was found to increase with the concentration of Ni^{2+} and decrease with increasing current density. The nickel content of the bath also affected tungsten content in the deposit. While the W-content was in the range 30–35 at.% for both the 0.05 M Ni^{2+} and the 0.10 M Ni^{2+} baths (containing 0.4 M tungstate and 0.5 M citrate) at all current densities, it increased significantly to 67 at.% for the 0.01 M Ni^{2+} bath. However, this was associated with very low FE and reflects a poorly adhering, thin deposit, rather than the good Ni-W deposits obtained at higher concentrations of Ni^{2+} in solution and, correspondingly, lower tungsten concentration in the alloy. For comparison, Younes and Gileadi[105] previously reported a maximum FE of 11% and a maximal tungsten content of about 67 at.% in two different baths containing different concentrations of Na_2WO_4. An increase in the concentration of Ni^{2+} was also reported to cause a significant increase in the FE.[92,124] Furthermore, it was reported that an increase in the concentration of Ni^{2+} results in an increase in the rate of both Ni and W deposition,[125,126] but the tungsten content in the alloy decreases.[93]

The *concentration* of citrate ions also affects the FE and W-content of the deposit. The effect was found to depend on current density and on the concentration of nickel ions in the bath. In general, the FE was found to decrease as the concentration of Cit^{3-} was increased. At a low current density (5 mA cm^{-2}), the tungsten content increased from 10 to 33 at.% as the concentration of Cit^{3-} was increased from 0.25 M to 0.60 M. At a higher current density (15 mA cm^{-2}), however, the trend reversed and the tungsten content increased from 31 to 60 at.% as the concentration of Cit^{3-} was decreased. The effects of the citrate ion concentration on the FE and on the tungsten content in the deposit were studied in solutions of very low concentration of Ni^{2+} ions, and the FE was also found to decrease with increasing Cit^{3-} concentration.

It is well known that the tungsten content of the alloy depends, among others, on the type of the complexing agent used in the bath. Citrate baths were reported to yield higher tungsten content than those containing tartrate or malate. However, an increase in the concentration of Cit^{3-} was sometimes found to result in a decrease in both the FE and the tungsten content.[78,93,125] Huang,[127] on the other hand, reported that addition of diammonium citrate to a sulfamate bath resulted in an increased concentration of tungsten in the alloy, but the residual stresses were also increased. In a bath containing 0.10 M Ni^{2+} and 0.10 M WO_4^{2-}, Younes and Gileadi observed[92] that the tungsten content of the alloy increased when the concentration of Cit^{3-} was increased up to 0.50 M, but then started to decrease. The FE decreased dramatically when the concentration of citrate exceeded 0.20 M, i.e., when the molar concentration of citrate exceeded the sum of concentrations of nickel and tungstate ions. The deposition potential shifted in the negative direction with increasing citrate concentration, first sharply and then moderately. The decrease of FE and the shift of potential indicate that the main reaction taking place in the presence of excess Cit^{3-} is hydrogen evolution, and this side reaction has a dominant effect on the potential. It was concluded that citrate, being a strong complexing agent for Ni^{2+}, tends to sequester this ion in a stable complex of $[Ni(Cit)_2]^{4-}$, decreasing the availability of Ni^{2+} to form a mixed Ni-WO_4-Cit complex, which is essential for the induced codeposition of tungsten (cf., Section 2.1.4). The relative effect on nickel was found to be stronger up to a concentration of 0.50 M citrate and weaker at higher concentrations, leading to a maximum tungsten content at a Cit^{3-} concentration of about 0.50 M.

The effect of nickel sulfamate, saccharin and sodium chloride as *additives* was studied by Eliaz et al.[104,108] The effect of adding nickel sulfamate to a bath containing 0.10 M Ni^{2+}, 0.40 M WO_2^{2-} and 0.50 M Cit^{3-} was to increase the FE. This observation was more pronounced at high than at low current densities. The increase in the FE was accompanied by a slight decrease in the tungsten content of the alloy. The morphology of the coating, observed by SEM, did not change significantly as a result of nickel sulfamate addition. However, metallographic cross-sections did show a remarkable increase in the thickness of the coating at a high current density. The addition of saccharin (14.6 mM) increased the FE at high current densities and decreased the tungsten content of the alloy at low current densities.

Adding sodium chloride (2–5 mM) stabilized the FE at a nearly constant value, independent of current density. In contrast, the FE dropped significantly at high current densities when chloride was absent. The addition of chloride ions did not change the composition of the alloy significantly. It should be noted, however, that excess chloride might be harmful to brightness and leveling.

Temperature is another important variable in the operation of W-Ni plating baths. The effect of this variable depends on the composition of the bath. When the bath contained tartaric and boric acids, both the tungsten content and the FE increased with increasing temperature in ammonia-containing baths,[128] whereas the effect of temperature on the tungsten content was small in ammonia-free baths.[120] Younes and Gileadi similarly observed that in ammonia-free solutions, almost no effect of temperature on the tungsten content exists, while the FE increases with temperature.[105] Krishnan et al.[113] also reported an increase in FE with increasing temperature, while Yamasaki et al.[123] described temperature effects both on the tungsten content and the ductility of the deposit. Atanassov et al.[112] observed an increase in the tungsten content with increasing temperature, both in unstirred and in stirred baths, and explained this behavior in terms of favorable conditions for tungsten transport toward the cathode surface. This argument can be questioned, however, since (a) it was not shown that mass-transport limitation played a role in the deposition of tungsten, and (b) it is not clear why the rate of mass transport of tungstate to the surface would be enhanced by increasing temperature more than the rate of transport of Ni^{2+} ions.

Eliaz et al.[104] also studied the effect of temperature within the range of 30°C to 70°C in an ammonia-free bath, containing 0.10 M Ni^{2+}, 0.40 M WO_4^{2-} and 0.50 M Cit^{3-}. As the bath temperature was increased, the FE increased. The tungsten content of the deposit did not change much with temperature, and showed a rather irregular behavior, with perhaps some tendency to decrease with increasing temperature. These results are in good agreement with the previous work of Younes and Gileadi,[105] keeping in mind that the latter was conducted at a current density of 15 mA cm^{-2} only. It was observed, rather unexpectedly, that in the higher temperature range (50–70°C), a Ni_4W phase could be formed by electrodeposition.

Increasing the *current density* in the same bath composition led to a decrease of the FE, whereas the tungsten content either increased or passed through a maximum at 10 mA cm^{-2}. At high current densities, where the FE is low, hydrogen evolution becomes more dominant and causes additional agitation in solution. Moreover, because Ni is deposited easily compared to W, higher FE is expected to result if the concentration of Ni in the alloy is higher. This implies some degree of mass-transport limitation, as will be discussed below (cf., Section 2.1.4). The effect of current density in the W-Ni system was also studied by others. Brenner et al.[119] observed a significant increase in tungsten content with increasing current density in ammonia-citrate bath. Yamasaki et al.[123] reported a similar trend. Atanassov et al.[112] noted a linear increase in tungsten content with increasing current density, when vigorous stirring was applied. On the other hand, a maximum was observed at 50–70 mA cm^{-2} in the absence of stirring. At current densities higher than 20 mA cm^{-2}, the FE was 15–40% higher in the stirred bath, where hydrogen evolution was less pronounced, compared to an unstirred bath. Krishnan et al.[113] also monitored a decrease in the FE with increasing current density. Finally, Huang et al.[129] reported an increase in residual stresses with increasing current density.

X-ray diffraction (XRD) studies have shown that the structure of electrodeposited tungsten alloys is different from that of thermally prepared tungsten alloys.[78] Several structures have been observed in the W-Ni system, depending on the operating conditions and the chemical composition of the resulting deposit.[91,93,104,105,108] Below 20 at.% tungsten, a solid solution of W in Ni, in the form of an fcc phase, is formed (as shown in Fig. 5a). An amorphous Ni-W phase was observed[91,93,105] when the tungsten content fell within the range of 20–40 at.% (Fig. 5a). This metastable phase cannot be predicted from the equilibrium binary phase diagram, according to which dual crystalline phases should form within this concentration range. Younes et al. also observed the formation of an orthorhombic NiW phase when the concentration of tungsten in the deposit exceeded 40 at.%,[91,105] as shown in Fig. 5a. It was argued that formation of equal amounts of Ni and W in the alloy may be regarded as evidence, albeit circumstantial, for the existence of a mixed nickel-tungstate-citrate complex. Later,[93] Younes et al. also reported the first plating of a body-centered tetragonal NiW_2 phase. It was found that the NiW and NiW_2 phases can be formed when the sum of the molar concentrations of nickel and tungstate ions exceeds slightly the molar concentration of the citrate ion, and the ratio of concentrations of WO_4^{2-} to Ni^{2+} ions is much larger than unity. It should be noted, however, that the concentration of citrate should not be much lower than the sum of concentrations of tungstate and nickel ions, otherwise precipitation will occur. At higher citrate concentrations, the amorphous Ni-W phase forms at room temperature and different current densities. On the other hand, under conditions of low current density (5 mA cm^{-2}) and higher temperature (50°C and above), the body-centered tetragonal Ni_4W phase can be formed electrochemically in a reproducible manner,[108] as shown in Fig. 5b.

Several studies were published in recent years on electrodeposition of W-Co alloys. Donten and Stojek[130] used pulse electrodeposition to increase the tungsten content in amorphous Co-W alloys. These alloys contained, in addition, small amounts of boron or phosphorous. They showed that, if a symmetrical current pulse was used, the tungsten content in the alloys reached a maximum value of 41.4 at.%, which is higher than in the case of direct current deposition. However, when using any asymmetrical

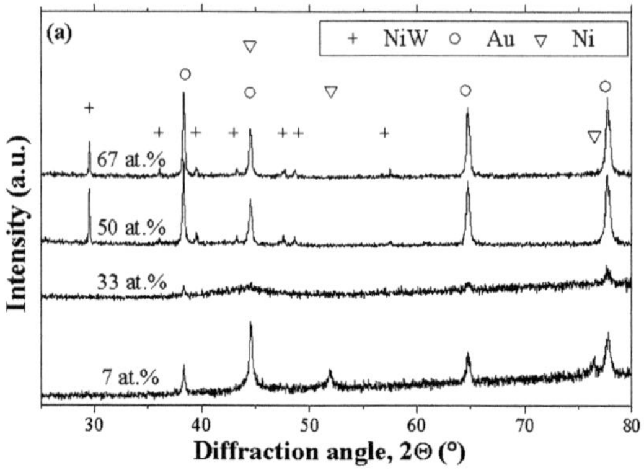

(a)

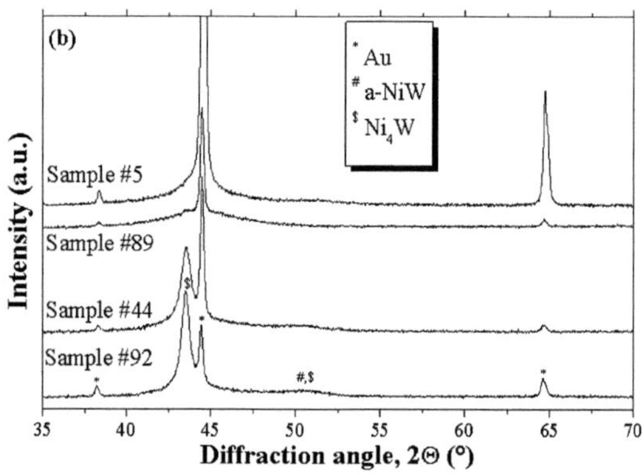

(b)

Figure 5. X-ray diffraction spectra, demonstrating the effect of bath temperature and operating conditions on the structure of Ni-W electrodeposits. (a) A solid solution of W in Ni at 7 at.% W with fcc structure, an amorphous Ni-W phase at 33 at.% W, and an orthorhombic Ni-W phase at 50 and 67 at.% W. Reprinted from Ref. 93, Copyright (2003) with permission from Elsevier. (b) The top two spectra demonstrate the deposition of amorphous Ni-W phase on gold. The lower two spectra show the (211) strongest reflection of the bct Ni_4W phase. Reprinted from Ref. 108, Copyright (2005) with permission from Elsevier.

current wave form, the tungsten content in the alloy decreased. The authors offered no explanation for this difference. Wikiel and Osteryoung[131] used anodic stripping voltammetry for monitoring in-situ the concentration of cobalt ions during electrodeposition of Co-W alloys on platinum micro-disk electrodes. A mechanism of electrodeposition of a Co-W alloy using cyclic voltammetry was suggested by Aravinda et al.[132] These authors assumed that the deposition of W-Co alloys occurs from a cobalt-tungstate complex, while the deposition of Co^{2+} occurs via a stepwise reaction involving Co^{+} ion as an intermediate. Although the present authors would agree tentatively that the precursor for deposition of a Co-W alloy is indeed a mixed-metal complex (based on the similarity of this system to the Ni-W system), it is not clear how this conclusion could have been reached based on measurements of cyclic voltammetry alone, considering that the current observed represents the sum of at least three different reactions occurring in parallel (cf., Section 1.2.2).

Several theories were proposed to explain the mechanism of induced codeposition of tungsten, but none could be confirmed experimentally. Here, we summarize a few of these theories. Firstly, it was assumed that tungsten-containing ions cannot be discharged on a pure tungsten surface due to the *self-polarizing* nature of tungsten, and that the presence of the iron-group metal alters the characteristics of the surface, e.g., increases the overpotential for the hydrogen evolution reaction.[78] However, this mechanism was later discredited, when it was observed that tungsten could not be deposited even on mercury, although the rate of hydrogen evolution on this metal is known to be very low.[133] Secondly, it was suggested that initially an oxide of tungsten is formed on the cathode. This oxide is subsequently reduced by atomic hydrogen, a reaction that is catalyzed by the presence of the iron-group metal.[134] Thirdly, it was suggested that the deposition potential of tungsten is too negative to be attained in aqueous solutions, but becomes more positive as a result of formation of a solid solution of tungsten with an iron-group metal.[123,128,135,136] One of the drawbacks of this mechanism is that measurements of the heat of solution of the two metals showed that it was too small to explain the shift in potential.[78] Finally, it has been postulated that a complex ion, containing both tungsten and an iron-group metal, forms in solution and that the discharge of this ion enables the simultaneous deposition of both metals.[78]

(*iv*) *New Interpretation of the Mechanism of Ni-W Codeposition*

With respect to the last type of mechanism, Gileadi and coworkers recently studied the induced codeposition of W-Ni from moderately alkaline baths containing citrate as a complexing agent, with or without ammonia.[91-93,105-107] The range of bath compositions and operating conditions is shown in Appendix A. The working hypotheses in these studies were that:

1. A tungstate-citrate complex reacts with a nickel-citrate complex, to form a mixed-metal complex, of the type $[(Ni)(HWO_4)(Cit)]^{2-}$.
2. This mixed-metal complex is the precursor for the deposition of the W-Ni alloy.
3. Nickel can also be deposited from its complexes with either citrate (e.g., $[NiCit]^-$) or NH_3 complexes of the form $[Ni(NH_3)_n]^{2+}$, where $n = 1–6$.

Although there has been no direct evidence for the formation of the above mixed-metal complex, there are ample experimental observations supporting the assumption that it is indeed formed and serves as the precursor for deposition of the Ni-W alloys. These indications include:

1. The mutual synergistic effect of nickel and tungsten on each other. Thus, the partial current density for deposition of W is increased as the concentration of $NiSO_4$ is increased. Similarly, increasing the concentration of Na_2WO_4 in solution results in an increase in the partial current density for deposition of Ni.
2. The W-content in the alloy increases markedly upon removal of ammonia or ammonium salts from the bath.
3. The W-content in the alloy decreases markedly as the pH exceeds 8.
4. An unexpected mass-transport limitation of the partial current density for deposition of W is observed.

In Fig. 6b the partial current density for deposition of tungsten is shown to increase by a factor of about 10 with increasing concentration of nickel in a solution containing 0.1 M Na_2WO_4 and 0.6 M citrate. In the same range of concentrations of Ni^{2+} ions, the concentration of W in the alloy decreases (by a factor of about 7.5). This is by no means surprising, and is consistent with the fact

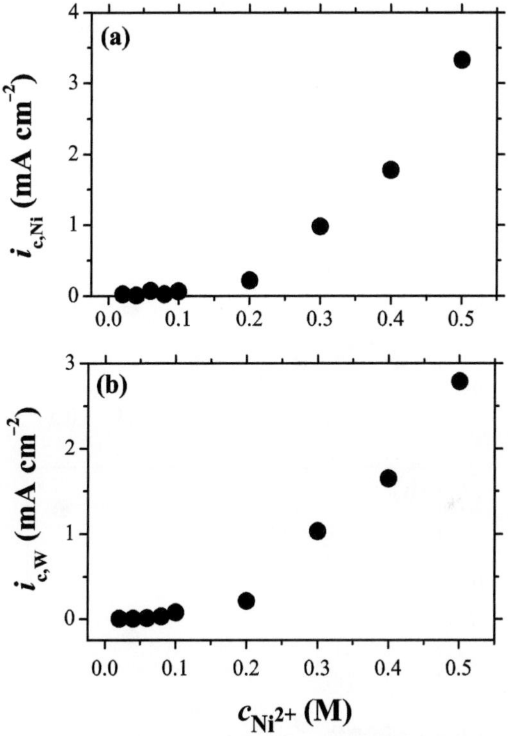

Figure 6. The effect of Ni^{2+} ion concentration on the partial current densities of nickel (a) and tungsten (b). All baths contained 0.1 M WO_4^{2-} and 0.6 M Cit^{3-}. Reprinted from Ref. 93, Copyright (2003) with permission from Elsevier.

that nickel can be deposited from several complexes in solution, in addition to its being deposited as an alloy with tungsten. Thus, the total increase in the partial current density for deposition of nickel upon increasing its concentration in solution (Fig. 6a) far exceeds the increase in the rate of deposition of tungsten. For the same reason, the FE is found to increase dramatically with increasing rate of deposition of nickel.[92,106]

A similar synergistic effect of addition of WO_4^{2-} on the rate of deposition of Ni is shown in Fig. 7 for solutions containing 0.1 M

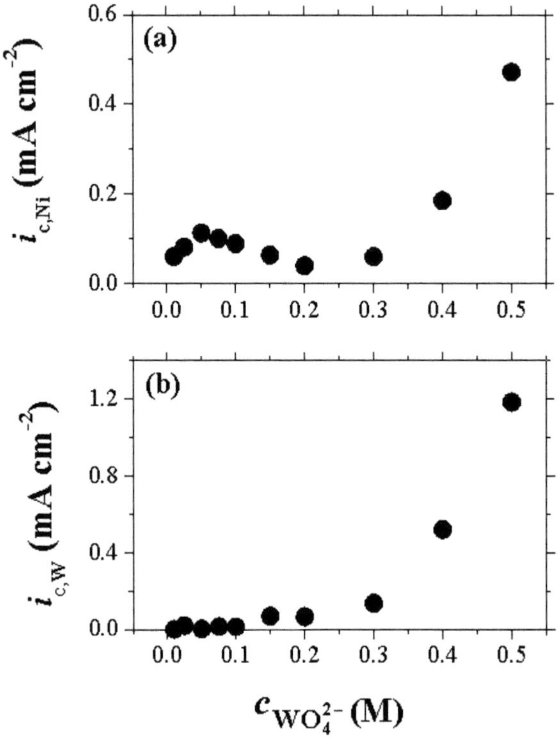

Figure 7. The effect of WO_4^{2-} ion concentration on the partial current densities of nickel (a) and tungsten (b). All baths contained 0.1 M Ni^{2+} and 0.6 M Cit^{3-}, pH = 8.0. Reprinted from Ref. 93, Copyright (2003) with permission from Elsevier.

$NiSO_4$ and 0.6 M citrate. As the concentration of WO_4^{2-} is increased above 0.3 M, the partial current density for W deposition increases sharply, while the partial current density for Ni deposition also increases. This is probably not due to any direct effect of WO_4^{2-} on the rate of deposition of Ni, but to the fact that an increasing proportion of Cit^{3-} ions is complexed by WO_4^{2-}. Hence, for WO_4^{2-} concentrations of 0.3, 0.4 and 0.5 M, the ratio of *free* Cit^{3-} (i.e., citrate not complexed with tungstate) to Ni^{2+} changes from 3 to 2 to 1, respectively, making it gradually easier to deposit Ni, as its predominant species in solution changes from the highly charged $[Ni(Cit)_2]^{4-}$ to the simpler $[NiCit]^{-}$, and it is well known that the complex containing two citrate ions sequesters the Ni^{2+} ions. In addition, increasing the concentration of WO_4^{2-} in solution, at a constant concentration of all other components of the bath, can increase the rate of formation of the mixed-metal complex, from which both tungsten and nickel are deposited.[93]

The effect of ammonia is discussed below, in conjunction with the effect of pH. As described in Section 1.2.2 and in Fig. 1a, citric acid can exist in different forms, depending on pH. The pK_a values corresponding to the three acidic groups and the alcohol group are 2.96, 4.38, 5.68, and 10.82, respectively (see Table 1). The complexes of tungstate with anions of citrate are of the form

$$[(WO_4)_k(Cit)_l(H)_m]^{-(2k+3l-m)} \tag{31}$$

depending on the pH of the solution and the relative concentrations of citrate and tungstate.[137] Several protonated forms, which can be represented by the equation $[(WO_4)(Cit)(H)_m]^{-(5-m)}$ exist, where m can assume values of 1–3. The corresponding $\log(\beta_n)$ values are given in Table 1. Figure 8 shows the calculated concentration distributions for tungstate-citrate complexes as a function of pH. According to this figure, in the range of pH = 6.8–10.2, the predominant species is the complex containing only one proton (m = 1), which can be notated as $[1,1,1]^{4-}$ for brevity. Since Cit^{3-} is not protonated at this pH, it may be assumed that the hydrogen atom originates from the partially protonated HWO_4^{-} ion.[92] It is assumed that the complex $[(WO_4)(Cit)(H)]^{4-}$ is the precursor for the formation of the mixed-metal complex with Ni^{2+}, $[(Ni)(HWO_4)(Cit)]^{2-}$. When the pH is further increased, the $[1,1,1]^{4-}$ complex is deprotonated. The resulting complex would have been $[(WO_4)(Cit)]^{5-}$, but this is destabilized by its high negative charge of –5, so that WO_4^{2-} is formed. As a result, the concentrations of the mixed-metal complex in solution and the W-content of the deposit are decreased. At pH lower than 7.0, the concentration of the $[1,1,1]^{4-}$ complex is decreased, and at pH 6.0 most of the tungstate is bound in the $[1,1,2]^{3-}$ complex. At pH lower than 4.0, the dominant complex in solution is the $[1,1,3]^{2-}$ complex,[91,92] but this range of pH is not relevant in the context of the studies discussed above.

The reduction of the mixed-metal complex in mildly alkaline solutions requires the transfer of eight electrons, and seven hydroxyl ions are formed, as shown in Eq. (32). Consequently, the evaluation of a detailed mechanism, including the sequence of transfer of protons and electrons, the possible adsorbed intermediates and the rate-determining step or steps, seem like an impossible task. This is true in particular in view of the fact (discussed in Section 1.2.2 above), that there are several reactions

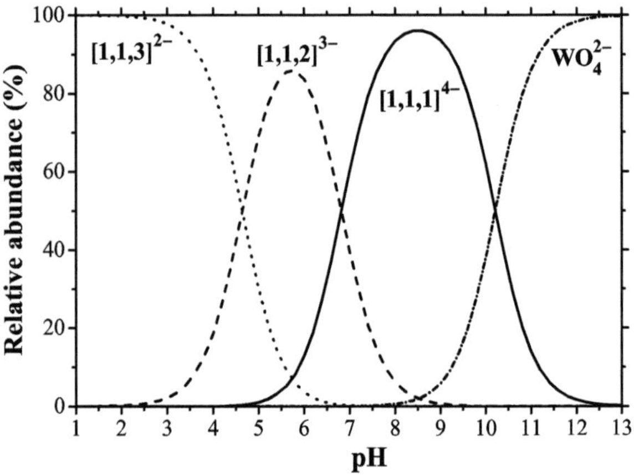

Figure 8. The relative abundance of the protonated tungstate-citrate complexes, $[(WO_4)(Cit)(H)_m]^{-(5-m)}$ as a function of pH (0.5 M Cit^{3-}, 0.1 M WO_4^{2-}). The abbreviated forms $[1,1,1]^{4-}$, $[1,1,2]^{3-}$, and $[1,1,3]^{2-}$ refer to values of $m = 1$, 2 and 3, respectively, in the above complex.

occurring in parallel, and measurement of current density only provides the sum of rates of all such steps.

All one can do is to write the overall reaction as:[105,106]

$$[(\mathrm{Ni})(\mathrm{HWO_4})(\mathrm{Cit})]^{2-} + 8e^- + 3H_2O \rightarrow \mathrm{NiW} + 7(\mathrm{OH})^- + \mathrm{Cit}^{3-} \quad (32)$$

The mixed-metal complex shown in Eq. (32) can be formed in the reaction

$$[(\mathrm{Ni})(\mathrm{Cit})]^- + [(\mathrm{HWO_4})(\mathrm{Cit})]^{4-} \rightarrow [(\mathrm{Ni})(\mathrm{HWO_4})(\mathrm{Cit})]^{2-} + \mathrm{Cit}^{3-} \quad (33)$$

Plotting the partial current densities for deposition of W as a function of the product of the concentrations of the two anions on the left-hand side of Eq. (33), a linear dependence was found for a wide range of solution compositions, supporting the proposed interpretation for the induced codeposition of W.[105,106] Unfortunately, the complex on the right-hand side of Eq. (33) has not been detected directly, either in solution or as an adsorbed species, even though its existence is strongly supported by all the experimental data discussed above. The rate of the reaction shown in Eq. (33) is expected to be quite low, in view of the fact that it involves an interaction between two negative ions, one of them highly charged, and requires the removal of a citrate ligand from one of the two complexes. This should lead to a low concentration of the mixed-metal complex at steady state, which could be the reason that this complex has not yet been detected. This is also consistent with the fact that the partial current density for deposition of tungsten is partially mass-transport limited, even though it is very low compared to the limiting current density calculated for the concentration of WO_4^{2-} in solution, as discussed below. The likelihood of a similar reaction involving the complex of nickel with two citrate ions, $[Ni(Cit)_2]^{4-}$, is low because of the electrostatic repulsion between the reacting anions, each having a charge of –4.[92]

In solutions where the ratio of concentrations of tungstate to nickel ions is very high, Eq. (33) could be followed by a reaction of the type

$$[(\mathrm{Ni})(\mathrm{HWO_4})(\mathrm{Cit})]^{2-} + [(\mathrm{HWO_4})(\mathrm{Cit})]^{4-} \rightarrow [(\mathrm{Ni})(\mathrm{HWO_4})_2(\mathrm{Cit})]^{3-} + \mathrm{Cit}^{3-} \quad (34)$$

The mixed-metal complex shown on the right-hand side of this equation could lead to an alloy having the composition of NiW_2. This was indeed observed in solutions containing 0.5 M Cit^{3-}, 0.4 M WO_4^{2-} and 0.01 M Ni^{2+}.[93]

In ammoniacal-citrate baths, Ni^{2+} can also form complexes with NH_3, the latter serving as a ligand, of the form $[Ni(NH_3)_n]^{2+}$, as shown in Fig. 2a, where $n = 1–6$, depending on the relative concentrations of the Ni^{2+} ion and ammonia. The $\log(\beta_n)$ values for these complexes are given in Table 1. The formation of these complexes can lead to a decrease in the concentrations of other complexes of Ni^{2+}, including the mixed-metal complex that is assumed here to be involved in the codeposition of W. Consequently, the rate of W deposition would be lowered. The fact that in ammoniacal-citrate baths the concentration of tungsten in the alloy could not exceed about 25 at.% W is easily understood if we remember that there are parallel paths through which Ni can be deposited alone, from its complexes either with citrate or with ammonia.[92,105]

The concentration of free NH_3 depends on pH in view of the equilibrium, given by the following equation:

$$\mathrm{NH_4^+} + \mathrm{H_2O} \rightleftarrows \mathrm{NH_3} + \mathrm{H_3O^+} \quad (35)$$

The acid dissociation constant, pK_a, corresponding to Eq. (35) is 9.25. As the pH is increased, Eq. (35) is shifted to the right, thus increasing the concentration of NH_3 in solution, as shown in Fig. 2b, allowing the formation of nickel-ammonia complexes:

$$\mathrm{Ni}^{2+} + n\cdot\mathrm{NH_3} \rightarrow [\mathrm{Ni}(\mathrm{NH_3})_n]^{2+} \quad (36)$$

This could explain the effect of pH on the composition of the alloy in solutions containing ammonium salts. However, a similar decrease in the tungsten content in the alloy was also observed when the pH was increased from 8.0 to 9.0 in citrate baths free of ammonia. This was explained by considering Fig. 8, according to which, above pH = 8.5 the complex $[(WO_4)(Cit)(H)_m]^{-(5-m)}$ starts loosing its last proton, thus forming the unstable $[(WO_4)(Cit)]^{5-}$ complex, which decomposes to form three tungstate ions. Since the mixed-metal complex cannot be formed by the reaction of nickel citrate with bare tungstate, a sharp decline of its concentration with increasing pH leads to a similarly sharp decline in the concentration of tungsten in the alloy, as discussed before.

The mechanism of induced codeposition of W must account for the effect of pH on other complexes or reactions too. Addition of ammonia to citrate baths lowers the tungsten content in the alloy significantly. This implies that the deposition of Ni from complexes with NH_3 is faster than from its complexes with citrate. This statement is also supported by the fact that both partial current densities for deposition of Ni and W are decreased with large excess of citrate.[106] Similarly, increasing the temperature should drive part of the NH_3 out of solution, allowing the formation of more of the mixed-metal complex, thus increasing the concentration of W in the alloy. Hence, the effects of both temperature and pH reported in the literature for ammonia-containing baths may be related to changes in the concentration of NH_3 in solution rather than to the effect of temperature or pH on the electrodeposition reaction per se.[105] In addition, because an effect of deaeration was observed only at high pH values, where NH_3 exists in solution, it is most likely that oxygen concentration in solution and the associated oxygen reduction reaction do not play an important part in the induced codeposition of W. Figure 9 shows the decrease in the W-content of the alloy upon the addition of ammonia.[92] On the other hand, the FE is seen to increase. Hence, adding ammonia could improve the properties of the plating bath if a high concentration of W in the alloy is not needed. It should be pointed out that the addition of ammonia should decrease the concentration of W in the alloy for any mechanism of induced codeposition, as long as there is a synergistic effect of Ni^{2+} in solution upon the deposition of W from aqueous solutions containing WO_4^{2-}, which is a well-established experimental fact.

Further support for the role of the mixed-metal complex in induced codeposition of W is the measured effect of mass transport. Gileadi et al. used rotating cylinder electrode (RCE) experiments to study this effect in two ways – by varying the rotation rate at a constant current density, or by changing the current density at a constant rotation rate. The limiting current density at a rotating cylinder electrode, operating in the region of turbulent flow, is given by:

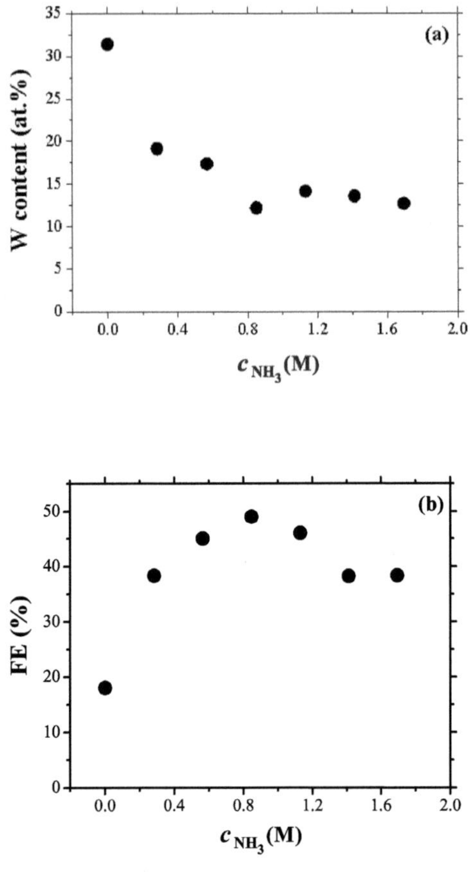

Figure 9. The effect of ammonia concentration in the plating bath on the W-content in the alloy (a), on the FE (b), and on the partial current densities for deposition of nickel (c) and tungsten (d). Plating conditions: 0.1 M $NiSO_4$, 0.5 M Na_2WO_4, 0.6 M Na_3Cit, pH = 8.0, i = 15 mA cm^{-2}, ω = 2,000 rpm. Reproduced with permission from Ref. 92. Copyright (2003) The Electrochemical Society.

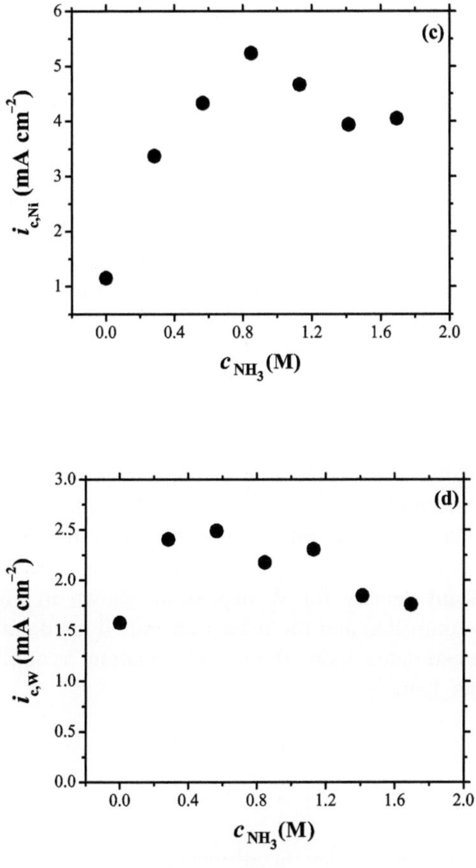

Figure 9. Continuation.

$$i_L = 0.079\, n\, F r^{0.4} D^{0.644} \nu^{-0.344} c_\infty \omega^{0.7} = B\omega^{0.7} \tag{37}$$

where r is the radius of the rotating cylinder (cm), ν is the kinematic viscosity (cm^2 s^{-1}), ω is the angular velocity (rad s^{-1}), and c_∞ is the bulk concentration of the electro-active species (mol cm^{-3}).

The use of baths with different ratios between the nickel and tungstate ions allowed concluding that the ion that is the minority component controls the rate of deposition of W. The current density calculated for deposition of tungsten was much lower than that expected, if WO_4^{2-} had been the electro-active species, indicating that either the formation of the mixed-metal complex or its activation-controlled rate of discharge is the slow process.[106]

The effect of rotation rate was studied in the range of 2,000 to 5,000 rpm, which represents a 90% (= $2.5^{0.7}$) increase in the rate of mass transport to a RCE. The effect of rotation rate on the deposition process is shown in Fig. 10. As the concentration of WO_4^{2-} is increased tenfold, from 0.04 to 0.40 M, the current density increases by a factor of only two. The limiting current density, calculated on the basis of the concentration of WO_4^{2-} in solution, is much higher than the partial current densities for deposition of this metal, so one would not expect a 40% increase of the rate of deposition of W with the increase of the rate of mass transport, as found experimentally. The explanation of these unexpected observations lies in the formation of the mixed-metal complex, as shown in Eq. (33). The concentration of this complex is low, and its rate of formation is also expected to be low. From the dependence of the partial current density for W deposition shown in Fig. 10a, the activation-controlled and the mass transport-limited current densities can be estimated, using the Levich equation, as applied to RCE experiments, namely

$$\frac{1}{i} = \frac{1}{i_{ac}} + \frac{1}{B\omega^{0.7}} \tag{38}$$

The concentration of the mixed-metal complex can now be obtained from the value of *B*. Neglecting the difference in diffusion coefficients (taking $D = 6.9\times10^{-6}$ cm^2 s^{-1}), this turns out to be 2.3 mM in a solution containing 0.04 M WO_4^{2-}. When the concentration of WO_4^{2-} is increased tenfold and that of Ni^{2+} is decreased by

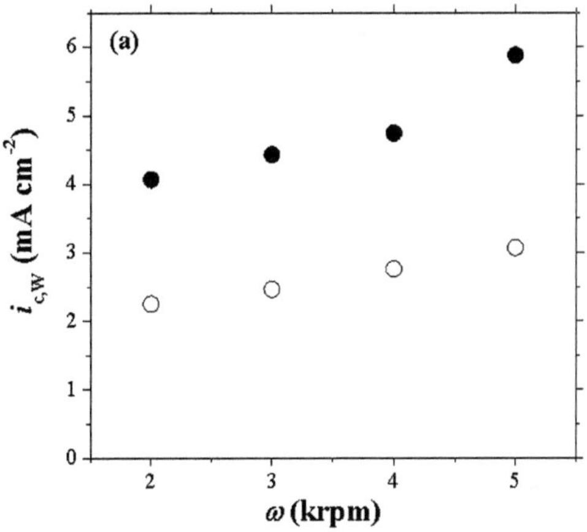

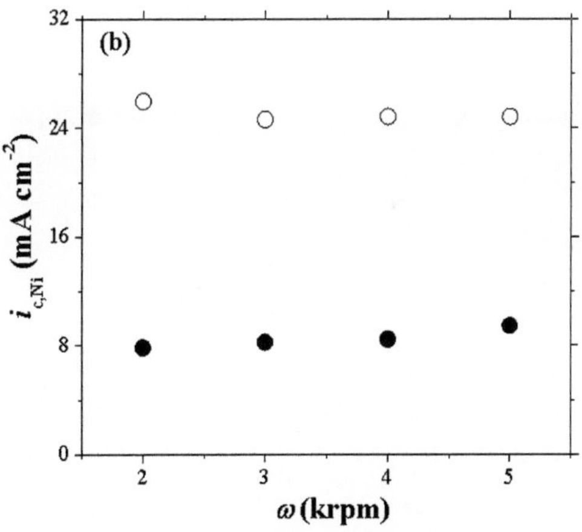

Figure 10. The effect of rotation rate on the partial current density for deposition of tungsten (a) and nickel (b), the W-content in the alloy (c), and the steady-state deposition potential (d). Plating conditions: excess of NH_4OH, 0.4 M Na_3Cit, pH = 8.0, i = 30 mA cm^{-2}. (•): 0.4 M Na_2WO_4, 0.04 M Ni-SO_4; (o): 0.04 M Na_2WO_4, 0.4 M $NiSO_4$. Reproduced with permission from Ref. 92. Copyright (2003) The Electrochemical Society.

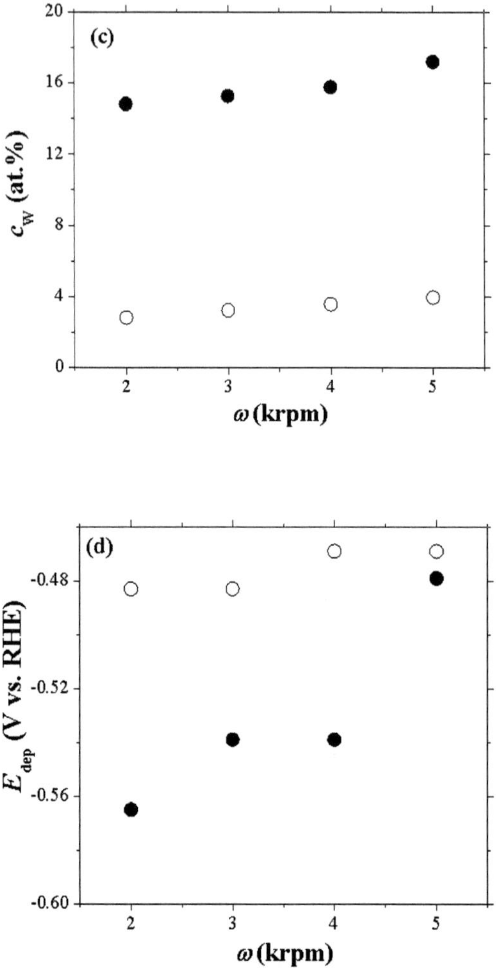

Figure 10. Continuation.

the same factor, both the activation-controlled current density and the limiting current density increase by a factor of 1.8 (from 5 to 9 and from 4.1 to 7.4 mA cm^{-2}, respectively, at 2,000 rpm). It may be concluded that the concentration of the precursor for the deposition of W increased by the same factor, to a value of 4.1 mM.

The partial current density for Ni deposition was found to be essentially independent of the rotation rate. An apparent anomaly is observed, however, even in this case, since increasing the concentration of Ni^{2+} in solution by a factor of 10 causes an increase of its partial current density only by a factor of 2–3, as seen in Fig. 10b. It should be recalled, however, that the Ni^{2+} ion can exist in solution in many different forms. There are six possible complexes with NH_3 and two complexes with Cit^{3-}, in addition to the mixed-metal complex with tungstate and citrate. The overpotential for deposition depends on the nature of each complex, and the relative abundance of the various complexes depends on the concentration of Ni^{2+}. Hence, the resulting partial current density for Ni deposition cannot be expected to depend linearly on concentration in such a complex system.[92,106]

In summary, even though the partial current density for deposition of tungsten $i_{c,W}$ is very small compared to the calculated limiting current density, $i_{L,W}$, a significant dependence of $i_{c,W}$ on the rate of mass transport is observed. This is taken as a strong indication that the electro-active species is indeed a mixed-metal complex of the type shown in Eq. (32), which exists in solution at a low concentration. On the other hand, $i_{c,Ni}$ is independent of rotation rate, indicating that this metal is also deposited by an independent parallel route from its complexes with either NH_3 or Cit^{3-}, which exists in solution at high concentrations.[93]

2. Molybdenum Alloys Containing Ni, Co and Fe

(*i*) *Properties of Molybdenum Alloys*

Molybdenum (Mo) is a silvery-white, hard, fairly ductile, refractory metal suitable for alloys that are required to exhibit a combination of high strength and rigidity at temperatures as high as 1,650°C. Selected properties of molybdenum are listed in Table 2, in comparison with the respective properties of tungsten and rhenium. Molybdenum is found in molybdenite (MoS_2) and wulfenite ($PbMoO_4$) ores. It is also recovered as a by-product of Cu and W mining. Molybdenum has the fifth highest melting point of all elements. Its electrical conductivity (30% IACS) is the highest of all refractory metals, about one third that of Cu, but higher than those of Ni, Pt and Hg. It has high-thermal conductivity—approximately 50% higher than that of Fe, steels or Ni-based alloys, and consequently finds wide usage as heat sinks. Its low thermal expansion coefficient plots almost linearly with temperature over a wide range, thus facilitating its use in bimetal thermocouples. Its mechanical properties include high tensile strength, high Young's modulus of elasticity, high hardness and toughness (it is softer and more ductile than W). Its high specific elastic modulus makes it attractive for applications that require both high stiffness and low weight. The high thermal conductivity, low coefficient of thermal expansion and low specific heat of Mo provide resistance to thermal shock and fatigue; these properties are also important in electronic applications. Molybdenum also exhibits good machinability and low vapor pressure at elevated temperatures. It is stable in a wide variety of chemical environments, including halogens, but oxidizes in air more readily than W at temperature higher than 760°C. Under these conditions, the oxide layer (MoO_3) sublimes, and the base metal is attacked. Therefore, Mo performs best in inert or vacuum environments. The corrosion resistance of Mo is very similar to that of W. It resists non-oxidizing mineral acids and reducing atmospheres containing hydrogen sulfide. It also offers excellent resistance to some liquid metals, including Li, Bi, Na and K, in the absence of oxygen.

(*ii*) *Applications of Molybdenum Alloys*

More Mo is consumed annually than any other refractory metal. The major use for Mo is as an alloying element in alloy steels, tool steels, stainless steels, Ni-based and Co-based superalloys. In these materials, it increases the hardenability, toughness, high-temperature strength, and corrosion resistance. Molybdenum is important in the missile industry, where it is used for high-temperature structural parts, such as nozzles, leading edges of control surfaces, support vanes, struts, reentry cones, radiation shields and heat sinks. In the electrical and electronic industries, Mo is used as cathode supports for radar devices, magnetron end hats, heat sinks for matching Si for semiconductor chip mounts, and sputtered layers for gates and interconnects on integrated circuit chips. Molybdenum has also been used in forging dies, rotating X-ray anodes in clinical diagnostics, furnace tubes, high-temperature solid lubricants (e.g., molybdenum sulfide), chemical processing equipment, flame retardants, light bulb filaments, inorganic paint pigments, chemical catalysts (e.g., in refining of petroleum), and scrubbers in flue gas desulfurization in coal-fired power stations. Finally, Mo alloyed with Re (e.g., Mo-41Re and Mo-47.5Re) is used in electronics, space programs and nuclear industries, while Mo-25Re alloys are used for rocket engine components and liquid-metal heat exchangers.

(iii) Electrodeposition of Molybdenum Alloys

A phenomenon of induced codeposition, similar to that discussed above for W, is observed when Mo is codeposited with iron-group metals. Similarly to tungsten, molybdenum cannot be deposited alone from aqueous solutions. Electrodeposition of Mo alloys exhibits similar dependencies on experimental variables as that of W. It should be noted that, although the two systems are very similar, some differences are found in the literature, as described bellow.

An early comprehensive review on electrodeposition of Mo alloys with iron-group metals was presented by Brenner.[78] The development of different baths, as well as the effect of operating conditions on the Mo content of the alloy is described in detail in that work. Electrodeposited alloys of Mo were claimed to be of limited practical value, because of their poor physical characteristics compared to the corresponding alloys of W.

The acid baths for electrodeposition of Mo alloys have been divided into two types: those that are wholly inorganic, and those that contain organic poly-acids as chelating agents. The addition of organic complexing agents, typically poly-hydroxy acids, considerably improved the quality and increased the Mo content of the electrodeposited alloys. This was attributed to complexing of the molybdate ion, the effectiveness being related to prevention of the formation and inclusion of partially reduced Mo compounds. The alkaline baths, which are considered to be preferential, were divided into several types, namely: ammoniacal baths, pyrophosphate, carbonate and caustic baths containing organic complexing agents. Representative bath compositions and operating conditions are listed in Appendix B. While Mo-Ni alloys were found to be the soundest among those of Mo-iron-group-metal alloys, their Mo content was the lowest.[78]

Ammoniacal baths have been found most suitable for codeposition of Mo alloys. The presence of free ammonia in the baths was claimed to be essential; similar to W codeposition – ammonia-containing baths had a higher FE. In contrast to W plating baths, there appeared to be no advantage of operating the Mo plating baths at elevated temperatures. Similar to W codeposition, the concentration of Mo in the deposit was higher than its metal ion concentration in the bath. For Mo-Ni, Mo-Co and Mo-Fe ammoniacal-citrate baths within the pH range of 2 to 12, the maximum Mo content of the deposit and minimum FE were observed around pH = 8. An increase in current density within the range of 10–200 mA cm^{-2} yielded a decrease in Mo concentration in Mo-Ni and Mo-Co alloys, in contrast to the trend reported for W alloys deposited from ammoniacal baths. As in the electrodeposition of W alloys, the current density-potential curves of the deposition of Mo-Ni and Mo-Co alloys occurred at less negative potentials than the curve for the iron-group metal alone.[78]

Newer studies have demonstrated that the properties of Ni-Mo alloys can be improved by application of pulsed electrodeposition.[145,146] Compared to DC electrodeposition, a higher Mo content in the alloy was obtained, which was accompanied by reduced residual stresses, improved mechanical properties and improved corrosion resistance.

According to Brenner, most of the theories proposed to explain the mechanism of induced codeposition of W can be applied also to induced codeposition of Mo, and vice versa. However, a theory that was found satisfactory to explain the induced codeposition of Mo from pyrophosphate bath, but would not be applicable to W codeposition, is that of Myers.[141] According to this model, a layer of hydroxide of Mo and the iron-group metal forms on the cathode. The hydroxide of the iron-group metal supposedly alters the permeability of the film so that it can be penetrated by molybdate ions. Otherwise, only hydrogen is discharged. Ernst and Holt suggested another theory.[147] According to this theory, Mo was first deposited as an oxide that, in the presence of the iron-group metal, could be reduced to metal by hydrogen. The number of unpaired electrons in the iron-group metal was considered as an index of its ability to form a bond with hydrogen and, thereby, reduce the Mo oxide.[78] However, the theoretical basis for this correlation could be questioned.

In recent years, several new papers have been published, attempting to explain the mechanism of induced codeposition of Mo with the iron-group metals. Podlaha and Landolt[95,96,148] studied the codeposition of Mo-Ni alloys from ammoniacal citrate solutions. They examined the influence of operating conditions on the composition of the alloy by means of rotating cylinder Hull (RCH) cell and RCE. The RCH cell is actually a cell with a rotating-cylinder working electrode, in which the anode and the cathode are concentric, and a third concentric cylinder, made of an electrically insulating material, is placed between them as shown in Fig. 11. This configuration forces a non-uniform current distribution on the cathode, thus allowing the evaluation of the properties of the elec-

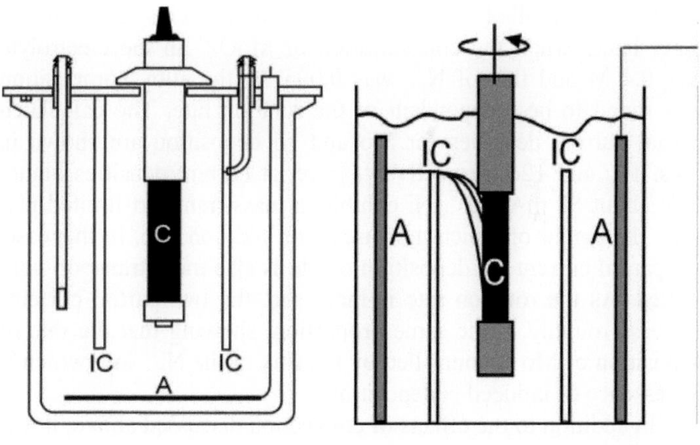

Figure 11. Illustration of two alternative designs for the rotating cylinder Hull (RCH) cell, which allows the study of non-uniform current distribution on the cathode, under controlled mass-transport conditions. A: anode, C: cathode, IC: insulating cylinder. Reproduced from Ref. 150 with kind permission of Springer Science and Business Media, and with permission from Ref. 95, Copyright (1996) The Electrochemical Society.

troplated film over a wide range of current densities in a single experiment. The RCH, developed by Landolt et al.[149-151] represents a significant improvement over the regular Hull cell, in that it allows measurements under controlled conditions of mass transport. In the cell used by these authors for the study of Mo deposition[95-97, 148] the radius of the rotating cylinder was 0.75 cm and the inner radius of the inert cylinder was 2.75 cm, thus leaving a space of 2 cm between them. The latter value is between two and three orders of magnitude larger than the thickness of the Nernst diffusion layer, ensuring a uniform rate of mass transport to the surface of the cylindrical cathode, while the current density on the cathode is non-uniform, of course. When the concentration of the Ni^{2+} ion in solution was much larger than that of MoO_4^{2-} (namely, 1.0 M $NiSO_4$, 0.005 M MoO_4^{2-}), the Mo-content in the alloy increased with rotation rate. This shows that a partial mass-transport limitation of the molybdate ion was involved, thereby lowering the Mo content in the deposit. From the measured alloy composition and film thickness, the partial current densities for Mo and Ni deposition were calculated. The results, shown in Fig. 12, demonstrate a rotation rate-dependent current plateau for Mo (Fig. 12a), and a largely activation-controlled reaction behavior for Ni (Fig. 12b). On the other hand, when the concentration of MoO_4^{2-} in the electrolyte was 0.4 M and that of Ni^{2+} was 0.005 M, the alloy composition was found to be independent of the rotation rate. The calculated partial current densities for Mo and Ni deposition are shown in Figs. 12c and 12d, respectively. Here, at current densities higher than about 50 mA cm^{-2}, Ni exhibits a mass-transport-limited plateau, the height of which increases with rotation rate. In this case, the partial current for deposition of Mo is also mass-transport controlled. As the rotation rate is increased, the two partial currents increase roughly in the same proportion, showing that the rate of deposition of Mo is controlled by the flux of the Ni^{2+} ion, which is the essence of induced codeposition.

In addition to the effects of convection discussed above, it was found[95] that increasing the concentration of the MoO_4^{2-} ion, the concentration of Cit^{3-} (up to a limit of 0.75 M) or the temperature resulted in an increase in the Mo-content in the deposit over a wide range of applied current densities. The concentration of ammonia was also found to have a significant effect—as it was increased

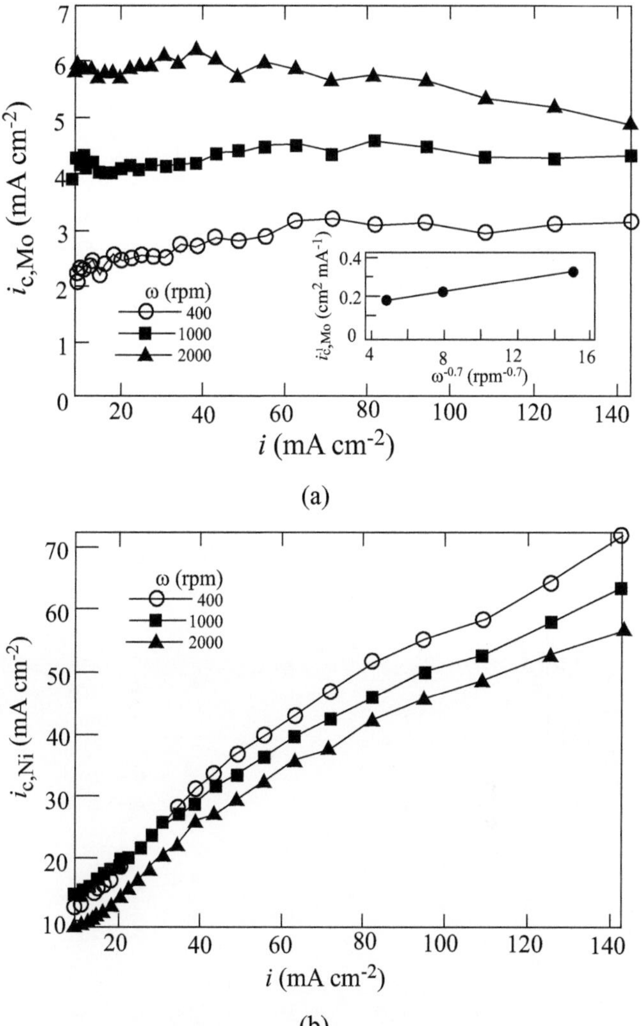

Figure 12. Partial current densities of Mo (a, c) and Ni (b, d) as a function of the approximate total current density for different rotation rates. The inset in (a) shows the inverse Mo partial current vs. $\omega^{-0.7}$. Electrolyte: 1.0 M Ni^{2+} and 0.005 M MoO_4^{2-} in (a) and (b); 0.005 M Ni^{2+} and 0.4 MoO_4^{2-} in (c) and (d). Reproduced with permission from Ref. 95, Copyright (1992), The Electrochemical Society.

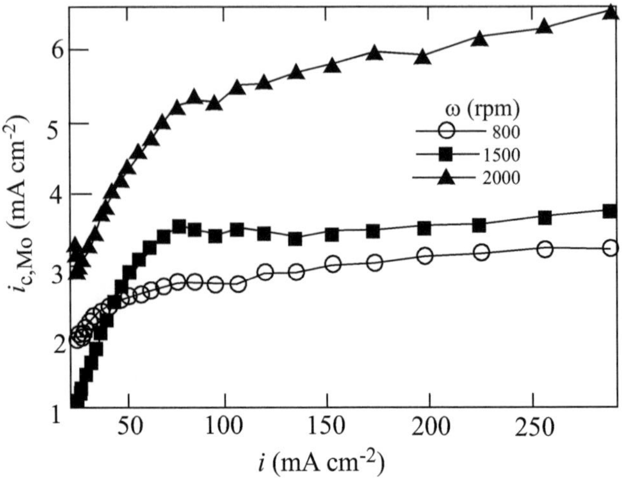

(c)

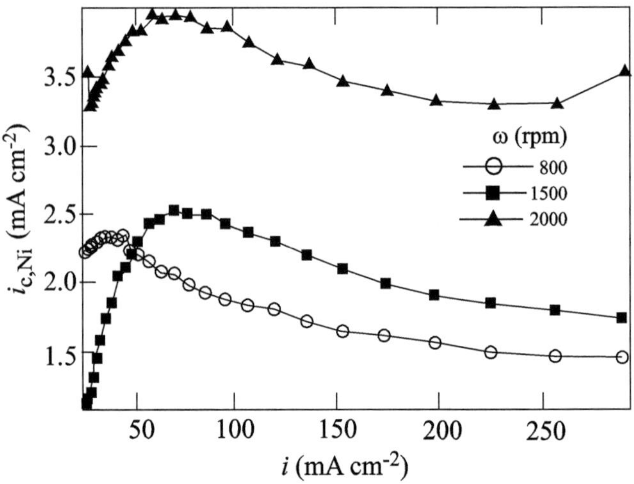

(d)

Figure 12. Continuation.

from 0 to 1 M, the FE increased, but the Mo-content in the deposits decreased.

On the basis of the observations described above, it was postulated that the precursor for the deposition of the Mo-Ni alloy was an adsorbed intermediate mixed-metal complex of the form $[\text{NiCit}(\text{MoO}_2)]^-_{\text{ads}}$. This intermediate can be reduced, thus allowing Mo deposition. The Ni^{2+} ion, complexed either with Cit^{3-} or with NH_3 (as discussed earlier for W codeposition), can be reduced in parallel with Mo, following the simple equations

$$[\text{NiCit}]^- + 2\,e^-_{\text{crys}} \rightarrow \text{Ni} + \text{Cit}^{3-} \tag{39}$$

$$[\text{Ni}(\text{NH}_3)_n]^{2+} + 2e^-_{\text{crys}} \rightarrow \text{Ni} + n\cdot\text{NH}_3 \tag{40}$$

The hydrogen evolution reaction is considered to be a side reaction, leading to a reduction of the FE to varying degrees, depending on the composition of the bath and the conditions of plating. In neutral or alkaline solutions this reaction can be written as

$$\text{H}_2\text{O} + e^-_{\text{crys}} \rightarrow \frac{1}{2}\text{H}_2 + \text{OH}^- \tag{41}$$

The reaction sequence leading to induced codeposition of Mo with Ni (as well as with Co and Fe) was assumed to be

$$[\text{NiCit}]^- + \text{MoO}_4^{2-} + 2\,\text{H}_2\text{O} + 2\,e^-_{\text{crys}} \rightarrow [(\text{NiCit})(\text{MoO}_2)]^-_{\text{ads}} + 4\,\text{OH}^- \tag{42}$$

$$[(\text{NiCit})(\text{MoO}_2)]^-_{\text{ads}} + 2\,\text{H}_2\text{O} + 4\,e^-_{\text{crys}} \rightarrow \text{Mo} + [\text{NiCit}]^- + 4\,\text{OH}^- \tag{43}$$

Thus, in solutions containing high $[\text{MoO}_4^{2-}]/[\text{Ni}^{2+}]$ ratio, the formation of the Ni-Mo intermediate is limited by transport of

$[NiCit]^-$ to the cathode surface. Because the deposition rate of the two metals is coupled, the alloy composition does not vary with rotation rate. In contrast, if the concentration of Ni^{2+} in solution is comparable to or higher than that of MoO_4^{2-}, the rate of formation of the Ni-Mo intermediate is limited by the transport of molybdate, while Ni can be deposited in parallel—its rate of deposition being independent of the rate of mass transport. Increasing either the rotation rate or the molybdate concentration—the partial current density of Mo deposition will also increase, while that of Ni deposition will not be affected.[95] Thus, it was proven beyond any doubt that the induced codeposition of Mo with Ni and other iron-group metals was dependent on the existence of the iron-group metal ions in solution.

However, the scheme proposed to explain the effect of Ni^{2+} ions on the codeposition of Mo (Eqs. 42 and 43) leaves some unanswered questions. Equation (39) shows the interaction between two soluble ions, leading to the two-electron reduction of one of them, and the formation of a mixed-metal complex adsorbed on the surface. All these processes cannot happen in a single step, but the authors did not specify what actually is assumed to happen: Is $[NiCit]^-$ first adsorbed on the surface and then acts as the catalyst for the reduction of MoO_4^{2-} to MoO_2? Or, perhaps, the two ions form a mixed-metal complex in solution, which is subsequently adsorbed and reduced?

Furthermore, $[NiCit]^-$ can readily be reduced to metallic nickel (c.f. Eq. 39) at the potential where the Ni-Mo alloy is formed. Yet, no such reduction is shown in Eqs. (42) and (43), in which Mo is first reduced from the hexavalent molybdate to the tetravalent molybdenum dioxide, and then further to metallic molybdenum. A solution, in which the concentrations of Ni^{2+} and Cit^{3-} were equal, was used and it was assumed that all citrate ions were bound to Ni in $[NiCit]^-$.[95] This assumption ignores the possible existence of complexes of the type $[(MoO_4)(Cit)(H)_m]^{-(5-m)}$ (written for short as $[1,1,m]^{-(5-m)}$, where m is the number of protons in the complex), in spite of the fact that such complexes are well known (cf., Fig. 13a). In addition, detailed calculation of the distribution of species in a Ni^{2+} and Cit^{3-} mixture shows that, when the ratio of the concentrations of the two ions is unity, there are still a few percents of free Ni^{2+} and Cit^{3-} ions not bound in the complex, as seen if Fig. 1b.

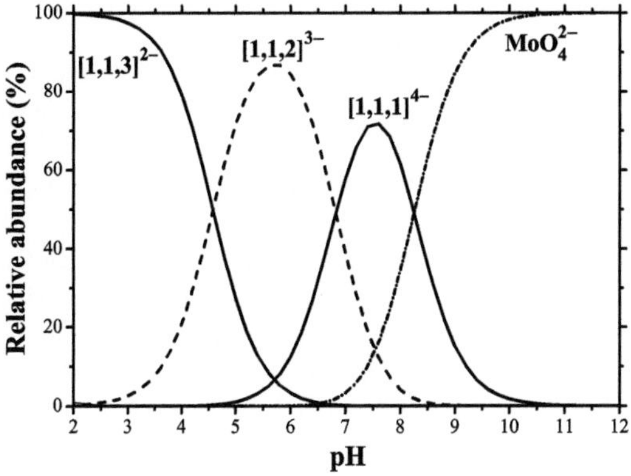

(a)

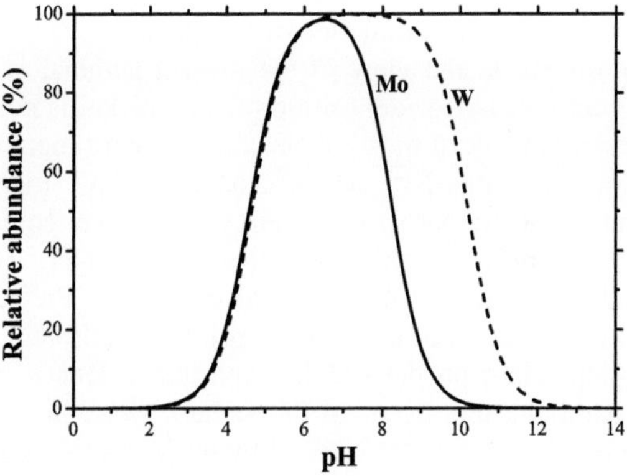

(b)

Figure 13. (a) The relative abundance of the protonated $[(MoO_4)(Cit)(H)_m]^{-(5-m)}$ complexes as a function of pH. The abbreviated forms $[1,1,1]^{4-}$, $[1,1,2]^{3-}$, and $[1,1,3]^{2-}$ refer to values of m = 1, 2 and 3, respectively, in the above complex. (b) Comparison between the sum of the concentrations of the complexes $[1,1,1]^{4-}$ and $[1,1,2]^{3-}$ of WO_4^{2-} and MoO_4^{2-} with Cit^{3-}, as a function of pH. The calculations are based on the data in Figs. 8 and 13.

In a following publication[96] a mathematical model was developed that was found to be in fairly good agreement with the experimental results. This, however, was based on a large number of adjustable parameters and several assumptions. Hence, this numerical agreement could at best be considered as consistent with the mechanism proposed, not proving the validity of the model. This model was extended by the same authors for the case of Mo alloys with Co and Fe.[97]

The shortcoming of the mathematical model in confirming the validity of the suggested physical model could best be demonstrated by the comparison of deposition of molybdenum ions with different iron-group metals.[97] In order to reduce the number of freely adjustable parameters, the authors plotted the partial current densities for Mo and Ni (or Co or Fe) as a function of potential, to obtain experimental values of respective Tafel slopes. The values obtained were very high, in the range of 0.36–0.59 V decade^{-1}, compared to typical values around 0.12 V decade^{-1} reported in the literature for deposition of the iron-group metals. No explanation was offered for this discrepancy and no mechanism was proposed that would explain such high Tafel slopes, which were subsequently used in the numerical simulation of the model.

The problem, in the view of the present authors, is that the partial current density for deposition of, say, nickel is determined from the total amount of nickel deposited per unit time. However, in a solution containing Ni^{2+}, MoO_4^{2-}, NH_3 and Cit^{3-}, there can be as many as nine different species from which nickel could be deposited (six complexes with 1–6 molecules of NH_3, two with citrate, and one adsorbed mixed-metal complex). The reversible potential for deposition of nickel is, in principle, different for each complex (depending on the stability constants). Hence, although all these parallel paths occur at the same *applied potential*, the *overpotential* is different for each of them. Moreover, there is no basis to assume that the exchange current densities or the Tafel slopes would be the same. If the observed Tafel plot would, nevertheless, be linear over at least two decades of current density, it could be argued that one of these parallel paths for deposition of nickel happens to be predominant. However, in the work quoted here, the apparent linearity of the Tafel plots extends only over a factor of about three in current density, namely over half a decade (cf., Fig. 4a in Ref. 97).

The difficulty in attempting to determine the mechanism of alloy deposition from the current-potential relationship observed in complex solutions, which sometimes contain more than one ligand, was alluded to in the introduction to this chapter (cf., Section 1.2.2). The comments made here are not meant to criticize the experimental work presented in these papers in the field of induced codeposition of Mo with iron-group metals. It is only given to show the limits of validity of mathematical models, particularly when the solution is complex and the number of freely adjustable parameters is large.

Figure 13a shows the calculated concentration distributions for molybdate-citrate complexes as a function of pH. This figure is analogous to Fig. 8 for tungstate-citrate complexes. The two figures are similar, but not identical. In particular, it is noted that the $[1,1,1]^{4-}$ species is predominant in the case of tungstate in the range of pH = 6.8–10.2. For molybdate, it reaches a maximum at pH = 7.6, but even there it only constitutes about 72% of all the species in solution. In Fig. 13b, the sum concentrations of $[1,1,1]^{4-}$ and $[1,1,2]^{3-}$ are compared. In studies of induced codeposition of Ni-W alloys it was found that the W-content of the alloy, measured as a function of pH, followed rather closely the variation of the relative abundance of the above sum of concentrations with pH, with the highest W concentration obtained at pH = 7 and a sharp decline above pH = 9.0. [93] If the same argument holds for induced codeposition of Mo, then the highest Mo concentration should be obtained at pH = 6.6, with a sharp decline at about pH = 8.3.

Gómez et al.[152] electrodeposited Co-Mo magnetic alloys from a sulfate-citrate bath on carbon electrodes. Although the focus of their paper was not on elucidating the mechanism of induced codeposition, it was suggested that hydrogen could not be responsible for the deposition of Mo in the Co-Mo system, because its concentration was fairly low and because another mechanism should explain the need for citrate or polycarboxylate anions in solution. The deposition process was found to be favored when molybdate was present in solution, even at very low concentrations. Hence, the authors adopted the model of Podlaha and Landolt, according to which a mixed-metal complex of cobalt(II), citrate and molybdenum dioxide is adsorbed on the surface and promotes Mo reduction.

Another mechanism for induced codeposition of Mo was suggested by Chassaing et al.[153] for electrodeposition of Mo-Ni alloys from citrate-ammonia electrolytes. Electrochemical impedance spectroscopy (EIS) measurements were carried out in order to better understand the different reactions occurring on the electrode surface during deposition. The proposed mechanism is based on a multi-step reduction of molybdate species. A MoO_2 layer is formed via reduction of molybdate ion as in Eq. (42). Then, if free Ni^{2+} is present in solution, this oxide can first combine at low polarization with Ni, following the reduction reaction:

$$MoO_2 + 4\,Ni^{2+} + 8\,e^-_{crys} \rightarrow MoO_2Ni_4 \tag{44}$$

A 25 nm-thick layer with a Ni/Mo atomic ratio of 4:1 was indeed detected by AES and EDS. The mixed Ni-Mo oxide is claimed to catalyze the HER. Next, when the polarization is increased, the mixed oxide is further reduced to a surface compound that inhibits the HER and is more catalytic for reduction of molybdate and nickel ions, forming an alloy according to the reaction:

$$MoO_4^{2-} + 3\,Ni^{2+} + 8\,e^-_{crys} + \;4 + n\cdot H_{ads} \rightarrow MoNi_3 + 4\,OH^- + n\cdot H_{abs} \tag{45}$$

The HER is claimed to take place in this case due to the alcohol functional group of the citrate ion, which is an unlikely reaction in solutions at pH 9.5. In the third stage, at still higher polarization, the Mo discharge is limited by diffusion, due to the low molybdate concentration of the electrolyte, thus decreasing the Mo content in the deposited alloy.

Several arguments against the last theory may be raised. First, the proposed mixed oxide contains nickel atoms that are fully reduced. If so, what makes this oxide stable? Secondly, no direct observation was provided for the reduction of this oxide into an alloy. Thirdly, the origin of the adsorbed hydrogen atoms included in Eq. (45) is not clear. Finally, it should be obvious that Eqs. (44) and (45) cannot be considered as elementary steps in the reaction sequence. Four Ni^{2+} ions could not be possibly reduced simultaneously. Neither could the eight-electron reduction of MoO_4^{2-} + $3Ni^{2+}$ occur in one step. Thus, there seem to be absolutely no basis to claim that the above sequence represents the mechanism of formation of the Mo-Ni alloy.

The formation of an intermediate, which is then reduced to form Mo—with either Ni or Fe acting as a catalyst, was also claimed, based on in-situ Raman spectroscopy studies.[154,155] Although the exact composition of the intermediate was not identified in these studies, it was argued that at low cathodic polarization, the main species on the electrode surface were polymolybdates, that could be reduced to Mo(IV) at a higher cathodic polarization. The species of Mo(IV) could be further reduced to Mo atoms only when cations of the iron-group metal were present in the electrolyte.

Crousier et al.[156] examined the role of hydrogen evolution in the process of deposition of Mo-Ni alloys on different substrates (glassy carbon, Ni and Pd). It was found that on carbon and Ni substrates, bright and smooth deposits were formed, while on Pd no alloy was formed. This observation was related to easy absorption and diffusion of atomic hydrogen into Pd, which prevented its availability for the alloy codeposition process. Hence, it was concluded that hydrogen plays an important role in the codeposition of the alloy. This conclusion of the authors is, however, not convincing. Firstly, it is known that hydrogen atoms can also permeate into Ni to some extent. Secondly, unsuccessful attempt to deposit Mo-Ni alloys on Pd may also be attributed, for example, to kinetic limitations.

3. Rhenium and its Alloys

(*i*) *Properties of Rhenium and its Alloys*

Rhenium (Re) differs from the other refractory metals (Nb, Ta, Mo and W) in that it has an hcp structure, and does not form carbides. Because it does not have a ductile-to-brittle transition temperature, Re retains its ductility from subzero to high temperatures. In addition, it can be mechanically formed and shaped to some degree at room temperature. It also has a very high modulus of elasticity that, among metals, is second only to those of Ir and Os. Compared with other refractory metals, Re has superior tensile strength and creep-rupture strength over a wide temperature range.

Among all elements, except W and C, Re has the highest melting point. It also has high electrical resistivity over a wide temperature range, and less volatilization in vacuum compared to W. At elevated temperature, Re resists attack in hydrogen and inert atmospheres. It is resistant to hydrochloric acid and seawater corrosion, and to the mechanical effects of electrical erosion (i.e., extensive destruction of an electrode in some forms of electrical discharge). Rhenium is nowadays available mainly as perrhenic acid ($HReO_4$), ammonium perrhenate (NH_4ReO_4), and metal powder.

The major drawbacks of Re include high density (that is exceeded only by those of Os, Ir and Pt), limited availability (that delayed its discovery until 1925), high cost of the raw material, and the need for intricate processing methods (e.g., because of its very high strain-hardening rate), which results in high fabrication costs. Rhenium is oxidized catastrophically in moist air at temperatures above 600°C; oxidation occurs as a result of the formation of rhenium heptoxide (Re_2O_7). In order to protect rhenium from oxidation at high temperatures, iridium (Ir) is often used as an oxidation-resistant coating.[157-161]

(*ii*) *Applications of Rhenium and its Alloys*

The scarcity and high cost of Re have limited its use. Nevertheless, its unique properties can be useful in important, albeit special, applications. Rhenium alloys are used in nuclear reactors, semiconductors, electronic tube components, thermocouples, gyroscopes, miniature rockets, electrical contacts, thermionic converters (i.e., direct producers of electric power from heat by thermionic electron emission), etc. Platinum-rhenium reforming catalysts are probably still the major end-use products of Re. They are used for the hydrogenation of fine chemicals and for hydrocracking, reforming and disproportionating of olefins, including increasing the octane rating in the production of lead-free petroleum products.[157,158] Rhenium has several properties of potential interest to the electronics industry, particularly its very high melting point and low volatility. In the electrical contact industry, Re is used because of its wear resistance and its ability to withstand arc erosion,[158] and since the oxides formed on its surface are relatively good conductors.[162] Thermocouples made of Re-W alloys are used for measuring temperatures up to 2,200°C, and Re igniter wires are used in photoflash bulbs for photography. Rhenium alloys are also used in grid heaters and cathode cups. The metal is used as a filament material for mass spectrographs and ion gages, because of its high electrical resistivity and low vapor pressures at high temperatures.[158] Tungsten-rhenium alloys are used to coat the surface of Mo targets in X-ray tube manufacture. Other applications of Re include heating elements, metallic coatings, alloying of W- and Mo-based alloys as well as of superalloys, fuel cell electrodes, and coordination compounds in radiopharmacy.[158,163,164]

Since the 1990s, much work has been focused on developing Re-based parts for aerospace applications. For example, Re has been considered for use in divert and attitude propulsion subsystems. Recession rates of Ir-lined Re thrust chambers have been estimated to be 0.42 $\mu m\ h^{-1}$ at 2,200°C.[165] In addition, Re wets nicely carbon and does not form carbides, which makes it attractive as a coating material for carbon-carbon composites.[166] In the absence of a protective atmosphere, a protective coating (e.g., of Ir, Pt or Rh), to inhibit oxygen attack, can be applied on top of Re.[157,158] A combustion chamber material system composed of a Re substrate and an Ir coating has already been proven to operate for longer lifetimes at higher temperatures of up to 2,200°C (compared to the state-of-the-art material system for radiation-cooled rockets, which is a C103 niobium alloy coated with R512E fused silica). The added thermal margin afforded by Ir-coated Re allows the virtual elimination of film cooling, leading to higher performance and cleaner spacecraft environments.[167]

Most of the published reports to-date deal with fabrication of Re-based items by chemical vapor deposition (CVD), as compared to any other coating process. Reed et al.[167] claimed that CVD is the only established process, by the year 1998, for fabricating Ir/Re combustion chambers. However, delamination-related failures have been reported on Re-coated components made by CVD. Fortunately, electroplating at near-room temperature may become a successful alternative and allow for fairly uniform Re coatings on such complex shapes. Although Re may be deposited as pure metal, binary Re-Rh or Re-Ir coatings should possess higher oxidation resistance and thermal stability. The second metal can also aid in healing of micro-cracks that form in the Re coating during electrodeposition or service (as discussed in the next section).

(*iii*) *Electrodeposition of Rhenium and its Alloys*

Compared to tungsten and molybdenum, much less information is available on electrodeposition of rhenium, possibly due to the confidentiality of many related projects worldwide. Furthermore, most studies have focused on the technological aspects, properties and applications of Re electrodeposition, while overlooking its electrochemistry. Rhenium belongs to a group of metals that are difficult to obtain by electrolysis of their aqueous solutions. This behavior has been attributed mainly to its very low overpotential for hydrogen evolution. For example, in a 2 N sulfuric acid solution, at a cathodic current density of 5 mA cm^{-2}, the overpotential for hydrogen evolution on Re was estimated to be no more than 90 mV.[160] In addition, because no simple Re(VII) salt is soluble in water without hydrolysis, it was further suggested that Re^{z+} ions are unlikely to exist in solution, or can exist only in very low concentrations.[162]

Figure 14 shows the Pourbaix equilibrium diagram for the Re-H_2O system at 25°C. In this figure, the lines for hydrogen and

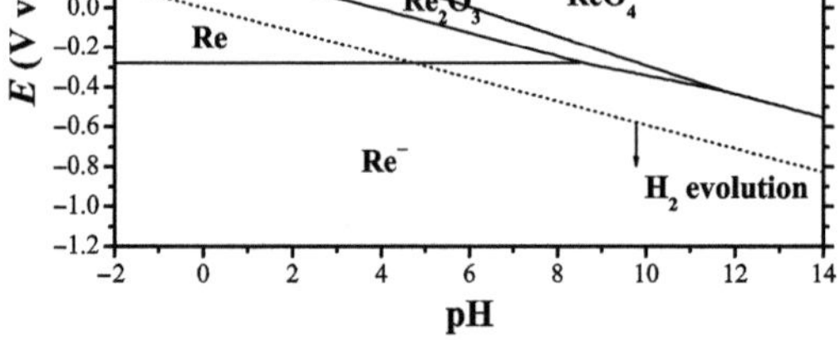

Figure 14. Potential-pH equilibrium diagram for the rhenium-water system at 25°C.

oxygen evolution are also shown. Between these two lines, water is stable. Figure 14 comprises an anodic corrosion domain (corresponding to the formation of ReO_4^-), a cathodic corrosion domain (formation of Re^-), an immunity domain (formation of Re), and a passivation domain (formation of more or less hydrated Re_2O_3). This figure is valid only in the absence of substances with which Re can form soluble complexes or insoluble salts. Metallic Re, shiny-white in the compact state and gray-black in the dispersed state, is a relatively base metal, having a very small zone of stability in common with that of water. Rhenium can be obtained, possibly mixed with some lower oxides, by the electrolysis of near-neutral or acidic solutions of perrhenates or perrhenic acid. It is obtained in a finely divided gray-black form when using current densities between 10 and 100 mA cm^{-2}. At higher current densities, lead-colored tree-like growths are formed at the cathode; these are thought to be compounds of metallic Re with brown hydrated lower oxides. Rhenium can also be produced by chemical reduction, but only with difficulty. More powerful reducing agents than zinc, such as stannous chloride ($SnCl_2$) and hydriodic acid (HI), reduce heptavalent rhenium only to the tetravalent state. Rhenium is easily oxidized in alkaline solutions, with the formation of various oxides (e.g., Re_2O_3, ReO_2, and ReO_3), as well as perrhenic acid and soluble perrhenates. In addition, it is easily reduced with the formation of soluble rhenide (Re^-). In the presence of moist air, Re is oxidized to perrhenic acid, even at room temperature. This oxidation is very rapid in the case of finely divided Re, which is occasionally pyrophoric. The oxidation of Re, even when compact, is rapid in the presence of solutions of hydrogen peroxide, alkaline or acid oxidizing solutions, solutions of nitrites/nitrates or alkali metal peroxides, and nitric acid solutions. Concentrated sulfuric acid also dissolves Re, but more slowly.[168] From Fig. 14 it may be concluded that a fine control of both potential and pH is required in order to form metallic rhenium without significant hydrogen evolution, which might result in hydrogen embrittlement and cracking.

Electrolysis has been applied primarily to extract metallic Re from solutions, and to produce Re coatings. In electrolytic extraction, the metal can be obtained in the form of a bright deposit or a black powder, depending on the conditions of electrolysis.[160] The overall reaction in a potassium perrhenate ($KReO_4$) solution was represented as:[162]

$$4\,HReO_4 + 2\,H_2O \rightarrow 4\,Re + 4\,KOH + 7\,O_2 \tag{46}$$

Different mechanisms have been proposed for the electrodeposition of Re, involving the stepwise reduction of the ReO_4^- ion, the role of adsorbed hydrogen atoms, following initial reduction to ReO_3, and even formation of the highly charged Re^{7+} ion as an intermediate.[160,162] No solid evidence was given for any of the reduction routes proposed; however, it is evident that the seven-electron reduction of ReO_4^- and the parallel removal of four oxygen atoms must involve many consecutive steps, even if the nature of the intermediates formed, some in solution and some perhaps adsorbed on the surface, is unknown. It should be noted, however, that a mechanism postulating the reduction of a heptavalent cation (Re^{7+}) can be rejected out of hand, since such ions are not known to exist in aqueous solutions. Moreover, the solvation energy of ions in aqueous solutions is roughly proportional to the square of the charge (namely, 5, 20 and 50 eV for Me^+, Me^{2+} and Me^{3+}, respectively). Extrapolating to Me^{7+} would lead to a solvation energy of 200–250 eV. Thus, if a Re^{7+} ion did exist in solution, it would be energetically impossible to remove its solvation shell and deposit metallic rhenium from it.

An early, relatively short, review on electrodeposition of Re alloys with iron-group metals was presented by Brenner.[78] In an acid perrhenate solution, the overall reaction can be written as

$$ReO_4^- + 8\,H^+ + 7\,e^-_{crys} \rightarrow Re + 4\,H_2O \tag{47a}$$

with a standard potential of +0.34 V vs. SHE,[169] which is more noble than that of hydrogen. In alkaline solutions, on the other hand, the overall reaction can be written as

$$ReO_4^- + 4\,H_2O + 7\,e^-_{crys} \rightarrow Re + 8\,(OH)^- \tag{47b}$$

and the corresponding standard potential is –0.604 V vs. SHE.[169] This value is approximately +0.21 V vs. RHE in a 1 M NaOH solution (pH 13.8). On the basis of the standard potentials, it should be possible to reduce Re from either aqueous or alkaline solutions. Yet, in practice the plating process takes place with a very low FE, of the order of 10%. Hence, Brenner concluded that the deposition of Re is subjected to some restraints, similar (but to a lesser degree) to those observed in deposition of W and Mo.

Typical bath compositions and operating conditions for electrodeposition of Re are listed in Appendix C. An acid sulfate solution, based on potassium perrhenate, was suggested by Fink and Deren.[170] Netherton and Holt[171] worked with similar baths, to which they added either citric acid or ammoniacal citrate.* Sligh and Brenner used more concentrated solutions of perrhenate, but still could not increase the FE significantly. Their deposits exfoliated when produced thicker than 10 μm, and oxidized rapidly upon exposure to air. Therefore, it was hypothesized that the electrodeposited metal was not pure, but contained oxide inclusions.[78]

Several recommendations regarding the electrodeposition process were later drawn by other authors. According to Lebedev,[160] the most widely used baths contain potassium perrhenate and free sulfuric acid, whereas Pt is the commonest anode material. One limitation in the use of potassium perrhenate is its low solubility in water (0.037 M at 21.5°C, compared to 0.215 M and 3.45 M at 20°C for ammonium and sodium perrhenates, respectively). However, its solubility is increased in acidic solutions. The deposition potential of Re was found to depend on the material of the cathode. On a Re cathode, the deposition potential was the less negative, with the lowest overpotential for hydrogen evolution. Additions of ammonium sulfate {$(NH_4)_2SO_4$} and gelatin to the bath resulted in an increase in the FE. In contrast to Lebedev's claim, a preliminary work of Treska et al.[166] recently demonstrated that the use of ammonium perrhenate, instead of potassium perrhenate, could result in fewer surface cracks, a higher metallic Re content in the coating, with no traces of undesired potassium contamination. Colton[159] claimed, on the basis of previous studies, that sulfuric acid solutions yield the best Re coatings. In addition, bet-

---

* It should be noted that at low pH citric acid exists in its fully protonated, uncharged form (cf., Fig. 1a). In this form, it is not a good complexing agent.

ter deposits were obtained if only a thin film of Re was deposited, followed by annealing at high temperature. Additional deposits of Re could then be added in a similar manner, forming a heavy deposit that had good chemical and mechanical stability. Meyer[162] suggested that the deposit properties could be improved by adding magnesium sulfamate {$Mg[SO_3(NH_2)]_2$}, which reduced the residual stresses, thus permitting the formation of thicker deposits at high current densities. It was also shown that addition of ammonium sulfate increased the conductivity and FE, particularly at low current densities. Lowering the pH to 1.0–1.5, by addition of sulfuric acid, resulted in higher FE and brighter deposits. The dependencies of FE on pH, current density and temperatures were also studied. While, at i = 100 mA cm^{-2} and T = 60°C, the deposits were bright and reasonably ductile in the range pH = 0.5–2.0, at higher pH the deposits were darker and with higher level of residual stresses. It was realized that Re has a very low hydrogen overpotential, which is comparable to that of electrodeposited Pt. It was also found that Re could be easily codeposited with several other metals, forming thicker, low-stressed alloy deposits at a much higher FE (as high as 80%, instead of 15% for the pure metal). This behavior was related to a decreased rate of hydrogen evolution. Such alloy coatings typically possess higher oxidation resistance and thermal stability, with a much lower crack density. Although Meyer himself claimed to be able to deposit Re of good quality, he criticized that previously-electrodeposited Re had been highly stressed, brittle, and of poor appearance – particularly when the thickness of the deposit exceeded 1 μm.[162]

Root and Beach[172] also found that the as-deposited Re was not stable in air and moisture, because of the inclusion of rhenium hydride impurities. The quantity of hydride depends on the pH of the bath and the cathode potential.[172, 173] Rhenium hydride is unstable in aqueous solutions and in moist air, and gradually transforms into rhenium oxides. After a time, the hydride-containing surface layer could contain some fine, powdery rhenium oxides, along with the metal. This could hinder the ability of the coating to act as a suitable oxidation barrier. In addition, residual stresses and cracks might form due to the presence of the brittle hydride phase per se. If small oxide impurities exist in the coating, they can readily be transformed into a tarnish-resistant metallic Re by annealing in pure hydrogen at 900–1,000°C for at least 15 min,[172] but preferably for 1–1.5 h.[173] Temperatures lower than 800°C and other annealing environments were found ineffective. This treatment above the hydride decomposition temperature, however, might introduce some cracks due to shrinkage of the coating during conversion of residual hydride to metallic Re. These cracks can allow oxidizing gases to gain access to the substrate. One way to heal these cracks is to fill them with additional metal, e.g., Ir or Rh. The best way to eliminate (or, at least, control) this problem is to establish optimum bath composition and operating conditions, so that the as-deposited coating would be hydride-free (or near hydride-free).[173] Figure 15 shows the transformation from an amorphous hydride phase into crystalline Re, following hydrogen firing.[166] In order to produce Re deposits thicker than 0.5 mil (12.7 μm), Root and Beach[172] had to employ alternate plating and hydrogen firing. This inability to deposit thick Re coating in one step was related to a lower overpotential of hydrogen on rhenium hydride, as compared to that on other metal surfaces. Hydrogen-fired Re deposits were found to be disconti-

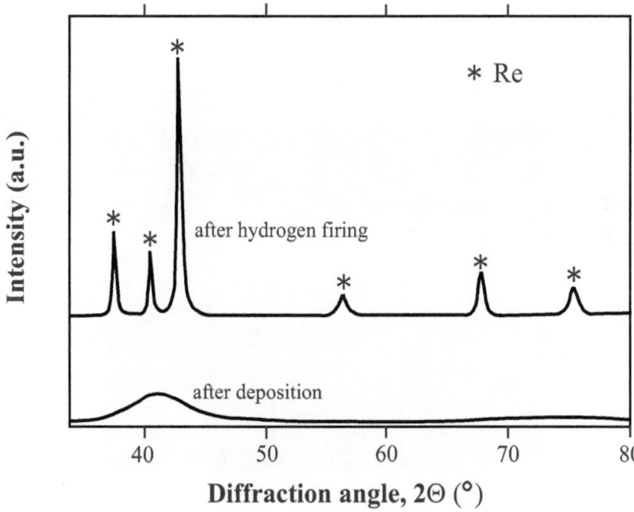

Figure 15. Phase transformation from amorphous rhenium hydride to crystalline Re following hydrogen annealing (T = 900-950°C, t = 1–1.5 h). Reprinted from Ref. 166, Copyright (1997) with permission from the Institute of Materials, Minerals and Mining.

nuous. This characteristic was related to shrinkage of the electrodeposited rhenium hydride, forming rhenium and hydrogen. Since the plating was postulated to involve the deposition of rhenium hydride, it was suggested that conditions, which minimize hydrogen absorption in the cathode, would reduce the FE. The FE was found to decrease due to periodic reverse current plating and to increase as a result of either addition of ammonium nitrate to the bath or increase of bath temperature.

One of the authors of this chapter (Eliaz) was involved for a short period of time in a related project that focused on electrodeposition of pure Re on parts made from graphite or carbon-carbon composites for aerospace applications. It was shown that, in the absence of fine pH and potential control, hydride formation led to significant cracking—mainly around carbon fibers (see Fig. 16). Preliminary runs of potentiodynamic polarization experiments were made to demonstrate that fine control of the potential, to within the immunity domain in the Pourbaix diagram, may aid in reducing the absorption of hydrogen and related cracking. However, neither optimization of the bath chemistry and operating conditions nor attempts to form Re-Rh alloy codeposition were made in that part of the research. Yet, it was found that better qual-

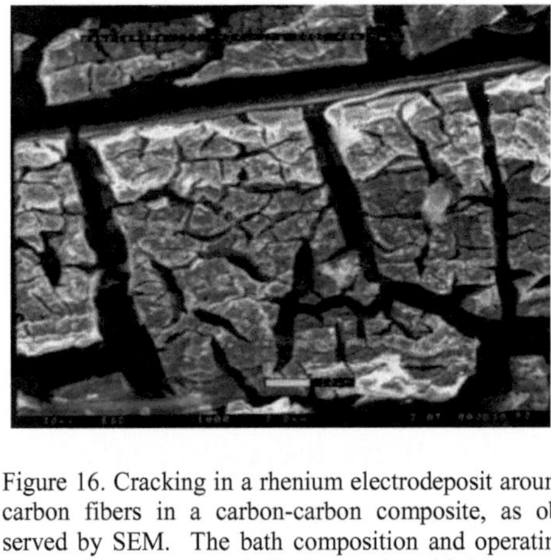

Figure 16. Cracking in a rhenium electrodeposit around carbon fibers in a carbon-carbon composite, as observed by SEM. The bath composition and operating conditions were not optimized; therefore, high contents of hydrogen resulted in cracking.[173]

ity coatings could be formed at high current densities. Pulse shapes that typically yielded reasonable results included 6-8 pulses forward (negative, plating, 140 mA each, 1.2 ms on, 4 ms off) and 1 reverse pulse (positive, metal-bridge dissolution, 10-mA peak, 9 ms on). Complex shapes (including wedge, throat and sphere) were coated, yielding nice visual appearance. The infiltrated and coated outer surfaces of these shapes were subsequently stabilized by an elevated temperature hydrogen anneal.[173]

Pemsler et al.[174,175] earlier introduced the concept of carbon fiber reinforced internal barrier, where diffusion barrier metals are infiltrated electrochemically into matrix-free carbon fibers in a specially oxidized surface ply of C-C. Infiltration was carried out successfully with Re, Rh and Ir. The rhodium coatings were found non-cracked, adherent, and survived thermal cycles to as high as 1,600°C.

Paris[176] described a method for Re electroplating of the hot plate used for contact ionization of barium in a Q-plasma source. In this case, the reason for coating the tungsten hot plate with rhenium was to increase the probability of ionizing barium (W and Re have work functions of 4.52 and 4.8–5.1 eV, respectively, whereas the ionization potential of Ba is 5.21 eV). Paris discussed some of the technical issues involved in the electrodeposition process, including the effectiveness of a slow rotation of the plate during deposition and a gentle removal of the bubbles from the solution by means of stirrers in producing more uniform coatings, short current reversal to remove roughness of non-uniform plating, and post-treatment in distilled water—followed by drying in an oven at 50–60°C for at least one hour—to prevent peeling of the Re deposit due to exposure to humid air.

Since the early 1990s, Bakos, Horányi, Szabó et al. have had a major contribution on the way to understanding the electrochemistry of Re.[177-185] The electrodeposition of metallic Re from aqueous perrhenate solutions on Au was found to take place at potentials within the hydrogen evolution region.[178] Next, potential oscillations during galvanostatic electrodeposition of Re from ReO_4^- species dissolved in perchloric acid ($HClO_4$) supporting electrolyte were reported.[179] No similar behavior was observed when sulfuric acid was used as the supporting electrolyte. Therefore, the oscillatory phenomenon was related to the reduction of ClO_4^- ions. In a following paper,[180] the electrodeposition of Re from ReO_4^- species dissolved in $HClO_4$ and H_2SO_4 supporting electrolytes was studied by coupling electrochemical and radiochemical (namely, backscattering of β^- radiation) methods. The rate of Re deposition at a given potential was very low in sulfuric acid compared to that in perchloric acid. It was suggested that deposition of metallic Re took place by at least two fundamental steps:

Step 1: formation of an oxide (or oxidized) layer, possibly ReO_2; and
Step 2: reduction of this layer to metallic Re.

The ratio between the rates of the two steps was expected to be potential and time dependent; the more negative the potential, the lower the ratio of the rate of Step 2 to that of Step 1.

The next paper dealt with Re deposition on Pt by reduction of perhhenic acid with methanol.[181] It had been suggested earlier[182] that methanol could act as the reducing agent, instead of preadsorbed hydrogen, in Re deposition on Pt surface, in the absence of electric polarization. In Ref. 177, pure Re was said to have been deposited via ReO_4^- reduction reaction with adsorbed hydrogen, which was one of the products of decomposition of methanol. It was concluded that this method of Re deposition resulted in the same adsorbed rhenium species as in reduction of ReO_4^- ions either with adsorbed hydrogen or with electric polarization. Almost the same amount of Re could be deposited via ionization of preadsorbed hydrogen as by reduction with methanol. It was later suggested[183] that during deposition of Re on Pt, ReO_2 adsorption took place prior to adsorption of atomic hydrogen, and consequently the hydrogen overpotential was decreased. After a monolayer had been formed, bulk phases of ReO_2 and ReO_3 could also be observed.

In a subsequent paper,[184] electrodeposition of rhenium species from sulfuric acid solutions of perhhenic acid onto polycrystalline Pt and Au surfaces was carried out both in underpotential and overpotential regimes. Metallic Re could be obtained by applying relatively high cathodic current densities, whereas Re(IV) compounds were more likely to form in reduction of ReO_4^- at low current densities. Lastly, the deposition of rhenium species on Au from $Re_2O_7^-$ containing sulfuric acid solutions was studied.[185] It was shown that the chemical nature of electrodeposited rhenium species depends on the concentration of H_2SO_4 in the supporting electrolyte, from which deposition took place. From H_2SO_4 solutions at pH ≈ 0.9, metallic rhenium was deposited and later oxidized at 0.67–0.7 V vs. RHE, which is far above the standard potential of +0.34 V vs. SHE for Eq. (47a).[169]

Zerbino et al.[186] used in-situ ellipsometry to study the initial stages of deposition of Re, comparing the effect of the substrate (Au and Pt). It was concluded that, on Au, a monolayer of Re was formed in parallel with hydrogen evolution. On Pt, a monolayer could be deposited in the region of formation of adsorbed atomic hydrogen (at +0.1 V vs. RHE). When the potential was shifted to –0.1 V vs. RHE, layers of metallic Re as thick as 5–30 nm were deposited.

Schrebler et al.[187] studied the nucleation and growth mechanisms for Re deposition on polycrystalline Au electrodes, from a bath containing 0.75 mM perrhenic acid and 0.1 M sodium sulfate at pH = 2. The potentiostatic step technique was simultaneously employed with measurements of mass changes in an electrochemical quartz-crystal microbalance. The mass vs. time transients were fitted with equations deduced from the current versus time relationships of the conventional nucleation and growth models. It was concluded that electrodeposition of Re started with progressive nucleation and two-dimensional growth, followed by two other contributions:

1. progressive nucleation, and three-dimensional growth under diffusion control,
2. progressive nucleation and three-dimensional growth under charge-transfer control, which was observed at longer times.

From these three contributions, the progressive nucleation and two-dimensional growth corresponded to the charge of a monolayer, and were attributed to two-dimensional nuclei of Re produced by the reduction of adsorbed perrhenate. The three-dimensional growth under diffusion control was the most important contribution, and represented 70–80% of the mass increase. The FE for the electrodeposition process was in the range of 12–18%. The nature of the adsorbed layer, however, was not identified in this study.

The last paragraphs of this chapter shall deal with electrodeposition of Re-based alloys. Based on the relatively positive reversible potential of Re, one might expect it to codeposit readily with other metals, in particular in alkaline solutions, where its deposition potential could be shifted in the negative direction. However, Brenner noted the unexpectedly low FE as an indicator to the difficulty of codeposition.[78] Fink and Deren[170] stated that Re could be codeposited with Co and Ni. Later, Netherton and Holt reported the successful codeposition of Re with Ni,[188] Co and Fe.[189] The iron-group metals were introduced into the bath in the form of simple salts, such as chloride or sulfate. Brenner[78] postulated that the citrate ion, that was contained in most of the baths, complexed the iron-group metal, thereby shifting its deposition potential cathodically and, consequently, favoring the deposition of Re. The presence of the citrate ion in the bath considerably increased the Re-content in the deposit and lowered the FE. The considerably more positive reversible potential of Re in acid solutions, compared to those of the iron-group metals, allowed for production of alloys with Re content that was significantly higher than its ionic content in the bath. Variation of the pH in the range 3 to 8 did not affect much the Re content in alloys with iron-group metals. The FE, on the other hand, increased considerably from pH 3 to 4, reached a maximum around pH = 5.5, and then decreased mildly. In any case, the FE for alloy deposition was much higher than for pure Re deposition. The Re-content in the deposit decreased with increasing current density, as commonly observed for the more noble metal in alloy deposition. This effect was most pronounced in baths containing a low concentration of ReO_4^-. An increase in bath temperature resulted in a considerable increase of the FE and in the concentration of Re in the Re-Ni alloy, but had little effect on the composition of the Re-Co alloy.

Meyer[162] claimed that both Ni and Co seem to stabilize the presence of ReO_4^- anions near the cathode. He proposed that there was a catalytic effect of Ni on the decomposition of ReO_4^-. Sadana and Wang[190] studied the effects of bath composition, pH, temperature, stirring, current density and pulsed current on the characteristics of Au-Re deposits, which contained 0.25–63.4 wt.% Re. The solution consisted of citric acid and potassium perrhenate. The Re-content of the deposit was found to increase markedly as a result of stirring, increase in current density, decrease in bath pH and temperature, and the use of pulsed current. In addition, the as-deposited alloys exhibited XRD patterns of supersaturated solid solution of Re in Au for Re contents lower than 21.7 wt.%, and XRD patterns of a mixture of Au fcc phase and an amorphous Re-Au phase for Re contents between 23.4 and 63.4 wt.% (ca. 24.4–64.7 at.%).

Kvokova and Lainer[191] electrodeposited pure Re and Re-Cr alloy on Mo substrate. For the deposition of Re itself, two baths were used: one containing perhhenic acid, and the other containing potassium perrhenate. In the first bath, the discharge of hydrogen ions was enhanced. The authors attributed the low overpotential of hydrogen on Re to its lattice parameter (a = 2.758 Å). However, a justification to this theory has not been proposed. For both deposition of Re and Re-Cr alloy, the concentration polarization was found to be insignificant compared to the activation polarization.

In the sections dealing with electrodeposition of W- and Mo-based alloys, the role of citrate as a complexing agent was described in detail (see, for example, Figs. 8 and 13). In an attempt to better understand the similarities and differences between electrodeposition of Re and that of W or Mo, a literature survey was conducted on electroreduction of perrhenate in the presence of citrate. Only one relevant, well-written, paper was found. Vajo et al.[192] applied polarography and controlled-potential coulometry to study the electroreduction of perrhenate in acidic solutions (pH = 1.6-5.4) of perhhenic acid with citrate (or oxalate). The presence of citrate was found to markedly enhance the reduction of perrhenate through reversible formation of a 1:1 complex, $[(ReO_4)(H_2Cit)]^{2-}$. This complex was sufficiently stable, that it yielded a diffusion-limited current. The very low value of the limiting current allowed for concluding that the equilibrium concentration of ReO_4^- was nearly equal to its analytical concentration. For a constant concentration of ReO_4^- (2.03 mM) and a varying excess concentration of citrate (from 0.05 to 0.4 M), the limiting current increased linearly with increased concentration of citrate, indicating that the equilibrium concentration of ReO_4^- indeed remained essentially equal to its analytical concentration independent of the citrate concentration, and that the electroactive complex contained one citrate moiety. Similarly, the linear increase of the limiting current density with increased concentration of ReO_4^- indicated that the electroactive species contained one Re-atom. When assuming that the citrate species reacting with ReO_4^- was H_2Cit^-, and taking into account pK_a = −1.25 for perrhenic acid, an excellent agreement was observed between the theoretical and experimental values of limiting current density versus pH. Hence, the stoichiometry of the formation of the perrhenate-citrate complex was claimed to be as follows; below pH = 3:

$$ReO_4^- + H_3Cit \rightarrow [(ReO_4)(H_2Cit)]^{2-} + H^+ \qquad (48a)$$

and above pH = 3:

$$ReO_4^- + H_2Cit^- \rightarrow [(ReO_4)(H_2Cit)]^{2-} \qquad (48b)$$

These two reactions are in accordance with the stepwise deprotonation of citric acid (c.f. Fig. 1a). The oxidation state of Re generated by electroreduction of the perrhenate-citrate complex was found to be Re(V), which could easily be reduced further, forming Re(IV). The enhanced reduction of the perrhenate-citrate complex was ascribed to expansion of the Re coordination sphere from 4 to 6, through formation of chelated structures by a concerted process, in which the incoming ligand transferred protons to coordinated oxo-groups.

In conclusion, it would seem that the chemistry of Re is very different from that of W and Mo. First, Re has a 7-valent form, while the other two do not exceed the 6-valent state. In addition, while the perrhenate is stable in strong acids, the tungstate and molybdate tend to hydrolyze and/or polymerize. Hence, it would be interesting to study further the mechanisms of electrodeposition of Re.

III. CONCLUDING REMARKS

It is pointed out above that metal deposition is different from outer-sphere charge-transfer reactions in that charge is carried across the metal/solution interface by the ions, not by electrons. Although this has been acknowledged by several noted electrochemists, a theory of charge transfer by ions, comparable in detail and depth to the theories of electron transfer, has yet to be developed. So far, metal deposition has been treated by the formalism developed successfully for electron transfer.

One of the unique features of metal deposition is that the surface is constantly being renewed. This is a mixed blessing: on the one hand it allows some reduction in the required purity level. On the other hand, the surface morphology and its roughness could change in the course of metal deposition, leading to a change of the real current density, although the applied current density i_{dep} is maintained constant.

The analysis of the kinetics of alloy deposition is complicated by the fact that at least two reactions occur in parallel. Consequently, the current-potential relation observed represents a combination of the contributions of two processes, each having its own overpotential, rate constant and potential dependence of the current density. Thus, any information obtained from the current-potential relation observed is of questionable value in evaluating the mechanism of the formation of the alloy.

The reduction of a large oxy-ion such as WO_4^- is a very complex process. It involves the transfer of six electrons and eight protons, and must proceed in several elementary steps. It is thus unrealistic to expect that one could determine this mechanism in detail, determine the nature of each step, the types of intermediates formed, the surface coverage by each of these intermediates, and so on. Nevertheless, there are certain aspects of this mechanism that can be studied in some detail, helping us to understand the fundamental factors that are critical for the operation of the plating bath. Such understanding is important in the general context of alloy plating, and could also help in systematic design of better plating parameters for industrial applications.

The unique feature of plating of tungsten (and, similarly, of molybdenum) from aqueous solutions is that they cannot be plated in the pure form. On the other hand, alloys of these metals can readily be deposited. The most common alloying elements are the iron-group transition metals (Ni, Co and Fe), but there are indications in the literature that other metals (e.g., Zn and Cd) could form similar alloys. The third metal discussed in this chapter is rhenium. Unlike W and Mo, Re can be deposited from aqueous solutions, albeit with great difficulty, but adding a nickel salt to the solution improves the process significantly, forming the Re-Ni alloy at a much higher FE, and producing much superior, smooth and crack-free deposits.

The above observation, called induced codeposition, which was first made about 70 years ago, presents a very interesting question: What is the mechanism by which the presence of ions of one of the iron-group metals in the plating bath enhances the deposition of W, Mo and Re, or indeed makes it possible? The purpose of this chapter is to answer this question, to the best of presently available knowledge. The section regarding alloys of tungsten relies heavily on work performed in our own laboratory, although a critical review of other work is also given, of course. The discussion of electrodeposition of molybdenum alloys is based heavily on the work of Landolt and his research group. The electrochemistry of rhenium deposition is much less understood, and the review represents what is known at present.

It should not be surprising that an ion such as WO_4^- cannot be reduced readily all the way to metallic tungsten. Indeed, it is surprising that there are certain conditions under which it can be reduced. Moreover, alloy deposition is often a complex, and quite unpredictable, process. In what is called *anomalous deposition* we classify processes that behave unexpectedly – the composition of the alloy cannot be predicted from the current-potential relation of the alloying elements studied each by itself. When forming a Ni-Fe alloy it seems that Fe^{2+} ions in solution inhibit the rate of deposition of nickel, while Ni^{2+} ions accelerate the rate of deposition of iron. In the deposition of a Ni-Zn alloy the situation is somewhat different. Here, one finds a complete synergistic effect: adding either ion to the solution enhances the rate of deposition of the other metal.

Anomalous alloy deposition is common in electroplating. Actually, it is so common that it is the rule rather than the exception. What could be the cause of this phenomenon? In general, it is postulated that one of the alloying elements forms a hydroxide on the surface, which inhibits the deposition of the other. This could explain the inhibiting effect, but it is much more difficult to explain the enhancement. Another explanation is that the two ions form a mixed-metal complex that, for some reason, is more readily reduced at the surface than each of the metal ions separately. Naturally, such complexes can only be expected to exist in solution containing a negatively charged complexing agent, which would hold the two positive metal ions together, for example:

$$Fe^{2+} + Cit^{3-} \rightarrow [FeCit]^{-} \tag{49}$$

$$Ni^{2+} + Cit^{3-} \rightarrow [NiCit]^{-} \tag{50}$$

followed by

$$[FeCit]^{-} + [NiCit]^{-} \rightarrow [FeNiCit]^{+} + Cit^{3-} \tag{51}$$

If the mixed-metal complex is the precursor for deposition of the alloy, then it can be seen that adding either of the two ions would increase the rate of deposition of the other. This simplistic interpretation could lead only to alloys having equal concentrations of the two elements. However, if the concentration of the two elements in solution is not equal, one could have a parallel reaction, in which one of the metals would also be deposited in parallel from its complex with citrate, giving rise to a whole range of alloy compositions.

This brings us to one of the main point made in this chapter, which is relevant both for anomalous alloy deposition and for induced codeposition: in order to understand the process, one should understand the chemistry of the solution, and particularly the distribution of species in plating baths that contain complexes. This type of analysis is shown in Figs. 1, 2, 8, 13 and 14, and has been used in our own work to explain the induced codeposition of tungsten.

Many publications were devoted over the years to explain the mechanism by which induced codeposition proceeds. None have been proven beyond doubt, and one has to look for the explanation that is consistent with the widest range of experimental observations.

For induced codeposition of Ni-W alloys, we concluded that the precursor for deposition of the alloy is a mixed-metal complex of the type $[(Ni)(HWO_4)(Cit)]^{2-}$. This complex is formed from a nickel citrate complex (cf., Eq. 50) and a tungstate citrate complex $[(WO_4)(Cit)(H)]^{4-}$. It may be somewhat surprising that the negative citrate ion forms a complex with a negative tungstate ion, but this is a well-established fact, and the stability constant of this complex is available in the literature. This should lead to a mutual synergistic effect of Ni on the rate of deposition of W and vice versa, which was confirmed experimentally. The partial current density for deposition of tungsten was found to be mass-transport dependent, although it was only a few percent of the limiting current density, calculated for the concentration of either Na_2WO_4 or $NiSO_4$ in solution. This indicates that the concentration of the above precursor is much smaller, and its rate of formation is low. The latter is not surprising, in view of the fact that it is formed by an interaction between two negatively charged ions, one of which having a high charge of –4. Alloys having a wide range of compositions were obtained, depending mostly on the composition of the bath. In most cases, the concentration of Ni in the alloy was higher than that of W, since there is a parallel path for deposition of Ni from its complex with citrate (or with NH_3, when present in solution). Nevertheless, conditions were found, under which an alloy with a 1:1 Ni:W ratio could be deposited. Different crystal structures were identified by X-ray diffraction. The structure was found to depend on the atomic composition of the alloy, not on the way it was prepared. The dependence of the alloy composition on pH was found to be consistent with the distribution of the relevant species in solution, as shown in Fig. 8. Finally, linear relationships were observed between the product of concentrations of the two complexes that form the mixed-metal complex and the partial current densities for deposition of tungsten.

Plating of alloys of Mo was studied intensively in recent years by Landolt and his co-workers. It was shown that in formation of Ni-Mo alloys, the rate of deposition of Mo (i.e., the partial current density for deposition of this metal) is controlled by the concentration of Ni in solution. This is consistent, of course, with the idea that the precursor for deposition of the alloy is a mixed-metal complex, as proposed for Ni-W alloys by Gileadi et al. It is also expected in view of the similarity of the chemistry of W and Mo ions in aqueous solutions. However, the mixed metal complex for Ni-Mo alloy deposition was assumed to be $[NiCit(MoO_2)]_{ads}^-$. The most important difference between the assumed mixed-metal complexes are that in the case of W the complex is in solution, while in the case of Mo it is assumed to be adsorbed on the surface. Moreover, in the precursor for deposition of Ni-W alloys, the mixed-metal complex contains the protonated tungstate anion HWO_4^-, while in the case of Ni-Mo alloys it is the neutral dioxide MoO_2, which has been formed from reduction of the MoO_4^{2-} ion in a previous step. In both cases, the authors have been able to deposit alloys with a range of concentrations of the refractory metals. However, in the case of W it was concluded that the mixed-metal complex was the source of deposition of an alloy of equal concentrations of Ni and W, while Ni was deposited in parallel from its complex with citrate. In the case of Mo it was assumed that the reaction proceeded by a four-electron reduction of the mixed-metal complex, while Ni was deposited independently from its complex with citrate.

The electrochemistry of rhenium is quite different from that of W and Mo, and has not been investigated in similar detail. To begin with, the most stable ion in solution is the perrhenate, ReO_4^-, from which metallic rhenium can be deposited directly, not only as an alloy. The hexavalent ReO_4^{2-} (similar to MoO_4^{2-}) does not exist in aqueous solutions. On the other hand, there are definite indications (which admittedly should be confirmed by further detailed studies) that forming alloys with Pt, and possibly with the iron-group transition metals, leads to higher-quality coatings with lower residual stresses, less cracking and lower hydride content.

In conclusion, it can be stated that alloys of the three refractory metals discussed in this chapter may be of significant importance for practical application, in view of their high corrosion resistance, stability at high temperatures and wear resistance. Some of these advantages are already being implemented, mainly for alloys of tungsten and of molybdenum, but only marginally for rhenium. Deeper understanding of the phenomenon of induced codeposition could lead to increasing the range of applications of such alloys in specialized applications.

ACKNOWLEDGEMENT

The authors wish to thank Prof. E. Kirowa-Eisner for her assistance in calculating the distribution of species shown in the relevant figures.

LIST OF ABBREVIATIONS AND SYMBOLS

ACD	Anomalous Codeposition
EIS	Electrochemical Impedance Spectroscopy
EMF	Electromotive Force
FE	Faradaic Efficiency
HAC	Hydrogen-Assisted Cracking
HE	Hydrogen Embrittlement
HER	Hydrogen Evolution Reaction
HIC	Hydrogen-Induced Cracking
HRC	Hardness Rockwell C
MEMS	Micro-Electro-Mechanical Systems
OCP	Open-Circuit Potential
OHP	Outer Helmholtz Plane
RHE	Reversible Hydrogen Electrode
SHE	Standard Hydrogen Electrode
ULSI	Ultra-Large-Scale Integration
UPD	Underpotential Deposition
VHN	Vickers Hardness Number

b	Tafel slope (V decade^{-1})
c	The subsurface solubility of a dissolved atom in a solid metal, expressed as solute-to-metal atom ratio
c_{bulk}	Bulk concentration of the electro-active species (mol cm^{-3})
$c_{Me^{z+}}$	Concentration of the metal cation (mol cm^{-3})
c_{surf}	Concentration of the electro-active species at the surface (mol cm^{-3})
D	Diffusion coefficient (cm^2 s^{-1})
E	Young's modulus of elasticity (GPa)
E^0	Standard (equilibrium) potentials (V)
$E^{0\prime}$	Standard potential in the presence of a complexing agent (V)
E_{corr}	Corrosion potential (V)
E_{dep}	Deposition potential (V)
E_{rev}	Reversible potential (V)
F	Faraday's constant (96,485 C equiv^{-1})
f	Fugacity (Pa)
I	Total current passed (A)

i	Current density (A cm^{-2})
i_{ac}	Activation-controlled current density (A cm^{-2})
$i_{c,i}$	Partial cathodic current density of the ith element in alloy deposition (A cm^{-2})
i_L	Limiting current density (A cm^{-2})
i_0	Exchange current density (A cm^{-2})
i_{pass}	Passivation current density (A cm^{-2})
K	Stability constant of a complex
K_S	Solubility (Sieverts') constant for dissociative adsorption of a diatomic gas, followed by absorption ($Pa^{-0.5}$)
K_{Ic}	Fracture toughness (MPa $\sqrt{m}$)
L	Ligand
l	Characteristic length
Me_i	Metal i
M_i	Atomic mass of the ith element (g mol^{-1})
n_i	The number of electrons transferred in the reduction of one atom of the ith element (equiv mol^{-1})
P	Partial pressure (Pa)
R	The ideal gas constant (1.987 cal $K^{-1}mol^{-1}$, or 8.314 J $K^{-1}mol^{-1}$)
R_{conc}	Mass transport resistance (Ω cm^{-2})
R_{ct}, R_F, R_{ac}	Charge transfer (Faradaic) resistance (Ω cm^{-2})
R_{soln}	Ohmic solution resistance (Ω cm^{-2})
r	Radius of the rotating cylinder electrode (cm)
r_{metal}	Atomic radius (Å)
T	Absolute temperature (K or°C)
T_m	Melting temperature (K or°C)
t_d	Deposition time (s)
W_a	Wagner number
w	The measured weight of a deposit (g)
x_i	The weight (or mole) fraction of an element in an alloy deposit
α	Linear thermal expansion coefficient ($^oC^{-1}$)
α	Transfer coefficient for the overall electrode reaction
β	Symmetry factor for an elementary charge-transfer step in the reaction sequence
β_n	Equilibrium constant for formation of complexes in the reactions Me + n·L → [MeL_n], usually given in its log form

δ	The thickness of the Nernst diffusion layer (cm)
Δ_{UPD}	UPD potential shift (V)
ε_f	Strain at fracture (%)
η	Observed overpotential (V)
η_{ct}	Charge-transfer (activation) overpotential (V)
η_{conc}	Concentration overpotential (V)
η_{iR}	Resistance overpotential (V)
κ	Specific conductance of solution (S cm^{-1})
κ	Thermal conductivity (W $m^{-1}K^{-1}$)
ν	Kinematic viscosity ($cm^2\ s^{-1}$)
ν	Poisson's ratio ($\equiv -\varepsilon_{zz}/\varepsilon_{xx} = -\varepsilon_{yy}/\varepsilon_{xx}$, where ε_{xx} is the principal strain in the x direction of the applied force, and ε_{yy} and ε_{zz} are the resulting principal strains in the orthogonal directions y and z, respectively)
ω	Angular velocity, or rotation rate (rad s^{-1}, or rpm)
ρ	Density (g cm^{-3})
ρ	Specific resistivity (Ohm·cm)
σ_u	Tensile strength (Pa)
σ_y	Yield strength (Pa)
θ	Partial surface coverage

APPENDICES

Appendix A

Typical bath compositions and operating conditions for electrodeposition of W-based alloys, cf., Table 3.

Appendix B

Typical bath compositions and operating conditions for electrodeposition of Mo-based alloys, cf., Table 4.

Appendix C

Typical bath compositions and operating conditions for electrodeposition of Re-based alloys, cf., Table 5.

Table 3
Typical Bath Compositions and Operating Conditions for Electrodeposition of W-Based Alloys

Bath composition									
$[WO_4^{2-}]$ (M)	Iron-group metal in bath (M)	Organic acid/salt (M)	Others (M)	pH	T (°C)	i (mA cm^{-2})	FE (%)	wt.% W in deposit	Ref.
0.03	Co: 0.60	Sodium citrate: 0.60	Ammonium chloride: 0.94	9.0	95	20	87	23	119
1.00	Co: 0.07	Tartaric acid: 0.13	Boric acid: 0.32	7.0	100	20	30	50	120
0.22	Co: 0.10	Citric acid: 0.65	Ammonium chloride: 0.94	8.2	50–55	150–200 (pulse plating)	20	65	121
0.27-0.54	Ni: 0.42	Rochelle salt: 1.4 or sodium citrate: 0.6 or hydroxyacetic acid: 2.6	Ammonium chloride: 0.94	9.0	95	20	85–93	10–20	119
0.15	Ni: 0.07	Citric acid: 0.34	—	8.0	70	70–150	45	35	122
0.40	Ni: 0.01–0.10	Tri-sodium citrate: 0.25–0.60	None or nickel sulfamate: 0.062–0.124 or saccharin: 0.0146 or sodium chloride: 0.002–0.005 or sodium sulfate: 0.176	8.0	RT–70	5–15	8–21	25–90	104 108

Table 3. Continuation

Bath composition				pH	T (°C)	i (mA cm^{-2})	FE (%)	wt.% W in deposit	Ref.
[WO_4^{2-}] (M)	Iron-group metal in bath (M)	Organic acid/salt (M)	Others (M)						
0.01-0.50	Ni: 0.01–0.20	Tri-sodium citrate: 0.10–1.0	Ammonia: 0–1.7	6.0–12.0	10–62	15	0.5–98	5–74	91–93 105
0.003–0.006	Ni: 0.38	Ammonium chloride: 0.93	With or without diammonium citrate: 0.013–0.044	Not stated	30–70	5–30	43–90	12–20 (at.% or wt.% not stated)	113
0.14	Ni: 0.06	Tri-sodium citrate: 0.3-0.5	Ammonium chloride: 0.5, sodium bromide: 0.15	8.5	60–90	0.5–2	Not stated	40–51	123
0.09	Ni: 0.051	Sodium citrate: 0.31	Sulfamate: 0.102	4.0–8.0	30–70	5–100	4–85	19–81 (at.% or wt.% not stated)	112
0.09–0.14	Fe: 0.27–0.43	Ammonium citrate: 0.40 or sodium citrate: 0.30 or ammonium malate: 1.5	Ammonium sulfate: 0.76	9.0	95	50	45–75	30–45	119

Table 4
Typical Bath Compositions and Operating Conditions for Electrodeposition of Mo-Based Alloys

Bath composition									
$[MoO_4^{2-}]$ (M)	Iron-group metal in bath (M)	Organic acid/salt (M)	Others (M)	pH	T (°C)	i (mA cm^{-2})	FE (%)	wt.% Mo in deposit	Ref.
0.02	Ni: 0.22	Citric acid: 0.31	—	4	25	50	32	17	138
0.52	Co: 0.17	Hydroxyacetic acid: 1.3	—	1.5–2.5	25	30–100	35	20–30	78
0.052	Ni: 0.22	Rochelle salt: 0.27	Sodium chloride: 0.12, ammonia	11	27	100	67	18	139
0.2	Ni: 0.30	Sodium citrate: 0.3	Ammonia	10.5	25	100	70	18	140
0.1	Fe: 0.30	Sodium citrate: 0.3	Ammonia	10.5	25	200	12	57	140
0.14	Co: 0.034	—	Sodium bicarbonate: 1.0, sodium pyrophosphate: 0.13	8.0	22	40	20	50	141

Table 4. Continuation

Bath composition									
$[MoO_4^{2-}]$ (M)	Iron-group metal in bath (M)	Organic acid/salt (M)	Others (M)	pH	*T* (°C)	*i* (mA cm^{-2})	*FE* (%)	wt.% Mo in deposit	Ref.
0.17	Fe: 0.033	—	Sodium pyrophosphate: 0.1, sodium bicarbonate: 0.9	8.0	50	150	44	61	142
0.52	Co: 0.17	—	Potassium carbonate: 4.7	11.0	100	10	75	25	143
0.067	Ni: 0.20	—	Sodium chloride: 0.05, ammonium hydroxide (25%): 5.71	10.7	30–40	100	41	17.5	144
0.005–0.05	Ni: 0.20	Sodium citrate: 0.25–0.95	Ammonia: 0-1.0	7.5–11.4	25–80	10-200	8–79 (partial range)	1–71	95
0.005	Ni: 0.7 or Co: 0.7 or Fe: 0.7	Sodium citrate: 0.70	Ammonia	7.4	25	35 (Mo-Co) 50 (Mo-Fe)	22–36 (Mo-Co), 7–21 (Mo-Fe), 61–95 (Mo-Ni)	31–43 (Mo-Co), 11–14 (Mo-Fe), 7–23 (Mo-Ni)	97
1.0–15.0	Co: 0.10	Sodium citrate: 0.20	—	6.6	25	—	—	max. 60	152
0–0.03	Ni: 0.2	Sodium citrate: 0.25	Ammoia	9.5	37	—	—	—	153

Table 5
Typical Bath Compositions and Operating Conditions for Electrodeposition of Re-Based Alloys

Bath composition								
$[ReO_4^-]$ (M)	Organic acid/salt (M)	Others (M)	pH	T (°C)	i (mA cm^{-2})	FE (%)	wt.% Re in deposit	Ref.
0.052	—	Ammonium sulfate: 0.030, sulfuric acid: 0.122, gelatin (4 g L^{-1})	0.9–1.0	85–90	100–150	—	—	160
0.035	—	Sulfuric acid: 0.1, ammonium sulfate: 0.89, magnesium sulfamate: 0.22	1.0–1.5	25–80	20–150	1.3–12	—	162
0.038	—	Sulfuric acid	1.0	RT–90	100–140	—	—	170
0.16	—	Sulfuric acid: 0.89, ammonium hydroxide: 0.71	0.9–1.1	21–88	108–2,800	—	—	172
0.048–0.052	—	Sulfuric acid: 0.122	1.0	50–70	5–150 (DC & pulse plating)	—	—	166 173
0.14	Sulfamic acid: 1.11	Magnesium sulfate: 0.22, ammonium sulfate: 0.76	1.0–1.5	—	—	—	—	176
0.0035	Boric acid: 0.49	Ni: 1.0	4.6	70	30	98	6	188
0.0035	Boric acid: 0.49, citric acid: 0.37	Ni: 1.0	2.3	70	30	35	45	188
0.035	Citric acid: 0.34	Co: 0.21, ammonium hydroxide	7-8	70	50	80	80	189
0.035	Citric acid: 0.34	Fe: 0.22, ammonium hydroxide	7-8	70	50	65	58	189

REFERENCES

[1]E. Raub and K. Müller, *Fundamentals of metal deposition*, Elsevier Publishing Company, Amsterdam, 1967.
[2]J. W. Dini, *Electrodeposition: the materials science of coatings and substrates*, Noyes Publications, New Jersey, 1993.
[3]E. Budevski, G. Staikov, and W.J. Lorenz, *Electrochemical phase formation and growth*, VCH, Weinheim, 1996.
[4]M. Paunovic and M. Schlesinger, *Fundamentals of electrochemical deposition*, John Wiley & Sons, New York, 1998.
[5]M. Schlesinger, M. Paunovic, (eds) *Modern electroplating*, 4th edn, Wiley- Interscience, New York, 2000.
[6]N. Kanani, *Electroplating: basic principles, processes and practice*, Elsevier, Oxford, 2004.
[7]D. Landolt, *J. Electrochem. Soc.* **149** (2002) S9.
[8]ASTM B 374 – 96: *Standard Terminology Relating to Electroplating*, American Society for Testing and Materials, Pennsylvania.
[9]H.H. Wan and H.Y. Cheh, *J. Electrochem. Soc.* **135** (1988) 643.
[10]O. Dossenbach, in *Theory and practice of pulse-plating*, Ed. by J.C. Puippe, F. Leaman, AESF, Florida, p. 73, 1986.
[11]D. Landolt, *Plat. Surf. Finish* **88** (2001) 70.
[12]P. J. Moran and E. Gileadi, *J. Chem. Education* **66** (1989) 91.
[13]H. L. F. von Helmholtz, *Ann. Physik*, **89** (1853) 211.
[14]H. L. F. von Helmholtz, *Ann. Physik* **7** (1879) 337.
[15]G. Gouy, *J. Phys. Radium* **9** (1910) 457.
[16]G. Gouy, *Compt. Rend. Acad. Sci.* **149** (1910) 654.
[17]D.L. Chapman, *Phil. Mag.*, **25** (1913) 475.
[18]O. Stern, *Zeit. Elektrochem.* **30** (1924) 508.
[19]E. Gileadi, in *Fifth international symposium on proton-exchange-membrane fuel cells*, Ed. by T. Fuller, C. Bock, C. Lamy, *Electrochem Soc. Trans.* 1:3, 2006.
[20]R. A. Marcus, *J. Chem. Phys.* **24** (1956) 4966.
[21]R. A. Marcus, *J. Chem. Phys.* **43**(1965) 679.
[22]R. A. Marcus, *Electrochim. Acta* **13** (1968) 955.
[23]R. R. Dogonadze, *Dokl. Akad. Nauk. SSSR* **133** (1960) 1368.
[24]R. R. Dogonadze and Y.A. Chizmadzhev, *Dokl. Akad. Nauk. SSSR* **144** (1962) 1077.
[25]R. R. Dogonadze, A. Kuznetzov, and Y. A. Chizmadzhev, *Zh. Fiz. Khim.* **38** (1964) 1195.
[26]V.G. Levich, in *Advances in electrochemistry and electrochemical engineering*, Interscience, Vol. 4, Ed. by P. Delahay, New York, 1966, p. 249.
[27]C. Grahame, *Ann. Rev. Phys. Chem.*, **6** (1955) 337.
[28]K. Vetter, *Electrochemical kinetics*, Academic Press, New York, p. 134, 1967.
[29]V. V. Losev, in *Modern Aspects of Electrochemistry*, Ed. by B. E. Conway, J. O'M. Bockris, Plenum Press, New York, 1972, Vol. 7, p 314.
[30]N. Sato, *Electrochemistry of metal and semiconductor electrodes*, Elsevier, Amsterdam, p. 289, 2002.

[31]E. Gileadi, *Interface*, **12** (2003) 10.
[32]E. Gileadi, *Chem. Phys. Lett.* **393** (2004) 421.
[33]E. Gileadi, in *Electrochemical Processing in ULSI and MEMS,* Ed. by H. Deligianni, T.P. Moffat, S.T. Mayer, and G.R. Stafford, *Proc. Electrochem. Soc. ECS*, Vol. 17, New Jersey, 2004, p. 3.
[34]E. Gileadi, E. Kirowa-Eisner, *Corros. Sci.* **47** (2005) 3068.
[35]E. Gileadi, *J. Electroanal. Chem.* **532** (2002) 181.
[36]E. Gileadi, *Electrode kinetics for chemists, chemical engineers and material scientists*, VCH, New York, 1993, pp. 159, 540.
[37]A. Vaskevich, M. Rosenblum, and E. Gileadi, *J. Electroanal. Chem.* **167** (1995) 383.
[38]D.M. Kolb, in *Advances in electrochemistry and electrochemical engineering*, Vol. 11, Ed. by H. Gerischerand, and C. W. Tobias, Wiley, New York, 1978.
[39]X. Zong and S. Bruckenstein, *J. Electrochem. Soc.* **146** (1999) 2555.
[40]B.E. Conway and J. S. Chacha, *J. Electroanal. Chem.* **287** (1990) 13.
[41]E. Herrero, L. J. Buller and H. D. Abruna, *Chem. Rev.*, **101**(2001) 1897.
[42]R. Adzic, in *Advances in electrochemistry and electrochemical engineering*, Vol. 13, Ed. by H. Gerischerand and C. W. Tobias, Wiley, New York, 1984.
[43]R. Adzic, E. Yeager, and B. D. Cahan, *J. Electrochem. Soc.* **121**(1974) 474.
[44]K. Engelsmann, W. J. Lorenz, and E. Schmidt, *J. Electroanal. Chem.* **114** (1980) 1.
[45]A. Hamelin, and J. Lipkowski, *J. Electroanal. Chem.* **171** (1984)317.
[46]J. Clavilier, J. M. Orts, J. M. Feliu, and A. Aldaz, *J. Electroanal. Chem.* **293** (1995) 197.
[47]H. Bort, K. Juttner, and W. J. Lorenz, *J. Electroanal. Chem.*, **90** (1978) 413.
[48]C. L. Faust, in *Modern electroplating*, 2nd edn. Ed. by F.A. Lowenheim, John Wiley & Sons, New York, 1963, p. 453.
[49]N. Eliaz, E. Moshe, S. Eliezer, and D. Eliezer, *Metall. Mater. Trans. A* **31** (2000)1085.
[50]N. Eliaz, D. Eliezer, E. Abramov, D. Zander, and U. Köster, *J. Alloys Compd.* **305** (2000) 272.
[51]N. Eliaz and D. Eliezer, *Metall. Mater. Trans. A,* **31** (2000)2517.
[52]A. S. Tetelman and W. D. Robertson, *Trans. Met. Soc. AIME* **224** (1962) 775.
[53]N. Eliaz, L. Banks-Sills, D. Ashkenazi, and R. Eliasi, *Acta Mater.* **52** (2004) 93.
[54]H. G. Nelson, in *Embrittlement of engineering alloys,* Ed. by C. L. Briant, S. K. Banerji, Academic Press, New York, 1983, p. 275.
[55]N. Eliaz, *J. Adv. Mater.* **34** (2001) 27.
[56]N. Eliaz, A. Shachar, B. Tal, and D. Eliezer, *Eng. Fail.Anal.*, **9** (2002) 167.
[57]W. J. Barnett, A. R. Troiano **209** (1957)486.
[58]R. A. Oriani, *Ann. Rev. Mater. Sci.* **8** (1978) 327.
[59]N. J. Petch, and P. Stables, *Nature* **169** (1952) 842.
[60]D. G. Westlake, *Trans. ASM* **62** (1969) 1000.
[61]H. K. Birnbaum, in *Hydrogen embrittlement and stress corrosion cracking*, ASM, Ed. by R. Gibala, and R. F. Hehemann, Metals Park, Ohio, 1984, p 153.
[62]H. K. Birnbaum, in *Hydrogen effects on material behavior, TMS*, Ed. by N. R. Moody, A. W. Thompson, Warrendale, Pennsylvania, 1990, p. 639.
[63]J. P. Hirth, *Metall. Trans.* **11A** (1980) 861.
[64]C. D. Beachem, *Metall. Trans.* **3** (1972) 437.
[65]S. P. Lynch, *Met. Forum* **2** (1979)189.
[66]J. J. Gilman, *Phil. Mag.* **26** (1972) 801.

[67]P. Bastien and P. Azou, in *Proceedings of the first world metallurgical congress*, ASM, Ed. by W. M. Baldwin, Metals Park, Ohio, 1951, p 535.
[68]A. N. Stroh, Adv. Phys. **6** (1957) 418.
[69]D. S. Shih, I. M. Robertson, and H. K. Birnbaum, *Acta Metall.* **36** (1988) 111.
[70]J. K. Tien, A. W. Thompson, I. M. Bernstein, and R. J. Richards, *Metall Trans.* **7A** (1976) 821.
[71]QQ-C-320 (1987) Federal Specification.
[72]ASTM F 519 (1993) Standard Test Method, West Conshohocken.
[73]F. C. Frank and J. H. van der Merwe, *Proc. Roy. Soc. London A* **198** (1949) 205.
[74]M. Volmer and A. Weber, *Z. Physik. Chem.*, **119** (1926) 277.
[75]I. N. Stranski and L. Krastanov,*Sitzungsber Akad Wiss Wien Math*, **146** (1938) 797.
[76]H. Liebscher, *Galvanotechnik* **88** (1993) 754.
[77]N. Kanani, U. Landau, and R. Kammel, *Erzmetall* **54** (2001) 369.
[78]A. Brenner, *Electrodeposition of alloys*, Vol. I & II. Academic Press, New York, 1963.
[79]H. Dahms and I. M. Croll, *J. Electrochem. Soc.* **112** (1965) 771.
[80]S. Hessami, C. W. Tobias, *J. Electrochem. Soc.* **136** (1989) 3611.
[81]I. M. Croll, L. T. Romankiw, in *Electrodeposition technology, theory and practice*, ECS, Ed. by L. T. Romankiw, New Jersey, 1987, p 285.
[82]M. Matlosz, *J. Electrochem. Soc.* **140** (1993) 2272.
[83]K. Y. Sasaki and J. B. Talbot, *J. Electrochem. Soc.* **142** (1995) 775.
[84]K. Y. Sasaki and J. B. Talbot, *J. Electrochem. Soc.* **145** (1998) 981.
[85]T. Erdey-Grúz, *Kinetics of electrode processes*. Wiley-Interscience, New York, 1972, p 25.
[86]L. J. Gao, P. Ma, K. M. Novogradecz, and P. R. Norton, *J. Appl. Phys.* **81** (1997) 7595.
[87]S. H. Liao, *IEEE Trans. Magn.* **26** (1990) 328.
[88] E. E. Castellani, J. V. Powers, and L. T. Romankiw, US Patent 4 102 756, 1978.
[89]N. Zech, E. J. Podlaha, and D. Landolt, *J. Electrochem. Soc.* **146** (1999)2886.
[90]N. Zech, E. J. Podlaha, and D. Landolt, *J. Electrochem. Soc.* **146** (1999) 2892.
[91]O.Younes, L. Zhu, Y. Rosenberg, Y. Shacham-Diamand, and E. Gileadi, *Langmuir* **17** (2001) 8270.
[92]O. Younes, E. Gileadi, *J. Electrochem. Soc.* **149** (2002) C100.
[93]O.Younes-Metzler, L. Zhu, and E. Gileadi, *Electrochim. Acta*, 48 (2003)2551.
[94]A. Brenner, D. E. Couch, and E. K. Williams, *J. Res. Natl. Bur. Stand.* **44** (1950) 109.
[95]E. J. Podlaha and D. Landolt, *J. Electrochem. Soc.* **143** (1996) 885.
[96]E. J. Podlaha and D. Landolt, *J. Electrochem. Soc.* **143** (1996) 893.
[97]E. J. Podlaha and D. Landolt, *J. Electrochem. Soc.* **144** (1997) 1672.
[98]Y. Shacham-Diamand, Y. Sverdlov, and N. Petrov, *J. Electrochem. Soc.* **148** (2001) C162.
[99]A. Kohn, M. Eizenberg, and Y. Shacham-Diamand, *J. Appl. Phys.* **92** (2002) 5508.
[100]Y. Shacham-Diamand and S. Lopatin, *Microelectron Eng.* **37/38** (1997) 77.
[101]A. Kohn, M. Eizenberg, Y. Shacham-Diamand, B. Israel, and Y. Sverdlov, *Microelectron Eng.* **55** (2001) 297.
[102]A. Kohn, M. Eizenberg, Y. Shacham-Diamand, and Y. Sverdlov, *Mater. Sci. Eng.*, **A302** (2001) 18.

[103]Cambridge Engineering Selector (CES), ver 3.1, Granta Design Limited, Cambridge, United Kingdom, 2000.
[104]N. Eliaz, T. M. Sridhar, and E. Gileadi, *Electrochim. Acta* **50** (2005) 2893.
[105]O.Younes and E. Gileadi, *Electrochem. Solid-State Lett.* **3** (2000) 543.
[106]O. Younes, *PhD thesis*, Tel-Aviv University, 2001.
[107]L. Zhu, O.Younes, N. Ashkenasy, Y. Shacham-Diamand, and E. Gileadi, *Appl. Surf. Sci.* **200** (2002) 1.
[108]T. M. Sridhar, N. Eliaz, and E. Gileadi, *Electrochem. Solid-State Lett.* **8** (2005) C58.
[109]S. Yao, S. Zhao, H.Guo, and M. Kowaka, *Corrosion*, **52** (1996) 183.
[110]PWA 36975: Electroplated NiW – Thin Deposit (Enloy Ni-500); http://www.enthone.com.
[111]P. Schloßmacher and T. Yamasaki, *Mikrochim. Acta* **132** (2000) 309.
[112]N. Atanassov, K. Gencheva, and M. Bratoeva, *Plat. Surf. Finish.* **84** (1997) 67.
[113]R. M. Krishnan, C. Joseph Kennedy, S. Jayakrishnan, S. Sriveeraraghavan, S. R. Natarajan, and P. G. Venkatakrishnan, *Metal Finish* **93** (1995) 33.
[114]L. Namburi, MSc thesis, Louisiana State University, 2001.
[115]N. Suliţanu, *J. Magnetism and Magnetic Mater.* **231** (2001) 85.
[116]N. D. Suliţanu, *Mater. Sci. Eng.* **B95** (2002) 230.
[117]C. G. Fink and F. L. Jones, *Trans. Electrochem. Soc.* **59** (1931) 461.
[118]L. N. Gol'tz and V. N. Kharlamov, *Zhur. Priklad. Khim.* **9** (1936) 640.
[119]A. Brenner, P. S. Burkhead, and E. Seegmiller, *J. Res. Natl. Bur. Standards* **39** (1947) 351.
[120]H. Offermanns and Mv. Stackelberg, *Metalloberfläche* **1** (1947) 142.
[121]T. P. Hoar and I. A. Bucklow, *Trans. Inst. Metal. Finishing* **32** (1955) 186.
[122]L. E.Vaaler and M. L. Holt, *Trans. Electrochem. Soc.* **90** (1946) 43.
[123]T. Yamasaki, R. Tomohira, Y. Ogino, P. Schloßmacher, and K. Ehrlich, *Plat. Surf. Finish* **87** (2000) 148.
[124]H. Cesiulis, A. Baltutiene, M. Donten, M. L. Donten, and Z. Stojek, *J. Solid State Electrochem.* **6** (2002) 237.
[125]M. Donten, Z. Stojek, and H. Cesiulis, *J. Electrochem. Soc.* **150** (2003) C95.
[126]W. H. Safranek, *The properties of electrodeposited metals and alloys*, 2nd edn. American Electroplaters and Surface Finishers Society, Florida, 1986, p 348.
[127]C. H. Huang, *Plat. Surf. Finish* **84** (1997) 62.
[128]T. F. Frantsevich-Zabludovskaya, and A. I. Zayats, *Zh. Prikl. Khim.* **30** (1957) 764.
[129]C. H. Huang, W. Y. She, and H. M. Wu, *Plat. Surf. Finish* **86** (1999) 79.
[130]M. Donten, and Z. Stojek, *J. Appl. Electrochem.* **26** (1996) 665.
[131]K. Wikiel, and J. Osteryoung, *J. Appl. Electrochem.* **22** (1992) 506.
[132]C. L. Aravinda, V. S. Muralidharan, and S. M. Mayanna, *J. Appl. Electrochem.* **30** (2000) 601.
[133]K. S. Jackson, A. S. Russell, and J. L. Merrill, *J. Chem. Soc.* (1929) 2394.
[134]A. G. Glazunov and V. Jolkin, *Chem. Listy.* **31** (1937) 309,322.
[135]G. L. Davis and C. H. R. Gentry, *Metallurgia* **53** (1956) 3.
[136]Z. A. Solovyerva and A. T. Vagramyan, *Izvest. Akad. Nauk. Otdel. Khim. Nauk.* **2** (1954) 230.
[137]J. J. Cruywagen, L. Krüger and E. A. Rohwer, *J. Chem. Soc. Dalton Trans.* (1991) 1727.
[138]H. J. Seim and M. L. Holt, *J. Electrochem. Soc.* **96** (1949) 205.

[139]T. F. Frantsevich-Zabludovskaya, A. I. Zayats, and K. D. Modylevskaya, *Zhur. Priklad. Khim.* **29** (1956) 1811.
[140]D. W. Ernst, R. F. Amlie and M. L. Holt, *J. Electrochem. Soc.* **102** (1955) 461.
[141]H. S. Myers, PhD thesis, Columbia University, 1941.
[142]L. O. Case and A. Krohn, *J. Electrochem. Soc.* **105** (1958)512.
[143]A. Brenner and P. Burkhead, US Patent 2 653 127, 1953.
[144]T. F. Frantsevich-Zabludovskaya, and A. I. Zayats, *Zhur. Priklad. Khim.* **31** (1958) 224.
[145]C. C.Nee, W. Kim, and R. Weil, *J. Electrochem. Soc.* **135** (1988) 1100.
[146]E. Chassaing, M. P. Roumegas, and M. F. Trichet, *J. Appl. Electrochem.* **25** (1995) 667.
[147]D. W. Ernst, and M. L. Holt, *J. Electrochem. Soc.* **105** (1958) 686.
[148]E. J. Podlaha and D. Landolt, *J. Electrochem. Soc.* **140** (1993) L149.
[149]C. Madore, C. West, M. Matlosz and D. Landolt, *Electrochim. Acta*,**7** (1992) 69.
[150]C. Madore, M. Matlosz and D. Landolt, *J. Appl. Electrochem.* **22** (1992) 1155.
[151]C. Madore and D. Landolt, *Plat. Surf. Finish* **80** (1993) 73.
[152]E. Gómez, E. Pellicer, and E. Vallés, *J. Electroanal. Chem.* **517** (2001) 109.
[153]E. Chassaing, K. V. Quang, and R. Wiart, *J. Appl. Electrochem.* **19** (1989) 839.
[154]Z. J. Niu, S. B. Yao, and S. M. Zhou, *J. Electroanal. Chem.* **455** (1998) 205.
[155]Y. Zeng, Z. Li, M. Ma, and S. Zhou, *Electrochem. Commun.* **2** (2000) 36.
[156]J.Crousier, M. Eyraud, J. P. Crousier, and J. M. Roman, *J. Appl. Electrochem.* **22** (1992) 749.
[157]J. B. Lambert, in *ASM handbook*, 10th ed., Vol. 2, ASM International, Ohio, 1990, p. 557.
[158]T. Grobstein, R. Titran and J. R. Stephens, in ibid, 1990, p 581.
[159]R. Colton, *The chemistry of rhenium and technetium*, John Wiley & Sons, London, 1965.
[160]K. B. Lebedev, *The chemistry of rhenium*, Butterworths, London, 1962.
[161]D. Mittendorf and G. A. West, *Mater. Manuf. Processes* **13** (1998) 749.
[162]A. R. Meyer, *Trans. Inst. Metal Finishing* **46** (1968) 209.
[163]G. Erickson, K. Harris, and R. Schwer, *Development of CMSX-5: a third-generation high strength single crystal superalloy.* Cannon-Muskegon Corp., Michigan, 1985.
[164]N. Czech, F. Schmitz, and W. Stamm, *Surf. Coatings Technol.* **68/69** (1994) 17.
[165]S. J. Schneider, *High temperature thruster technology for spacecraft propulsion*, NASA technical memorandum 105348, IAF-91-254, 1991.
[166]M. Treska, L. W. Hobbs, and J. P. Pemsler, in *Microscopy of oxidation 3*, Ed. by S. B. Newcomb, J. A. Little, Institute of Materials, London, 1997, p 720.
[167]B. D. Reed, J. A. Biaglow, and S. J. Schneider, *Mater. Manuf. Processes* **13** (1998) 757.
[168]M. Pourbaix, *Atlas of electrochemical equilibria in aqueous solutions*, NACE, Texas, 1974, p 300.
[169]A. J. Bard, R. Parsons, and J. Jordan, *Standard potentials in aqueous solutions*, Dekker, New York, 1985, p 444.
[170]C. G. Fink and P. Deren, *Trans. Electrochem. Soc.* **66** (1934) 471.
[171]L. E. Netherton and M. L. Holt, *Trans. Electrochem. Soc.* **95** (1949) 324.
[172]G. S. Root, J. G. Beach, in *Rhenium – Electrochemical Society Symposium*, Ed. by B. W. Gonser, Elsevier, New York, 1962, p. 181.
[173]N. Eliaz, M. Treska, L. W. Hobbs, Pemsler JP (2000) unpublished data.

[174]S. A. Sastri, J. P. Pemsler, R. A. Cooke, J. K. Litchfield, and M. B. Smith U.S. Patent 5, 495, 979, 1996.
[175]P. Pemsler, *Electrochemical coating of complex carbon-carbon parts for low mass divert propulsion systems*, SBIR Phase I Contract N00167-99-C-0058, Castle Technology Corp., Massachusetts, 2000.
[176]P. J. Paris, *Rev. Sci. Instrum.* **60** (1989) 2802.
[177]G. Horányi, I. Bakos, S. Szabó, and E. M. Rizmayer, *J. Electroanal. Chem.* **337** (1992) 365.
[178]I. Bakos, G. Horányi, S. Szabó, and E. M. Rizmayer, *J. Electroanal. Chem.* **359** (1993) 241.
[179]I. Bakos and G. Horányi, *J. Electroanal. Chem.* **375** (1994) 387.
[180]G. Horányi and I. Bakos, *J. Electroanal. Chem.* **378** (1994) 143.
[181]S. Szabó and I. Bakos, *React. Kinet. Catal. Lett.* **62** (1997) 267.
[182]S. Szabó, *Int. Rev. Phys. Chem.* **10** (1991) 207.
[183]S. Szabó and I. Bakos, *Stud. Surf. Sci. Catal.* **118** (1998) 269.
[184]S. Szabó and I. Bakos, *J. Electroanal. Chem.* **492** (2000) 103.
[185]S. Szabó and I. Bakos, *J. Sol. State Electrochem.* **8** (2004) 190.
[186]J. O. Zerbino, A. M. Castro Luna, C. F. Zinola, E. Méndez, and M. E. Martins, *J. Braz. Chem. Soc.* **13** (2002) 510.
[187] R. Schrebler, P. Cury, M. Orellana, H. Gómez, R. Córdova, and E. Dalchiele *Electrochim. Acta* **46** (2001) 4309.
[188]L. E. Netherton and M. L. Holt, *J. Electrochem. Soc.* **98** (1951)106.
[189]L. E. Netherton and M. L. Holt, *J. Electrochem. Soc.* **99** (1952) 44.
[190]Y. N. Sadana and Z. Z. Wang, *Surf. Coatings Technol.* **37** (1989) 419.
[191]I. M. Kvokova and V. I. Lainer, *Zashchita Metallov* **1** (1965) 515.
[192]J. J. Vajo, D. A. Aikens, L. Ashley, D. E. Poeltl, R. A. Bailey, H. M. Clark, and S. C. Bunce, *Inorg. Chem.*, 20 (1981) 3328.
[193]L. R. Sillén and A.E. Martell, *Stability constants of metal-ion complexes, supplement 1,* The Chemical Society, London, 1971.
[194]L. R. Sillén and A.E. Martell, *Stability constants of metal-ion complexes, supplement 17.* The Chemical Society, London, 1964.
[195]G. R. Hedwig, J. R. Liddle and R. D. Reeves, *Aust. J. Chem.* **33** (1980) 1685.
[196] J. J. Cruywagen and R. F. van de Water, *Polyhedron* **5** (1986) 521.

5

25 Years of the Scanning Tunneling Microscopy

20 Years of Application of STM in Electrochemistry

Marek Szklarczyk,* Marcin Strawski, Krzysztof Bieńkowski

Laboratory of Electrochemistry, Department of Chemistry, Warsaw University, ul. Pasteura 1, 02-093 Warsaw, Poland

I. INTRODUCTION

"for their design of the scanning tunneling microscope"
The Nobel Prize in Physics, 1986

In the spring of 1981 at the IBM Zürich Research Laboratory, Gerd Binnig, Heinrich Rohrer, Christoph Gerber and Edmund Weibel observed a tunneling current between a platinum sample and a tungsten tip. This phenomenon combined with the ability to scan the tip against the sample surface gave the birth of scanning tunneling microscope (STM). Since then this novel type of microscopy has been providing information about nanometer-scale properties of matter. The rapid development in the tunneling microscope has triggered off the invention of the whole family of

*Author to whom correspondence should be addressed.

Modern Aspects of Electrochemistry, Number 42, edited by C. Vayenas et al., Springer, New York, 2008.

scanning probe microscopies (SPM), which makes use of almost every kind of conceivable interaction between a sharp tip and a sample. In 1986, just only five years after the invention of STM, the Atomic Force Microscope (AFM) was presented by Gerd Binnig, Calvin Quate and Christoph Gerber. The invention of STM has dramatically influenced basic science and technology on the nanometer scale. The impact of this new technique on our perception of the nanoscale world, and the offered opportunity to observe the beauty of nature at the atomic level, led to Gerd Binnig and Heinrich Rohrer being awarded the Nobel Prize in 1986 for the STM invention.

The unique and powerful advantage offered by STM and all STM derived techniques is the ability to perform *direct* studies of a great number of different phenomena. It is a particularly powerful technique for revealing the local aspects of the atomic and electronic surface structures and dynamics. Studies of key energetic parameters, like structural, electronic, thermal, mechanical, and all other properties of materials can be done with extremely high spatial resolution, down to the atomic scale. Many other techniques such as e.g. electron diffraction, transmission electron microscopy, and X-ray diffraction provide only *indirect* information about sample structures, while offering to probe certain features at the atomic scale, inevitably average these properties over substantially larger areas.

At present, the microscopes founded on the STM idea are the most ubiquitous tools in surface science. They work from the micrometer down to the sub-nanometer scale. The studied phenomena include surface topography, adsorption, surface diffusion, surface potential and surface electric field, the strength of individual chemical bonds, electronic and vibrational properties, nanotribology: friction, lubrication, wear and adhesion, dielectric and magnetic properties. The range of phenomena studied by SPM techniques is continuing to grow. What is more, the use of these techniques in the molecular manipulation and rapidly growing nanotechnology fields is also increasing.

SPM can be applied in any environment and provides the ways in which theory, ultra high vacuum (UHV) and in situ observations can be linked directly. These possibilities have opened a wide variety of important fields for SPM application. Biology and electrochemistry prior to the development SPM had to infer some necessary information, e.g., structural one, from the spectroscopic or microscopic measurements gained under *ex situ* conditions, i.e. under UHV conditions. The use of SPM techniques allow to collect experimental data in the real space and in real time.

The application of the STM to study of structural problems at the electrified solid-liquid interface has been one of the most important advances in electrochemistry over the past two decades. STM data provide a hitherto unobtainable level of structural insight into surfaces in solution. It is because the charge transfer and capacitive charging event central to electrochemical reactivity occur within a few atomic diameters of the electrode surface in the inner Helmholtz plane, so the detailed arrangement of atoms and molecules at this interface strongly controls the corresponding electrochemical reactivity. Although classical electrochemical methods such as e.g., cyclic voltammetry have provided remarkable sensitivity for the characterization of submonolayer processes occurring at electrode/liquid interfaces, the scanning probe microscopy has particularly great potential to overcome the difficulty of heterogeneity, giving the possibility of discrimination of specific sites from the other sites in the atomic scale. The twenty years of application of *in-situ* STM in electrochemistry have contributed to a better understanding of structure and processes at the solid liquid interface. Several important developments altogether have brought about the emergence of in-situ electrochemical surface science.

Like in other fields of science, in the SPM domain a lot of effort is devoted to develop a reliable theoretical model of experiments, which can be used to interpret experimental images without resorting to direct simulation. The aim of the theory would be to provide an understanding of image contrast, and give an interpretation of particular experimental images. Several experimental and theoretical groups had been working together in order to arrive to sustainable models. However, it seems that this goal has been achieved only to some extent. This is because as the complexity of the experimental SPM techniques continues to increase, it becomes more difficult. There is no doubt that the SPM tip can influence the acquired image data, which introduces additional difficulties in the development of the theoretical model. The main difference between SPM modeling and, for example, modeling of adsorption or modeling of other surface phenomena is a very rare equilibrium of the SPM tip with the surface. Additionally, many experimentally observed phenomena are unique and are not subject to statistical averaging. Consequently, theories until recently have been concerned mainly with qualitative predictions. The development of quantitative models will require determination of parameters for comparison, formulation of criteria of agreement, and common calibration for theory and experiment. Currently, experimental research is increasingly focused on still more subtle effects. One reason for this is the continuing integration of chemistry, solid state physics, and biology. This poses unprecedented challenges of a precision for theoretical models aiming to reach a common framework of understanding.

Despite the experimental problems experienced daily, such as tip/cantilever preparation, scanned surface preparation, noise dumping, data interpretation, the SPM field is still developing rapidly and new SPM techniques are proposed.

Twenty five years after registering the first STM image and twenty years after first application of STM for studying electrochemical system, and also twenty years after the Nobel award for STM invention it is already impossible to present a complete review of all experimental and theoretical work which have been devoted to SPM field as well as to provide full reference list of thousands of papers which have already been published. This is not the aim of this chapter. There is a lot of excellent review papers and books devoted to SPM field as the whole, and to the particular SPM application (e.g., Refs. 1–15 and cited therein). Our intention is to give a compact description of any advance in overcoming successive threshold in the application of the STM technique, pointing to new possibilities and advantages of these techniques, and calling to mind older reports, which have a real impact on today's stage of the STM application in different fields. The description of the results will be summarized as landmarks in a tabularized form.

II. STM LANDMARKS

The experimental and theoretical work on tunneling current phenomena[16-25] and wave nature of microscopic particles[26] (cf. Table 1) finally resulted in the portentous night of 16 March, 1981, which can be considered as the date of birth of STM.[1] That night, under high vacuum conditions (UHV) the first experimental *log i-s* characteristics of the tunneling current, *i*, on the tip-surface separation, *s*, were obtained for a platinum sample.[27,28] However, the real breakthrough in the STM field came with the first atomic resolution image of the Si(111) 7×7 surface in 1982.[29,30] Some consider this achievement as the birth date of modern tunneling microscope, superior in comparison with all other vacuum tunneling instruments. The most advanced apparatus presented to this time has been a microscope built up by Young et al.[31,32] Due to the fact that in this instrument the tip sample separation was controlled by a field emission current, the lateral resolution was limited to only about 400 nm. In fact this microscope was the first one in the family of SPM to attract much interest among the scientific community due to its ultimate resolution power down even to the atomic scale, suggested by Young et al.[32] The other reports on observation of vacuum tunneling followed. E. C. Teague reported *i(s)* curves in his thesis,[33,34] and U. Poppe reported vacuum tunneling, but he did not measure *i(s)* dependence.[35]

The application of STM can be divided into different fields of investigation; solid-vacuum interface (UHV), solid-gas interface, solid-liquid interface (electrochemistry), biology, and nanotechnology. Some landmarks are shown in Table 1.

1. Solid-Vacuum Interface STM Investigations

Over the first ten years following its invention, STM has had its greatest impact on the field of surface science. Numerous structures of clean and adsorbate-covered surfaces have been obtained. The image of the Si(111)7×7 reconstructed surface showing two complete rhombohedral unit cells was the first historical atomic resolution STM image and it is presented in Fig. 1. This picture demonstrates that STM is able to clearly resolve individual atoms on a surface, less than 7-Å apart.[29,30] Furthermore, the analysis of this image shows that part of the rhombic cell lies higher than the other part, which was interpreted as an electronic inequivalence on the surface owing to a structural inequivalence in the underlying layers. The other materials studied by Binnig et al. were $CaIrSn_4$ and Au(110) single crystal.[36] The authors reported the observation of monoatomic steps for surfaces of these materials. The corruga-

Table 1
Landmarks in the STM Field

Tunneling	
• Observation of field emission from metals, J. E. Lilienfeld, 1922.[16] • Wave nature of microscopic particles, L. de Broglie, 1923.[26] • Theory of α decay phenomenon, G. Gamov, R. W. Gourney, E.U. Condon, 1928.[18,19] • Model of field emission, R. H. Fowler, L. Nordheim, 1928.[17] • Tunneling in p-n junctions, L. Esaki, 1958.[20] • Measurement of energy gap of superconductors, I. Giaever, 1960.[21] • Perturbation theory of tunneling, J. Bardeen, 1961.[22] • Tunneling of Cooper particles, B. D. Josephson, 1962.[23] • Experimental verification of Josephson effect, P. W. Anderson, J. M. Rowell, 1963.[24] • Inelastic tunneling spectroscopy, R.C. Jaklevic, J. Lambe, 1966.[25] • Three dimensional theory of the scanning tunneling microscopy, J. Tersoff, D. R. Hamann, 1985.[43]	
UHV imaging	**Imaging under air and liquids***
• Topografiner instrument and work on vacuum tunneling and field emission. (R. Young, J. Ward, F. Scire, 1971)[31,32] • Development of scanning tunneling microscope. The first atomic resolution image,Si(111) 7×7 surface. (G. Binning, H. Rohrer, Ch. Gerber, E. Weibel, 1982)[27,30]	• Imaging in air of biological sample bacteriophage ϕ29. (A.M. Baró, R. Miranda, J. Alamán, N. Garcia, G. Binnig, H. Rohrer, Ch. Gerber, L. Carrascosa,1985)[56] • The invention of AFM microscope (G. Binnig, C. F. Quate, Ch. Gerber, 1986)[62] • Imaging under water. (R. Sonnenfeld, P.K. Hansma, 1986)[58]

Table 1. Continuation

UHV imaging	Imaging under air and liquids*
• The first atomic resolution image of a metal surface, Au(110). (G. Binning, H. Rohrer, Ch. Gerber, E. Weibel, 1983)[37]	• Imaging in electrolyte solution with potential control. (R. Sonnenfeld, B.C. Schardt, 1986)[59]
• Atomic resolution image of adatoms, O, on surface of Ni(110). (A. M. Baró, G. Binnig, H. Rohrer, Ch. Gerber, E. Stoll, A. Baratoff, F. Salvan, 1984)[44]	• Atomic corrugation of metal surface in air, Au(111). (V. M Hallmark, S. Chiang, J. F. Rabolt, J. D. Swalen, R. J. Wilson, 1987)[67]
• Application of STM in biology, observation of DNA chains. (G. Binning, H. Rohrer, 1984)[45]	• First commercial STM unit. (1987, cf. text).
• Images of graphite structures. (P. K. Hansma, 1985)[46]	• Molecular manipulation in air, moving a di(2-ethylhexyl)phthalate molecule by ca. 50 nm) (J. S. Foster, J. E. Frommer, P. C. Arnett, 1988)[135]
• Beginning of lithography. (M. Ringger, H.R. Hedber, R. Schlogl, P. Oelhafen, H. J. Guntherodt, 1985)[60]	• Monitoring of Au(111) surface changes in dependence on electrode potential and chloride ions adsorption. (J. Wiechers, T. Twomey, D. M. Kolb, R. J. Behm, 1988)[125]
• Atom handling. (R. S. Becker, J.A. Golovchenko, B. S. Swartzentruber, 1987)[61]	• Minitoring of HOPG surface oxidation. (A. A. Gewirth, A. J. Bard 1988)[142]
• Atomic corrugation of a metal surface, Au(111). (V. M. Hallmark, S. Chiang, J. F. Rabolt, J. D. Swalen, R. J. Wilson, 1987)[67]	• Application of STM to programmed metal deposition within polymer film. (D. H. Craston, C. W. Lin, A. J. Bard, 1988)[153]
• Molecular resolution of benzene. (H. Ohtani, R. J. Wilson, S. Chiang, C. M. Mate, 1988)[80]	• First atomic corrugation images in air for an adsorbate, Pt(111)-($\sqrt{7}\times\sqrt{7}$)R19.1°-I adlattice. (B.C. Schardt, S-L. Yau, F. Rinaldi, 1989)[154]
• Resolution of the internal structure of copper-phthalocyanine molecules. (P. H. Lippel, R. J. Wilson, M. D. Miller, Ch. Wöll, S. Chiang, 1989)[81]	• First atomic resolution of metal surface under in situ condition of bulk deposited metal, Pb(100). (M. Szklarczyk, O. Velev, J. O'M. Bockris, 1989)[157]

Table 1. Continuation

UHV imaging	Imaging under air and liquids*
• The use of STM to move and position atoms; formation of IBM inscription of Xe atoms. (D. M. Eigler, E. K. Schweizer, 1990)[101] • Imaging of electronic standing waves. (M. F. Crommie, C. P. Lutz, D. M. Eigler, 1993)[98] • Imaging of dissociation of single molecules on both semiconductor[105] and metal[104] has been accomplished. (G. Dujardin, R. E. Walkup, Ph. Avouris, 1992;[105] B. C. Stipe, M. A. Rezaei, W. Ho, S. Gao, M. Persson, B. I. Lundqvist, 1997[104]) • Observation of rotation forced by STM tip of adsorbed molecules, Sb dimmers[117] and O_2 molecules[118] (Y.W. Mo, 1993;[117] B. C. Stipe, M. A. Rezaei, W. Ho 1998[118]) • Registering of inelastic tunneling spectra for an isolated acetylene, and distinguishing the C_2H_2 from C_2D_2 molecule. (B. C. Stipe, M. A. Rezaei, W. Ho 1998)[118] • The microscopic characterization of bond formation between CO molecule and Fe atom. (H. J. Lee, W. Ho, 1999)[106] • Controlling of the molecular dynamics of a single molecule through the localization of the electronic excitation inside the molecule. (M. Lastapis, M. Martin, D. Riedel, L. Hellner, G. Comtet, G. Dujardin, 2005)[112]	• *In situ* monitoring of semiconductor surface, n-TiO_2. (K. Itaya, E. Tomita, 1989)[245] • The first model of electron tunneling through water layers was proposed by W. Schmickler and D. Henderson 1990.[131] • High resolution atomic images of the Pt(111)-($\sqrt{3}\times\sqrt{3}$)R30°-I adlattice and Pt substrate. (S-L. Yau, M. Carissima, M. Vitus, B. C. Schardt, 1990)[171] • Atomic resolution of Au(100) and UPD deposited Cu. (O. M. Magnussen, J. Hotlos, R. J. Nichols, D. M. Kolb, R. J. Behm, 1990)[173] • Using of the STM to investigate the surface state formed in the electrochemical environment in semiconductor, p-Si(111). (M. Szklarczyk, A. González-Martin, O. Velev, J. O'M. Bockris, 1990)[248] • First nanofabrication of nanostructures on semiconductors, Si(100) and GaAs(100). (L. A. Nagahara, T. Thundat, S. M. Lindsay, 1990)[268] • Molecular resolution of organic molecules, guanine on HOPG. (R. Srinivasan, J. C. Murphy, R. Fainchtein, N. Pattabiraman, 1991)[226] • Direct imaging of SO_4^{2-} adsorbed on metal, Au(111). (O. M. Magnussen, J. Hageböck, J. Hotlos, R. J. Behm, 1992)[203] • Imaging of S^{2-} adlayer, Au (111) electrode. (X. Gao, Y. Zhang, M. J. Weaver, 1992)[224]

Table 1. Continuation

UHV imaging	Imaging under air and liquids*
• Fabrication and monitoring the *walking* molecular ensemble, molecules of 9,10-thioanthracene (DTA) on Cu(111) surface. (K-Y. Kwon, K. L. Wong, G. Pawin, L. Bartels, S. Stolbov, T. S. Rahman, 2005)[119] • Monitoring the ability to switch the charged state of an individual metal atom in the reversible way by voltage pulse. (J. Repp, G. Meyer, F. E. Olsson, M. Persson, 2004)[120]	• Atomic resolution of semiconductor surface, Si(111). (S. –L. You, F. –R. Fan, A. J. Bard, 1992)[253] • First report on nanofabrication of metal nanoclusters, silver and copper nanoclusters. (W. J. Li. J. A. Nirtanen, R. M. Penner, 1992)[259,260] • Imaging of self-assembly of inorganic anions, monolayer α-$SiW_{12}O_{40}{}^{4-}$ on Ag (111) surface. (M. Ge, B. Zhong, W. G. Klemperer, A. A. Gewirth, 1994)[225] • Resolution of the (3×3) adlayer structure of benzene on Rh(111) surface. (S. –L. Yau, Y. –G. Kim, K. Itaya, 1996)[182] • Resolution of internal structure of porphyrin adsorbed on Ag(111)-I surface. (K. Ogaki, N. Batina, M. Kunitake, K. Itaya, 1996)[233] • The use of STM to the identification of structurally similar molecules by probing resonant tunneling. (N. J. Tao, 1996)[241] • Fabrication of nanowires, (C. Z. Li. N. J. Tao, 1998)[269]

* If there is no other notice, studies were carried out under *in situ* conditions.

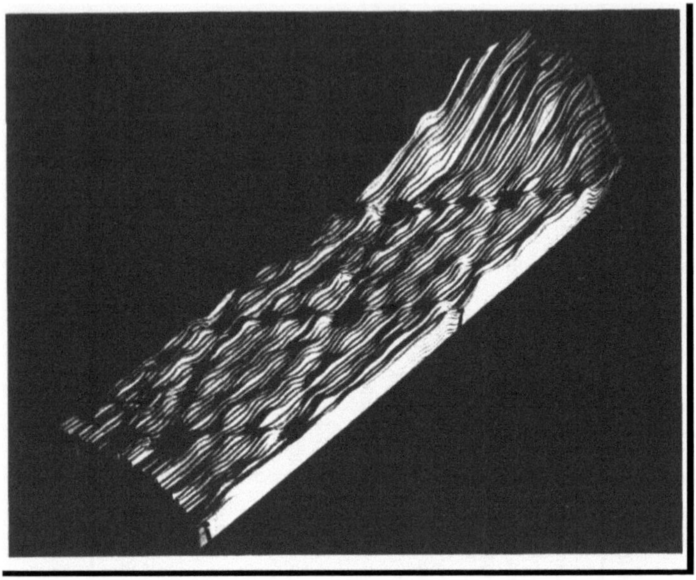

Figure 1. UHV-STM image of the Si(111) 7x7 surface showing two complete rhombohedral unit cells. Reprinted by permission from Ref. 29, Copyright (1982) Helvetica Phisica Acta.

tion in the (001) direction for a gold surface was described. The first report on atomic resolution of the metal surface was published by the same group in 1983[37] The authors succeeded in resolving the problem of the missing atomic row for the 1×2 reconstruction on the Au(110) surface (Fig. 2). Meanwhile islands of Au metal, on a Si semiconductor surface, were imaged by the same authors.[29,38]

The successful STM experiments accelerated the first theoretical efforts on a microscopic level. Tersoff and Hamann,[39] and Baratoff[40] applied Bardeen's formalism to the small geometries of STM tip and atomically corrugated surface. Garcia et al.[41] and Stoll et al.[42] worked out a scattering approach to tunneling in small geometries. In 1985 Tersoff and Hamann applied their model to test the structure sensitivity of the STM apparatus.[43]

STM is an excellent example of a practical exploitation of a strictly quantum mechanical phenomenon, i.e., quantum tunneling.

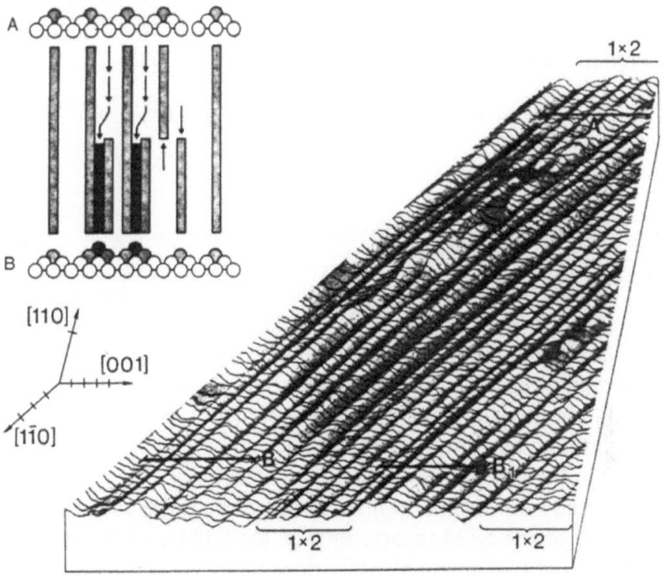

Figure 2. UHV-STM atomic resolution image of the Au(110) surface (structure A) showing 1×2 reconstructed ribbons (structure B). Reprinted from Ref. 37, Copyright (1983), with permission from Elsevier.

The effect which involves the penetration of a potential barrier by an electron wavefunction originating from the well and penetrating the potential barrier which is described by the Schrödinger equation with two components;[6] the first one, for tunneling distance $x < 0$, inside the well,

$$\boldsymbol{H} = -(\hbar^2/2m)(\mathrm{d}^2/\mathrm{d}x^2) \quad (1)$$

and the second one, for $x > 0$, inside the potential barrier.

$$\boldsymbol{H} = -(\hbar^2/2m)(\mathrm{d}^2/\mathrm{dx}^2) + V \quad (2)$$

where $\hbar$ is Planck's constant divided by 2π, m is the electron mass and V is the height of potential barrier.

The solutions of these equations, for electron of energy *E,* are:

inside the well,

$$\Psi = Ae^{ikx} + Be^{-ikx}, \text{ where } k = (2mE/\hbar^2) \quad (3)$$

and inside the barrier,

$$\Psi = CAe^{ik'x} + De^{-ik'x}, \text{ where } k' = (2m(E-V)/\hbar^2)^{1/2}. \quad (4)$$

The wavefunction, Ψ, inside the barrier has an imaginary part rising to infinity and is thus discounted and a real part which decays exponentially with distance inside the barrier. This result is important because where penetration is classically forbidden, for $E < V$, the quantum mechanical wavefunction is non-zero and the electron may enter and tunnel into the potential barrier. There is a finite probability that an electron will be found in the potential barrier and a finite current may be detected. These conclusions are the basis of the STM theory. In fact, the STM case involves consideration of two potential wells close together, i.e., separated be a potential barrier of finite thickness. The condition for the flow of tunneling current is a small separation distance *d* between wells. Under this condition the overlap of two wavefunctions might be sufficiently large to facilitate quantum mechanical tunneling and under an applied potential difference, the passage of a measurable current might be observed. This macroscopic current is a measure of the overlap of the two quantum wavefunctions. The magnitude of current in this one dimensional model is given by

$$i = \propto e^{(-2Ks)} \text{ where } K = (2m\phi/\hbar^2)^{1/2} \quad (5)$$

and *s* is tunneling barrier thickness, and ϕ is the local work function.

In practice, the above relation (5) is only a good approximation and is not really adequate for full description of STM. The electronic structures of the tip and the studied surface are different and their work functions are in general not equal. Furthermore, the density of states above the surface varies with its crystallographic structure and the decay of electron states depends on their location in the Brillouin zone. These complex state demands much more complicated a three dimensional treatment. The cited authors[39-43] have included these problems in their theoretical considerations. One of the most widely quoted treatment is due to Tersoff and Hamann.[43] The general expression for the tunneling current is:[43]

$$i = (2\pi e\phi/\hbar)e^2 V \sum_{\mu,\nu} |M_{\mu,\nu}|^2 \delta(E_\nu - E_F)\delta(E_\mu - E_F) \quad (6)$$

where E_F is the Fermi energy, E_μ is the energy of the state Ψ_μ in the absence of tunneling, and $M_{\mu,\nu}$ is the tunneling matrix element between states Ψ_μ of the probe tip and Ψ_ν of the sample.

For the spherical STM tip, low tunneling voltages and close work functions of a tip and a sample, i.e., for a real experimental conditions, Tersoff and Hamann showed that the expression for tunneling current is[43]

$$i = 32\pi^3 \hbar^{-1} e^2 V \phi_o^2 D(E_F) R^2 K^{-4} e^{2KR} \sum |\Psi_\nu(r_o)|^2 \delta(E_\nu - E_F) \quad (7)$$

where ϕ_o is the average work function, R is the tip radius, r_o is the position of the center of tip curvature determined by $(s + R)$, and $D(E_F)$ is the density of states at the Fermi level per volume of tip.

Equation (7) preserves important results which explain the astonishing resolution of the STM technique. It shows that the tunneling current depends exponentially on the distance between the STM tip and the studied sample. To a rough approximation, the tunneling current can decrease by an order of magnitude for every increase of 1 Å in the tunneling distance. Furthermore, Eq. (7) shows that the tunneling current is proportional to the local density of states at the Fermi level at the centre of the STM tip. This means that the STM can provide a direct image of electron states at the surface and, therefore, this equation provides a theoretical basis for the application of STM technique to the atomic-scale surface spectroscopy.

The successful theoretical efforts[39-43] on implementation of microscopic experimental details (e.g., tip diameter, tip-sample separation) into quantum theory have accelerated considerably the achievement of the top experimental STM results leading to the establishing the atomic-scale microscopy and spectroscopy and modern surface chemistry.

The realm of experimental surface chemistry was pioneered with the work of Baró et al.,[44] who observed an individual oxygen atom on a Ni(110) surface. Furthermore, the authors demonstrated that the oxygen overlayer was not irreversibly disordered by the scanning tunnel tip.

The first effort for the application of STM in biology was reported in 1984. Binnig and Rohrer reported that they could follow DNA chains lying on a carbon film deposited on a Ag-coated Si wafer.[45]

In the period of 1985–1987 a great number of papers were published. The authors reported results showing almost limitless possibilities of the STM technique. Hansma[46] and Binnig et al.,[47] presented atomically resolved graphite structures distinguishing energetically inequivalent carbon sites in graphite (Fig. 3).[47] Becker et al. described various reconstructions of Ge films deposited on Si(111),[48] Feenstra and Fein came up with cleaved GaAs(110) showing corrugations as small as 0.02 nm,[49] Quate's group resolved the 1×1 structure on Pt(100),[50] and Behm et al. monitored rotational domains of the hexagonal reconstruction on Pt(100).[51] Payne and Inkson[52] published theoretical and experimental studies

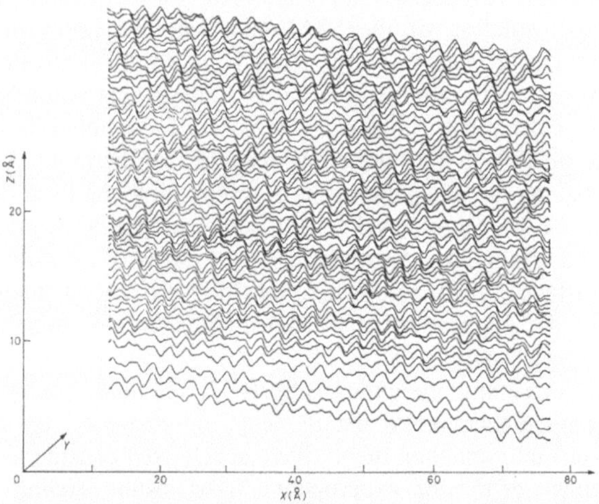

Figure 3. STM graph of a cleaved graphite surface as taken. Actual Y-distance from scan to scan is 0.246 ± 0.027 nm. Reprinted from Ref. 47, Copyright (1986) EDP Sciences.

on the effect of image forces on imaging parameters, voltage, current and tip-surface distance showing the influence of the image potential on the tunneling of electrons. Pashley, et al.[53] used an STM apparatus for measurement of the work function and tip-sample separation. The spectroscopic studies of single atoms showing that peaks in a plot of dlni/dlnV versus V correspond to the position of resonances in the densities of states of both sample and tip, have been published by Lang.[54] In 1987 this author published the first discussion on the apparent size of an atom in STM experiments,[55] and showed a strong dependence of the atomic size on bias voltage.

Meanwhile, the efforts for the application of STM in other environment than UHV were undertaken. Images of biological samples, bacteriophage ϕ 29, obtained in ambient-air pressure were reported by Baró et al. (Fig. 4).[56] STM was applied in cryo-

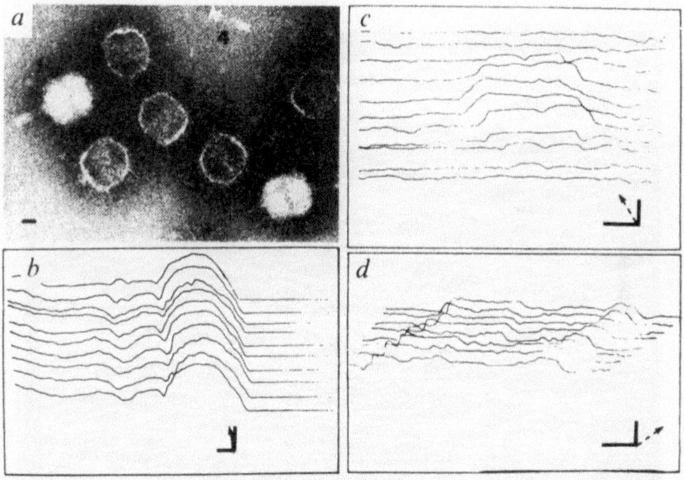

Figure 4. (a) Electron micrograph of various structures derived from bacteriophage ϕ 29. (b-d) STM images taken in air for structures derived from bacteriofage ϕ 29 drayed over graphite. Profiles in b were interpreted as a scan along head-tail axis, height of head *ca.* 20 nm, profiles in (c) were correlated with empty flattened structure an empty capsid without tail. Profiles presented in (b) were interpreted as scans along the head-tail axis of an empty phage particles with tail complex. The height of head in (c) and (d) *ca.* 7 nm. Scale bars 10 nm. Reprinted by permission from Macmillan Publishers Ltd: Ref. 56, Copyright (1985).

genic liquids by Coleman et al.,[57] under distilled water and in saline solution by Sonnenfeld and Hansma (Fig. 5),[58] and to the study of electrodeposition, by Sonnenfeld and Schardt.[59]

The other landmarks which happened in the middle of the 80's were a demonstration of the possibility of application of the STM apparatus for lithography (Fig. 6),[60] and for controlled atom handling.[61] The invention of AFM microscope[62] opened the route for the construction of several different force microscopies (SPM techniques). This way the use of STM/SPM techniques as a Feynman Machine[63] finally had been realized, the STM apprenticeship time came to an end and the time of travel began.

Initially, the application of STM to semiconductor surfaces was the most fruitful. Independently of the studies on surface geometry, research on surface chemistry was of great interest. The STM started to be used to study the early stages of oxidation,[64] adatoms adsorption,[65] and hydrogen forced surface reconstruction.[66]

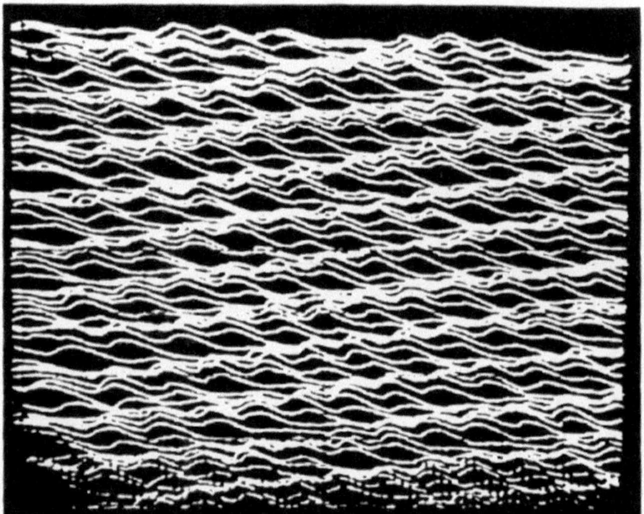

Figure 5. An STM image of a HOPG surface immersed in water. The graphite hexagonal lattice is clearly distinguished. Reprinted with permission from Ref. 58, Copyright © (1986) AAAS.

Figure 6. UHV-SEM image of $Pd_{81}Si_{19}$ surface showing three pairs crossed lines drown with STM tip. Reprinted with permission from Ref. 60, Copyright (1985) American Institute of Physics.

The first atomic resolution for metal surfaces came later because of the different surface electronic structure of semiconductors and metals. In the case of semiconductor, the energies

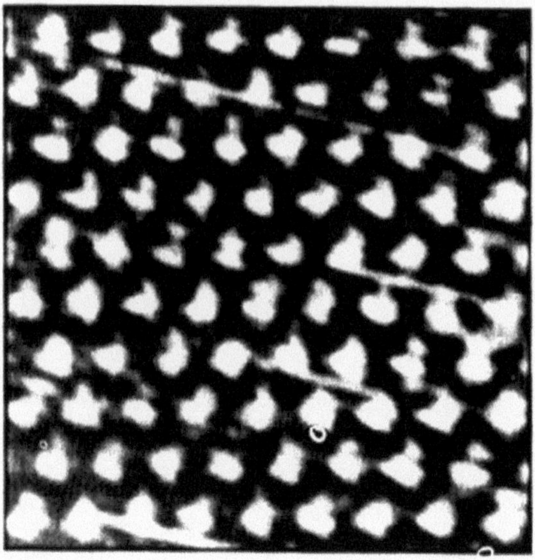

Figure 7. STM image of Au(111) surface taken in air. Reprinted with permission from Ref. 67, Copyright (1987) American Physical Society.

of dangling bonds with s or p surface electrons dominate the STM image. In the case of a metal, the electronic states can have s, p, d, f or hybrid character. The metallic states d and f are much more difficult to be observed due to their smaller amplitude in comparison with the dangling s or p electrons, which results from their stronger localization around atomic cores and which leads to an additional barrier for tunneling.[55] The ultimately available s and p states at metal surface have a delocalized character resulting in small energy corrugation amplitudes of the order 0.01 nm or less, which makes their imaging extremely difficult.

With the improvement of STM equipment and by collecting more experimental experience, the first observation of atomic corrugation on metal surface came in 1987.[67] Hallmark et al. presented the images of atomic corrugation on Au(111) registered in UHV and in air with the corrugation of 0.03 nm (cf. Fig. 7).[67] An Au(111) surface became a model surface for metals as it was in the case of Si(111) surface for semiconductors. Dynamic processes such as surface diffusion were observed by Jaklevic and Elie.[68] They observed the time dependent diffusion of surface marks made with the tunneling tip. Soon, atomic-resolution was reported for other metal surface. Wintterlin et al. showed the Al(111) surface.[69] They pointed out that a big corrugation observed, 0.03 nm at metal surfaces, must be due to the tip-surface interaction rather than to the electronic structure of a metal sample itself. Other metals surfaces were resolved also, e.g., Cu,[70,71] W,[72] and Pd.[73]

Right from the start of the STM research, the importance of the tunneling tip quality was obvious and the influence of the geometry and electronic structure of the tunneling tip were discussed. Tromp et al. reported the possibility of monitoring the empty or filled states on the sample by adsorbing an electronegative atom at the very end of a STM tip.[74] The effect of multiple geometry on registered images has been reported.[75–77] Sang-il Park et al. presented the image doubling phenomenon,[75] and the effect of imaging by two chemically different atoms adsorbed at the very end of the tip.[77]

The application of scanning tunneling microscopy to organic materials, ranging from small molecules to supramolecular assemblies, still has been a challenging task, much tougher than atomic resolution for solid surfaces. There is a variety of aspects to be taken into account. A suitable substrate with a surface roughness considerably less than the size of organic species to be imaged has to be available. Furthermore, the substrate has to be electrically conductive and chemically inert towards organics. Deposited organics have to be immobilized in some way to allow stable imaging. Although these problems have been solved, there is still considerable uncertainty about the interpretation of the obtained images. A detailed discussion of these problems is given in refs. 4 and 78. The early results by Baro et al.[56] have been followed by a wide range of studies on adsorbed molecules. Gimzewski et al.[79] reported monitoring individual molecules. The authors observed distinct topographic features on the surfaces, interpreted as copper-phthalocyanine molecules adsorbed on polycrystalline Ag. The low symmetry of the recorded images was interpreted in terms of molecular motion induced by the electric field gradients near the tip.[79] Ohtani et al. examined coadsorbed benzene and carbon monoxide on Rh(111) in 1988.[80] They reported molecular resolution

Figure 8. UHV-STM image of benzene molecules adsorbed on Rh(111) surface emphasizes the ring like structure of benzene molecule. Reprinted with permission from Ref. 80, Copyright (1988) American Physical Society.

for C_6H_6 (Fig. 8), while CO was not resolved. A further step in imaging organic molecules was made by Lippel et al. in 1989.[81] These authors presented high resolution images of copper-phthalocyanine molecules showing an internal atomic structure of the molecule of interest (Fig. 9). In 1993 Weiss and Eigler reported the dependence of the organic molecule shape, benzene, on the adsorption site.[82] They showed the dependence of the diameter and the height of the benzene molecule when it was adsorbed either at a top of a surface atom, or at the bridge site and at 3-fold hollow site (Fig. 10).[82] Despite the problems in the imaging of organics, there are a lot of papers which have been published with clear and beautiful images of different organic molecules, from simple ones (e.g., alkanes)[83] through molecular crystals,[84–86] DNA,[87–90] proteins,[91,92] to Langmuir-Blodgett films,[93–95] and the images are taken in any environment.

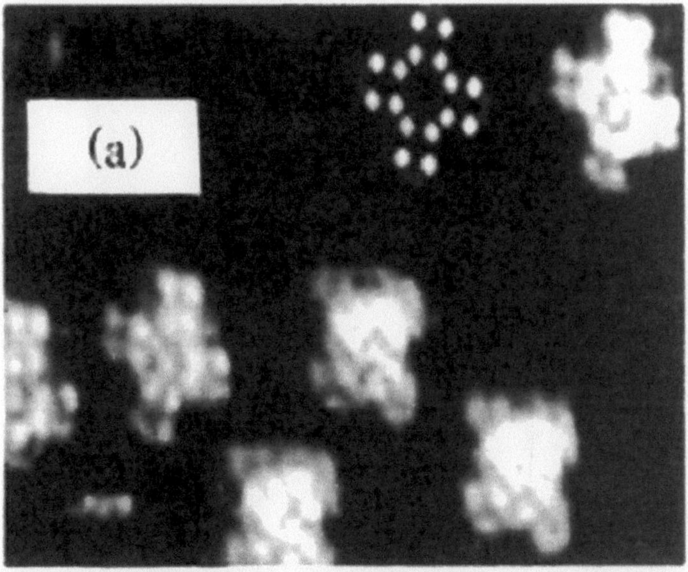

Figure 9. UHV-STM image of copper-phthalocyanine molecules showing fine structure of molecules. In the image is embedded HOMO calculated molecular scheme. Reprinted with permission from Ref. 81, Copyright (1989) American Physical Society.

The most tiny atom, hydrogen, was quoted to be imaged in 1993.[96] Wang et al. showed the images of Si(100) surface with adsorbed hydrogen. They show the dependence of density of structures identified as hydrogen with the hydrogen dosage to a vacuum chamber. The image of the hydrogen on metal surface, Pd(111) has been shown by T. Mitsui et al.[97]

In the same year the applications of STM to monitor electronic standing waves were reported in two well known works.[98,99] Crommie et al. described a method for confining electrons to artificial structures and the way of monitoring electronic waves formed in this way.[98] They presented the confinement property of the Fe adatoms on Cu(111) surface, and showed images of the secret resonance waves around one Fe atom and closed by 48 Fe atoms secret resonance waves (cf. Fig. 11).[98] To date, it seems that the images of structures presented in that work, which are of the

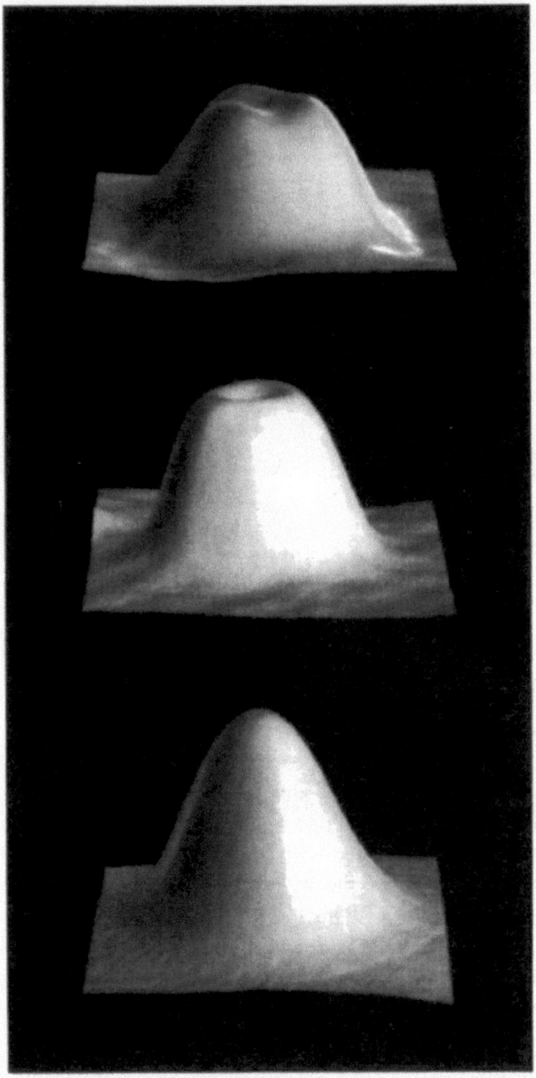

Figure 10. UHV-STM images, 1.5 × 1.5 nm, each showing a single adsorbed benzene molecule, a) assigned to hcp 3-fold hollow site, b) on top site, and c) on bridge site, respectively. Reprinted with permission from Ref. 82, Copyright (1993) American Physical Society.

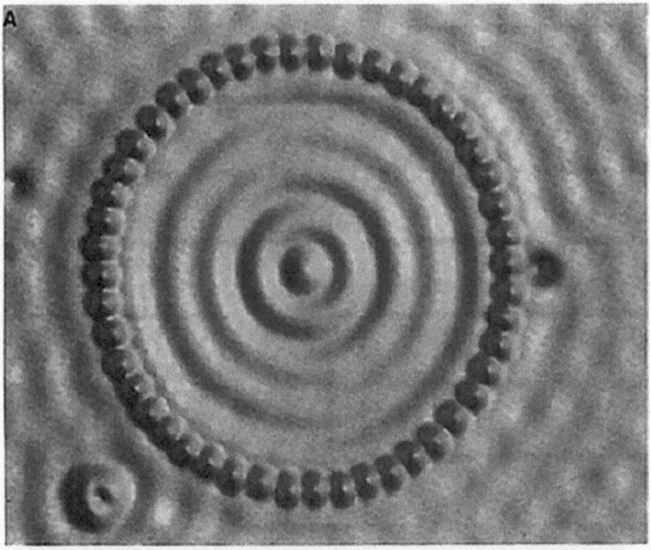

Figure 11. Spatial image of the eigenstates of a quantum corral made of 48 Fe atoms. Reprinted with permission from Ref. 98, Copyright (1993) AAAS.

order of de Broglie wavelength, thus, dominated by quantum mechanical effects, show the extreme of the STM possibilities. One can say that with the results presented in this paper the adolescence time of STM application was over and time of adulthood has started.

The ability of STM to control a surface at the spatial limits of individual atoms and molecules makes this technique enormously suitable for imaging chemical reactions at the level of single atoms or molecules. The following examples can be given on such applications of the STM apparatus.

A report on the early stages of thin-film growth has been given by Zhang and Lagally.[100] The authors described the early formation of a stable nucleus, fractal island growth, dendrite formation and kinetic phenomena, like island mobility and surface diffusion on an atomistic level. The scanning tunneling microscope has been used to move and position atoms.[101,102] and molecules.[103] Dissociation of single molecules has been accomplished on both

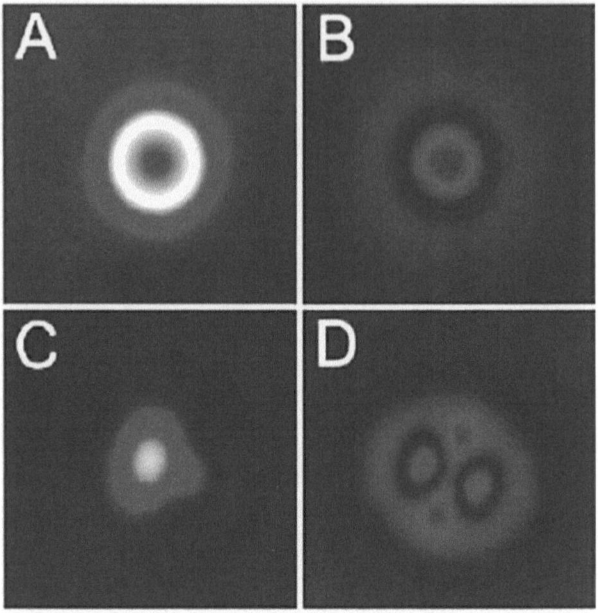

Figure 12. 2.5-nm by 2.5-nm topographic images for A) Fe, B) CO, C) FeCO and D) $Fe(CO)_2$. Reprinted with permission from Ref. 106, Copyright (1999) AAAS.

metals,[104] and semiconductor.[105] The microscopic characterization of bond formation between CO molecules and Fe atoms was reported in 1999, Fig.12.[106]

In 1966 it was discovered that vibrational spectra can be obtained from molecules adsorbed at the buried metal oxide interface.[25] In that experiment the tunneling current was measured as a function of voltage across the junction. In the experiment a peak d^2i/dV^2 dependence at each vibrational energy was observed. This method is known as inelastic tunneling spectroscopy (IETS). In the STM setup, this method is presently applied to study single molecules.[107] Stipe et al. have been shown inelastic tunneling spectra for an isolated acetylene molecule and they distinguished the C_2H_2 from the C_2D_2 molecules.[107] The scanning tunneling spectroscopy is presently used to monitor such quantized energy levels in carbon nanotubes,[108] to observe magnetic hysteresis at nanometer scales,[109] to measure single molecule resistance,[110] to probe the single atom spin flip,[111] and to control the molecular dynamics of a single molecule through the appropriate localization of the electronic excitation inside the molecule.[112]

It was conceivable from the moment of the invention of STM that this technique could become a principal technique on the way to increase the degree of miniaturization in different fields (electronics, medicine, biology). Nanotechnology applications started with the works of Ringger et al.[60] and Schneir et al.[113] The use of the STM tip as a nano-pen was presented by Albrecht et al.[114] and Staufer et al.,[115] who wrote the phrases "STANFORD UNIVERSITY APRIL 1989", and "I ♥ STM", respectively. After that, Eigler et al. demonstrated an atomic scale electronic switch [(116)]. The rotation forced by the STM tip on adsorbed molecules, Sb dimers and O_2 molecules, was presented in the '90.[117, 118]

The report on the *walking* molecular ensemble has been reported by Kwon et al.[119] With STM they manipulated the molecules of 9,10-thioanthracene (DTA) placing them on a Cu(111) surface. The molecules were attached to the surface by two thiol linkers, and when they activated the molecular ensemble by heating or the nudge of STM tip they observed the movement of DTA molecules along the Cu atoms rows of the substrate. Another surprising STM finding of great scientific and technological importance was described by Repp and coworkers.[120] They showed the ability to switch the charged state of an individual metal atom by voltage pulse, that is to convert it from the neutral state to a negatively charged ion. It was also shown that this process can be reversible. These results[120] can push the data storage devices to a limit never possible before.

The achieved results in atomic/molecular controlled manipulation by means of the STM apparatus show that Richard Feynman's extreme vision of miniaturization is within the reach.[63,121]

2. STM Investigations in Air and in Liquid Environment: 20 Years in Electrochemical STM Probing

Since the invention of the STM technique, it has been realized that this technique has the potential to be also employed for ambient and liquid environment studies. The needs of such applications, on one hand, gave the opportunity for the wide popularization of this technique, and on the other hand, it provided the chance to apply STM for the direct probing in real space and time with molecular resolution of such phenomena as corrosion, plating, dissolution, catalysis, and adsorption. The success in such a use of STM was expected to trigger off the development of new methods for industrial use and to provide molecular insight into basic electrochemistry. While the progress of STM results obtained under UHV condition has been rapid and enormous, the development of the application of tunneling microscope in air, and especially in electrolyte solution, has been much slower. In contrast to UHV condition there is no *clean* surface (without adsorbed species) in other environments. In air, it is always covered with an adsorbed species; simple gas molecules, organic impurities, and even a water layer, because of the ambient humidity. In the solution an electrical double-layer is formed, which influences conditions for the tunneling process in comparison with UHV condition. This phenomenon complicates the situation at the surface, creating two sets of problems; experimental ones, and theoretical ones.

The experimental difficulties are the consequence of the electrochemical phenomena taking place at the studied surface and at the metallic tunneling tip as well, which is inevitably immersed in the electrolyte and hence acts as an electrode, with electrochemical reactions taking place at the tip-electrolyte interface. Because of these phenomena, a faradaic current approaching several mA can flow through a tip, overwhelming the much smaller tunneling current of the order of nA. Another problem is the chemical stability of the tip in the solution which can dissolve or can start to grow because of the deposition of ions from the solution. Hence it was soon recognized, that it is necessary to control the electrochemical potential at the studied electrode surface and at the exposed part of the tip, which should be insulated to the very end. For this reason the concept of STM bipotentiostat was introduced by several research groups.[122–128] An extensive review of experimental problems and their solution was given by Colton et al.[9]

An example of a theoretical problem is the interpretation of the tunneling barrier in aqueous solution. Lindsay and coworkers[129,130] carried out a systematic study on the tunneling barrier in STM as a function of bias voltage and electrochemical potential, and they found a striking non-exponential dependence of the tunneling resistance on the tip-substrate distance. Such behavior in vacuum tunneling occurs only at an extremely small tip-substrate distance, in which the tunneling barrier vanishes. The first model of electron tunneling through water layers was proposed by Schmickler and Henderson[131] who treated water molecules in the tunneling gap as a dielectric medium and predicted a smaller tunneling barrier than in vacuum. A review of the theoretical problems met in wet STM was given by Tao and coworkers.[132] Another question that electrochemists have to consider, and which is not the case in under UHV condition, is what is imaged by STM, deposited atoms or adsorbed anions if deposition is accompanied by anionic coadsorption. The other problem which is tied up mainly with the STM electrochemical research is to what extent the information collected from a very tiny area is representative for the entire surface. In electrochemistry, it is necessary to deal much more often with a polycrystalline surface, than under UHV condition. The general question arising is the correlation between the integral electrochemical data and the STM-derived microscopic data.

These limitations, fortunately, only slightly influence the wide application of STM in electrochemistry and the number of users grows up continuously.

The first image of a biological sample in air was reported by Baro et al.,[56] who presented an image of a bacteriophage sample (cf. Fig. 4). The atomic corrugation for a graphite (HOPG) surface with a resolution smaller than 0.2 nm was reported by Sang-Il Park and C. F. Quate.[133] Morita et al. showed the possibility of imaging an insulator layer deposited on a metallic substrate.[134] In that paper the authors monitored electrochemically prepared anodic aluminum oxide with a Pt-Pd evaporated layer. Atomic corrugation at metal, Au (111), surface in air was reported by Hallmark et al.,[67] (see Fig. 7). Foster et al. focused their research on monitoring of organics. They have shown the possibility of molecular manipulation with STM apparatus in air, moving a di(2-ethylhexyl)phthalate molecule by ca. 50 nm. (cf. Fig. 13).[135] Then, they reported images of a liquid crystal, 4-n-octyl-4'-cyanophenyl.[136] They showed the periodicity of the interplanar packing and the periodicity within the planes. Images of polymer morphology have been reported by Albrecht et al.[137] Mate et al. have measured the forces acting during graphite imaging in air.[138]

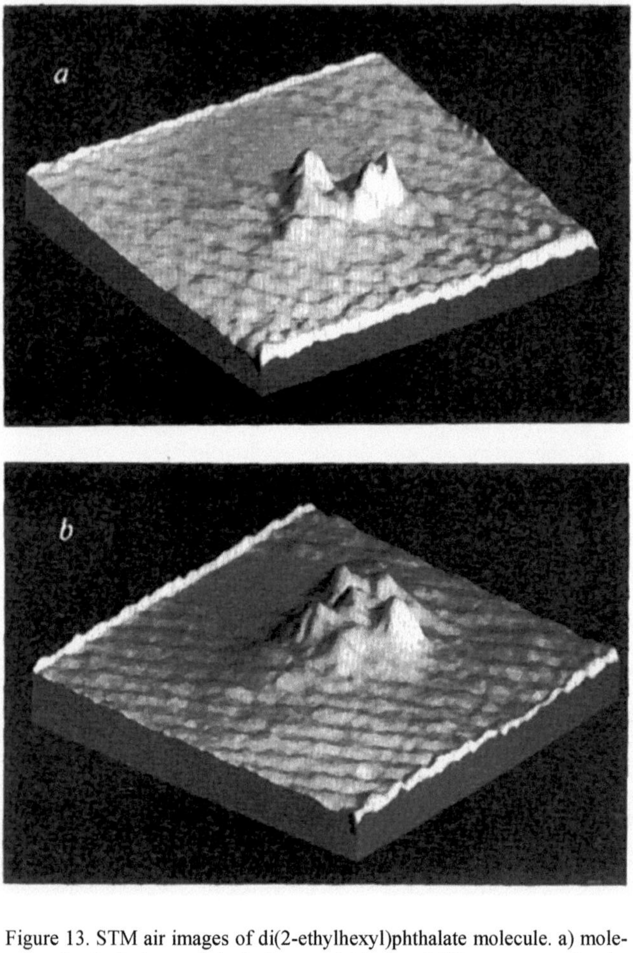

Figure 13. STM air images of di(2-ethylhexyl)phthalate molecule. a) molecule adsorbed at an original place, and b) placed 50 nm. away after by pinning it to tunneling tip by a voltage pulse and then pinning off by another voltage pulse. Reprinted by permission from Macmillan Publishers Ltd: Ref. 135, Copyright (1988).

They measured changes in the forces acting to maintain a constant current as the tip scans from one part of the graphite unit cell to another one.

Sonnenfeld and Hansma demonstrated first, twenty years ago in 1986, that a scanning tunneling microscope could be used in electrolyte solution, which opened up the way for the STM to become a major new tool for in situ electrode surface characterization.[58] These authors reported atomic corrugation of HOPG surface immersed in NaCl solution, Fig. 5. Subsequently, Liu et al. obtained 30 nm resolution scans of a Pt coated integrated circuit.[139] Schneir et al. compared results on graphite imaging obtained in air and water with theoretical calculations.[140] They obtained qualitative agreement of the experimental data with theory. The first report on the application of STM as an *in situ* technique to monitor electrochemical process was given by Sonnenfeld and Schardt.[59] These authors observed potential controlled electrodeposition of Ag films and islands on a graphite substrate, Fig. 14. An image of a graphite surface plated with gold, and images of electrochemical corrosion of Fe and Al taken in solution but with no potential control, were reported by Drake et al.[141]

Highly oriented pyrolytic graphite (HOPG) surface was and still is the test and standard surface in the *in situ* electrochemical studies due to the easy preparation of smooth and well defined surfaces. The electrochemical stability of HOPG was tested by Gewirth and Bard.[142] In this paper the authors first described the STM monitoring of the surface oxidation process, showing the formation of an amorphous surface oxide resulting from electrochemical potential cycling in the anodic potential range. They also reported atomic resolution for the HOPG surface under potential control.

The STM apparatus in the electrochemical studies has been used for

- monitoring of metallic electrodes surfaces and processes taking place at either anionic adsorption, or underpotential deposition processes (UPD), or bulk deposition processes;
- imaging semiconductive electrode materials;
- monitoring of organics adsorption.

Lev et al. reported images of a Ni foil.[123] They revealed flat terraces, 2–3 nm high. The authors noted the influence of the surface oxidation state on the tunneling current increase while the tip was engaged, and they explained this effect by the resistance of the oxide layer.

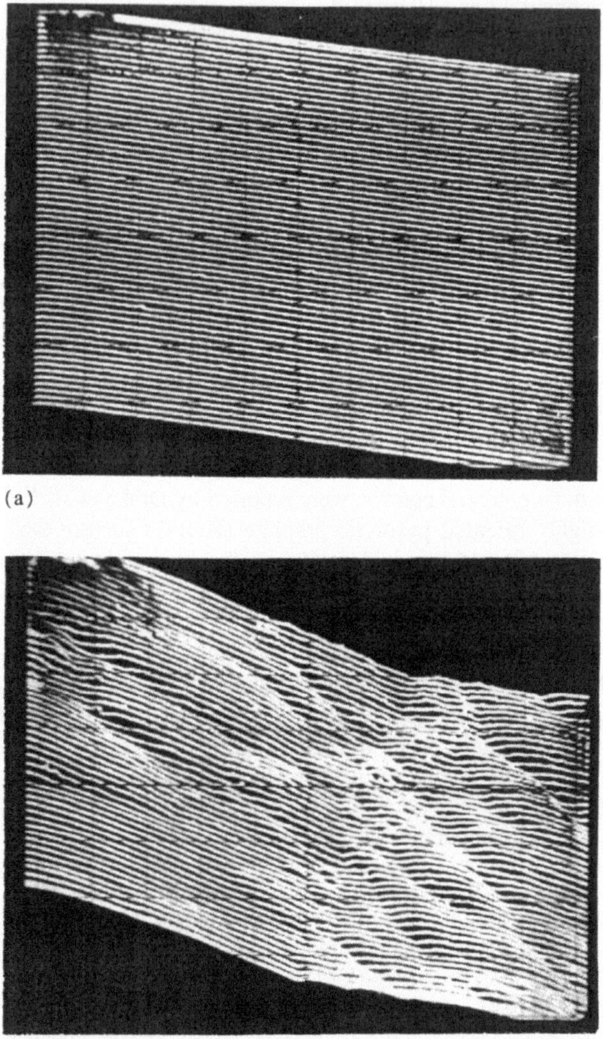

(a)

(b)

Figure 14. STM images registered *in situ*. (a) Image of HOPG surface before Ag plating, 90 nm × 150 nm. (b) Image of HOPG surface after Ag deposition. Reprinted with permission from Ref. 59, Copyright (1986) American Institute of Physics.

In 1988 an extended report on monitoring of the surface changes of the Au(111) surface in dependence on electrode potential and chloride ion adsorption was published by Kolb's group.[125] First this group had been monitoring the anion influence on the surface state. They observed atomically flat terraces for several nanometers with no atomic corrugation but of atomic height. Following addition of Cl^- ion to the solution, the authors obtained different images for the surface free of anions and with the adsorbed ones (Fig. 15).

Hottenhuis et al. studied Ag (100) surface. They made a comparison between *ex situ* images and *in situ* images.[143] To interpret their results they postulated a resolution increase in dependence on electrolyte concentration and its limitation to the Gouy-Chapman layer thickness.

A report on Pt microparticles dispersed electrochemically onto glassy carbon electrodes was published by Shimazu et al.[144] The authors made an interesting comparison. They examined Pt microparticles electrodeposited by both STM and SEM (Secondary Electron Microscopy) techniques. They found differences in the shape and dimensions of Pt microparticles depending on the image technique. As an explanation for this observation they proposed the finite dimensions and shape of the STM tip. In 1988, two other papers on imaging of Pt were published. Itaya and coworkers[145] reported Pt electrode surface changes due to the terraced surface restructuring forced by potential cycling in the potential range 0 - 1.5 V *vs.* NHE. Fan and Bard, were the first to report STM imaging of Pt electrode surface in organic solvent, n-heptane and they made an attempt to monitor adsorbed organics, i.e., ethyl acetate.[146] The species adsorbed were identified as surface local patches and it was concluded that organics adsorption may follow a nucleaction growth pattern, in a manner analogous to that frequently found for metal deposition.

Parallel with *in situ* STM electrochemical studies an *ex situ* microscopic studies of electrode surface were carried out. Arvia and coworkers reported studies on surfaces of platinum,[147,148] gold,[149,150] and palladium.[151] For the Pt electrode they reported a dependence of surface faceting on the applied electrochemical procedure. Furthermore, they tied up catalytic activity of the surface with its topography. In the case of Au electrode they studied the influence of Au morphology on lead UPD deposition. They

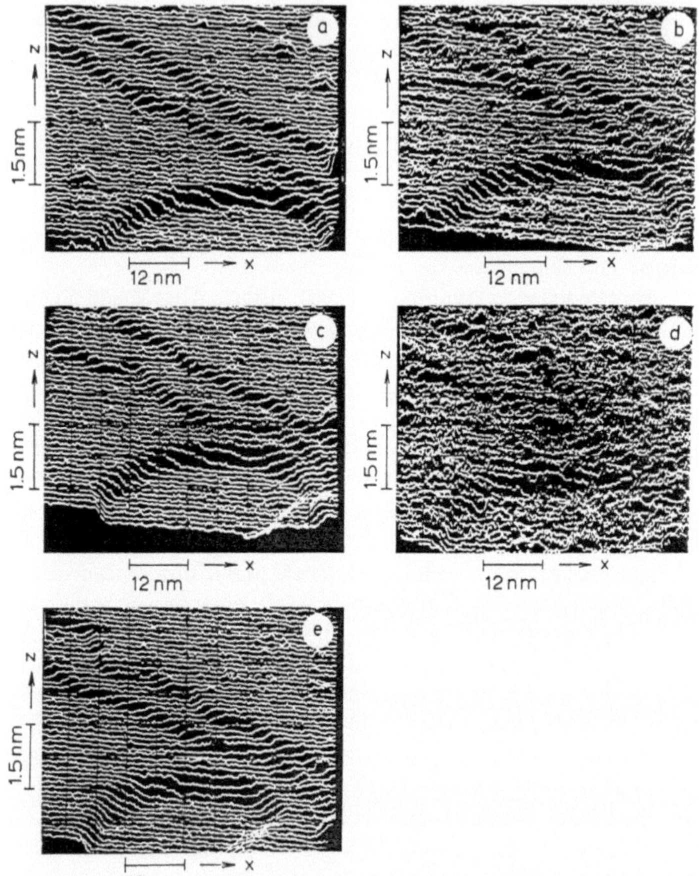

Figure 15. *In situ* STM images (60 nm × 150 nm) of Au(111) electrode surface in 0.05 M H_2SO_4 + 5 mM NaCl solution. (a) electrode potential 0.25 V, no chlorides adsorption; (b) electrode potential 0.70 V, chlorides adsorption; (c) electrode potential 0.25 V, no chlorides adsorption; (d) electrode potential 0.90 V, chlorides adsorption; (e) electrode potential 0.25 V, no chlorides adsorption. Potentials *vs.* Pd-H electrode. Reprinted from Ref. 125, Copyright (1998) with permission from Elsevier.

assigned voltammetric peaks for UPD deposition and desorption with terraces of given crystallographic structure. Additionally, they correlated STM and SEM images.

An original idea that would connect UHV and *in situ* application of STM apparatus, for simulation of electrochemical charge transfer was described by Sass and Gimzewski.[152] They proposed that an one atom sharp end STM tip can be compared with an anion/cation in solution because it can probe a single electron. They compared the electrical double-layer to adsorbed layers of solvent in vacuum. They assumed that potential drop across solvent layer can be induced by coadsorption of halide ions. For such a system they expected to detect significant differences in the tunneling current related to the charge transfer rate under *in situ* condition, depending upon where the tip is positioned laterally, relatively to the network of adsorbed solvent molecules. Unfortunately, they did not report any data obtained in this experiment.

A novel *ex situ* method for the deposition of metal structures within a polymer film based on the use of STM was described by Craston et al in 1988.[153] The authors presented the possibility of programmed pattern produced by manipulating the voltage of the STM tip scanning over the nafion film saturated with a cation solution. They deposited silver and copper atoms in the reduction reaction forced by the tip potential.

A state of art STM report on *ex situ* imaging of electrode surface with an adsorbate was published by Shardt et al. (cf. Fig. 16).[154] They were the first to obtain atomic corrugation images in air for an adsorbate. They were studying the iodine adsorption on Pt(111) surface and resolved the Pt(111)–($\sqrt{7}\times\sqrt{7}$)R19.1°-I adlattice with a very high resolution. By varying the tunneling current and tunneling voltage they were able to image the three iodine atoms expected to be in the ($\sqrt{7}\times\sqrt{7}$) unit cell.

Monitoring by the STM technique of the lead deposition either in a bulk deposition process (BDP) or in a UPD process was an important early example of STM application to monitor electrochemical processes in real time. Christoph et al.[155] studied bulk Pb deposition on a Ag(100) substrate. During the potential-controlled Pb phase deposition, they observed irreversible changes of the Ag(100) surface morphology by the Pb deposit. Furthermore, they reported a recrystallization phenomenon during repetitive Pb deposition/dissolution processes. Additionally, they determined the

Figure 16. STM *ex situ* images of Pt(111)-($\sqrt{7}\times\sqrt{7}$)R19.1°-I adlattice. Top view image (A) was recorded with V_t = 2.1 mV and i_t = 20.7 nA. Image (B) is 3-D view of image presented in in (A). Top view image (C) was recorded with V_t = 5.8 mV and i_t = 17.1 nA. Image (D) is 3-D view of image presented in (C). Reprinted with permission from Ref. 154, Copyright (1989) AAAS.

tunneling current vs. tunneling distance dependence. This dependence has showed that the change of the average tunneling barrier height due to Pb depositon (ca. 1 eV) can not be derived from the work functions of Pb and Ag.[156]

The first example of atomic resolution under *in situ* condition of the metal surface was reported by Bockris and coworkers in 1989.[157,158] They studied Pb bulk deposition on HOPG crystal in $NaClO_4$ solution and observed deposition of Pb crystallites. The Pb atomic lattice was detected over the C lattice of HOPG crystal

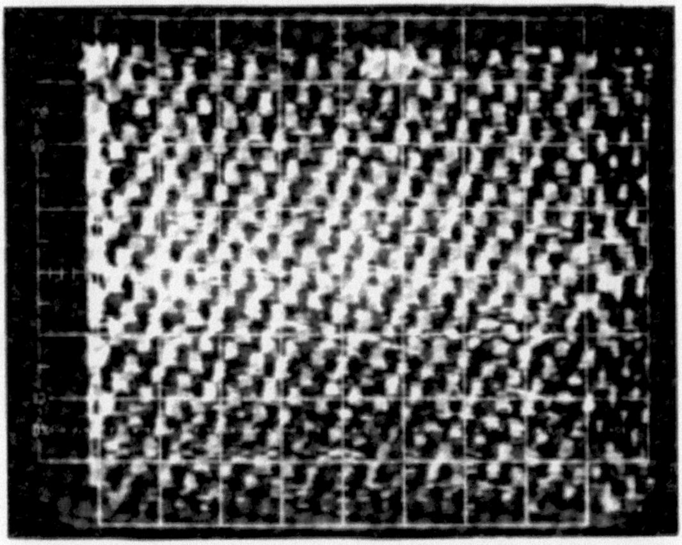

Figure 17. The *in situ* registered image of atomic corrugation of Pb deposit. Reproduced with permission from Ref. 157 Copyright (1989) The Electrochemical Society.

(Fig. 17) and identified as (100) plane which confirmed earlier electrochemical work on the BDP process of lead.[159] They reported the time dependent Pb surface reconstruction from (100) plane to (111). The observation of formation of (111) plane has supported Finch and Layton's view on lateral changes of Pb deposits.[160] They reported the same observation for anodic dissolution inducing remarkable surface morphology changes.

The early UPD lead deposition on Au(111) was reported by Green et al.[127,161,162] These studies indicated that a deposit developed a monoatomic high adlayer associated probably with Pb atoms in a (1×1) structure, and that during removal of this layer a series of surface pits were formed. Later on, Tao et al. showed that an (1×1) layer is formed, indeed, in the UPD process showing images of atomic resolution.[163]

The early studies on the one of the most popular process, copper UPD was presented by Uosaki and Kita.[128,164] They were studying Cu UPD process on Pt and Pd substrate. They did obtain terraced deposits of thickness dependent on time of deposition.

The application of STM technique spread out remarkably at the end of the eighties. The development of commercially available STM apparatuses by researchers of University of California, Santa Barbara, USA, in 1987, was the first step for the wide spreading of the technique. It resulted in wide accessibility of STM systems comparing to the former situation when only homemade systems were available. The work with the first systems was rather difficult, especially viewing data collection and their interpretation. In the mid eighties, the available data registering system was an oscilloscope. Fortunately, that was time of the fast growth and developments in computers and in computer software. The importance of computer acquisition and data processing for STM technique was understood early and several solutions were proposed.[165–169] At the end of the eighties, commercially available software was widely available. The computer procedures (filters, Fourier transform procedures, etc.) for data analysis helping to distinguish the real tunneling signal from noises, which is particularly important in the electrochemical environment, became available. It greatly helps to interpret and to present the collected STM data. Furthermore, at the late eighties, the techniques for preparation of single crystal electrodes became fairly well known, which resulted in the preparation of smooth and well defined electrodes surfaces.[170] The described innovations resulted in an enormous increase in the number of papers devoted to the application of STM in the electrochemistry field, and greatly increased the number of reports in which atomic corrugation registered under *in situ* conditions were presented.

(*i*) *Imaging of Metals and Metallic Deposits*

In 1990 Schardt and coworkers reported *in situ* results[171] similar to those obtained by this group in air.[154] They showed high resolution atomic images of the Pt(111)-($\sqrt{3}\times\sqrt{3}$)R30°-I adlattice and image of Pt substrate as well (cf. Fig. 18). The authors pointed out in their report that the iodine structure observed in solution differs from that imaged in air.

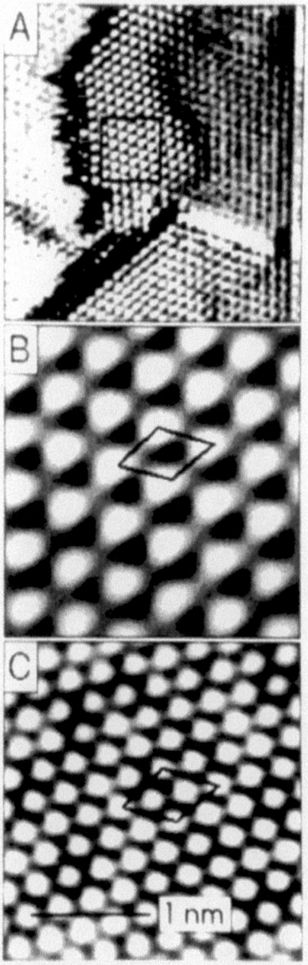

Figure 18. *In situ* images (A) of the Pt(111)-($\sqrt{3}\times\sqrt{3}$)R30°-I adlattice, scan size 12.5 nm.(B) Magnified image of the boxed region shown in (A), scan size 2.5 nm. (C) Pt(111) electrode atomic lattice showing a rhombic ($\sqrt{3}\times\sqrt{3}$)R30° unit cell, scan size 2.5 nm. Reprinted with permission from Ref. 171, Copyright (1990) American Chemical Society.

At the beginning of the nineties the process of the deposition of copper on gold electrode was one of the most extensively studied. Behm, Kolb and their coworkers used STM to study the changes in surface topography occurring on the flame annealed Au(100) and Au(111) surfaces.[172] They followed the surface oxidation process showing electrode roughening and they also reported atomic resolution for gold electrode surface and UPD deposited copper (cf. Fig. 19).[173] Moreover, they detected a quasi

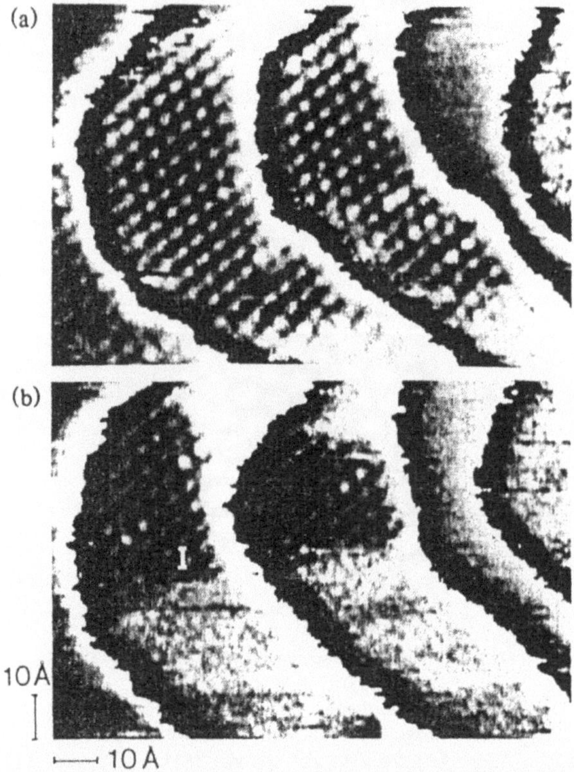

Figure 19. *In situ* STM images of Cu covered Au(111) surface taken in 0.05 M H_2SO_4 + 5 mM $CuSO_4$ solution. (a) electrode potential 0,215 V and (b) 0.20 V *vs.* SCE. Reprinted with permission from Ref. 173, Copyright (1990) American Physical Society.

hexagonal Cu adlattice, different from that formed under UHV conditions and explained this phenomenon by the structure determining role of coadsorbed anions in the electrochemical environment. Itaya and coworkers carried out coulometric and STM studies on the same system and reported formation of a $(\sqrt{3}\times\sqrt{3})R30^{o}$-Cu adlattice[174] related with coadsorbed sulfates.

In 1991–1996 several different metal electrodes surfaces and electrochemically deposited metals were imaged with atomic resolution. It proved that in an electrochemical environment, atomic resolution can be achieved as under UHV conditions independently of the problems involved in the electrochemical system.

Itaya and coworkers reported high resolution results for Cu deposited on Pt,[175] Cu deposited on Pt,[176] Ag deposited on Au,[177,178] Ag deposited on Pt,[179] Hg deposited on Au.[180] This group simultaneously presented atomic resolution images for bare electrodes surfaces, e.g., Pt,[181] Rh,[182] Ni,[183] and Co.[184]

Cadmium UPD deposition was studied by Bondos et al.[185] They monitored this process on Cu(111) and Au(111) surfaces and observed a (4×4) structure on the copper electrode, while on gold electrode they imaged a series of linear structures. They explained the formation of these structures by coadsorption of sulfate anions.

The important *in situ* STM work has tied up not only with resolving single crystal metal electrodes lattices, but also with studies on surface corrosion, roughening through oxidation processes, steps and domel structures formation. The earliest work in this field was reported on the oxidation of HOPG electrode surface.[142] The oxidation of Pt electrode was studied as early as in 1988 by Itaya and coworkers,[186] who observed roughening of Pt surface with repeated cycling to oxidation potentials. Uosaki and Kita reported stepping the electrode surface in the cathodic potentials.[164] Szklarczyk and Bockris observed potential dependent roughening of a Pt electrode during oxidation and then smoothing while the surface was reduced.[187]. The study of surface changes of the less noble metals started with the work of Lev et al.[123] They reported data for nickel oxidation and an interesting dependence of tunneling current magnitude on oxide thickness. Important work has also been concerned with Al surfaces. It has been shown that the initial stages of Al oxidation are associated with the formation of clusters or patches.[188] Another interesting application has been reported by Wang et al.[189] These authors correlated the surface morphology of glassy carbon electrodes with its electrochemical activity towards reaction with marker species like catechol and ferricyanide.[189]

The development of STM construction and microscopic software on the one hand, and the improvement of electrochemical tip preparation and its insulation against faradaic current flow, on the other hand, have resulted in improving the resolution in the direction perpendicular to the electrode surface. A corrugation as small as 0.03 nm was detected under *in situ* conditions (Fig. 20).[190] Such an experimental achievement made it possible to monitor phase changes for the bare electrode surface and for the electrodeposited films.[190–193] Two different phases were observed for Au (111) in dependence on electrochemical potential.[190] The transition from one phase to another has been observed for Cu deposited on Ag(100) in dependence on the amount of deposited Cu.[191] The comparison of surface structure of deposited film, Ni, on different substrates, Ag(111) and Au(111) was given by S. Morin et al.[192] In this report the authors made an attempt to correlate similarities and differences in surface structures of deposited Ni films with nucleation mechanisms proposed for Ag and Au substrates. Stickney and coworkers studied the dependence of deposited Se and Te surface structures on the thickness of the deposited layer and on the surface crystallographic structures, Au(100) and Au(111).[193] On Au(100), they observed the same structures for Se and Te deposits, which changes their symmetry with the increase of coverage. On Au(111), they observed a similar structure for both deposited elements, but with an increase in coverage the observed structures of

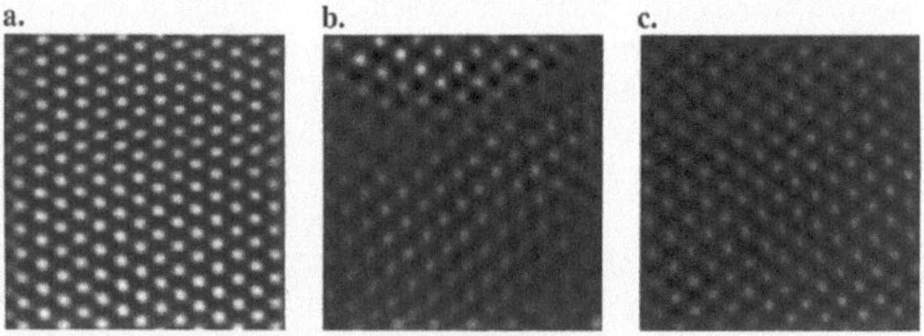

Figure 20. *In situ* STM Images of I adlattice on Au(111) electrode. (a) electrode potential –0,2 V *vs* Ag/AgI, (b) 0.3 V *vs* Ag/AgI, (c) 0.5 V *vs* Ag/AgI. Reprinted with permission from Ref. 190, Copyright (1995) American Chemical Society.

Se and Te deposits were different. They correlated these differences with the atomic diameters and preferred stable allotropes of the two elements.[193]

One of the most comprehensive reports on the application of the STM technique to the characterization of surface morphology as a function of sample preparation was presented by Kibler et al.[194] These authors used STM to study the influence of flame annealing in different atmospheres (air, N_2, H_2, CO and their mixtures), of the way of cooling, and of the electrochemical procedures on the surface morphology of Pt(111) and Pt(100). They reported the effect of adsorbed CO against surface contamination and that application of a CO atmosphere leads to clean well-ordered, unreconstructed surfaces.

(*ii*) *Imaging of Adsorbed Ions Adlattices*

Studies of the adsorption of ions and atoms on electrode surface have traditionally played a fundamental role in the characterization of the electrode surface. The species most often studied by STM have been halides, cyanides, sulfates and sulfides. Besides the basic theoretical aspects of such studies, the ionic interaction with metal surfaces has a huge practical importance because of electrodeposition, corrosion and dissolution.

Halides exhibit a strong interaction with many metals, and among them the iodine undergoes the strongest adsorption. The system platinum-iodine was studied as early as in 1990 (Fig. 16).[154] On platinum single crystal electrodes different iodine adlattices were observed and resolved.[195–199] The authors monitored overlayers structures on the Pt crystal plane, dependent on coverage and electrode potential.

Two different potential dependent structures of iodides were observed[200] on a gold Au(110) electrode. These authors presented images showing bare electrode surface and iodide adlayer in the same image frame of which particular parts were collected for different electrode potentials (cf. Fig. 21). They interpreted the absence of iodides on the STM image by their mobility at a given electrode potential.[200] More recently, Nagatami et al. determined iodine layers at Pt and Au electrodes.[198] These authors found that the presence of the iodine adlayer significantly decreases the tunneling barrier height relative to that found on the bare surface

Figure 21. *In situ* image showing potential induced formation of iodide layer. Lower part of frame presents the Au(110) pattern, and upper part of frame presents ordered iodide layer after potential change. Reprinted with permission from Ref. 200, Copyright (1995) American Chemical Society.

either in air or in vacuum. Furthermore, they found that the tunneling barrier height for Au and Pt electrodes covered with iodine is not significantly influenced by the electrode potential.[198]

Early reports showing the influence of halides on corrosion processes on atomic level were presented for Cu corrosion in chloride media by Suggs and Bard[201] and for Pd in iodide media by Sashikata et al.[202] For copper dissolution the most favourable corrosion sites were identified and the Cu(100){100} direction was found to be the most preferred corrosion direction.[201] In the case of Pd corrosion, it was found that corrosion is an iodine structure dependent reaction and that dissolution takes place anisotropically along a step aligned in the {100} direction without disruption of the iodine adlattice.[202]

Sulfate anions are among the most important anions in electrochemistry because their interaction with electrode surfaces forms the basis of several electrochemical studies. The first direct STM observation of SO_4^{2-} adsorbed on a metal, surface Au (111), was reported by Magnussen et al. (cf. Fig. 22).[203,204] Until today

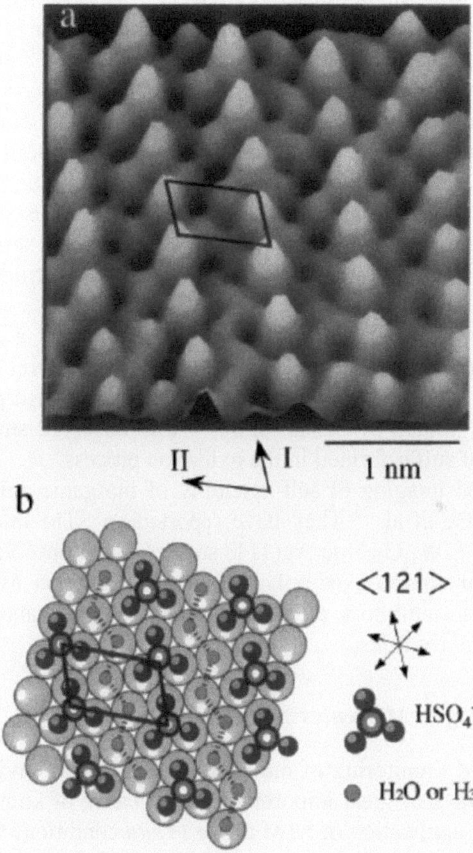

Figure 22. (a) High resolution STM *in situ* image recorded on Ir (111) electrode covered with HSO_4^- anions showing ($\sqrt{3}\times\sqrt{7}$) structure. (b) structural model for HSO_4^- anions and chains of H_3O^+ions or H_2O molecules. Reprinted from Ref. 215, Copyright (2001) with permission from Elsevier.

the adsorption of sulfate/bisulfate anions has been studied by means STM on several metals, Au(111),[203–206] Pt(111),[207–209] Rh(111),[210] Cu(111),[211–213] Ir(111),[214,215] Pd(111)[216] (the high resolution attained by the authors of this report allowed to distinguish the tunneling signal separated maxima due to water molecules,) Ru(111),[217] and Ag(111).[218–220] It is interesting, and rather surprising that the *in situ* high resolution images of adsorbed sulfates revealed the same highly ordered adlayers ($\sqrt{3}\times\sqrt{7}$) for all metals except for the Ag(111) surface. Other discussed matter is the type of adsorbed species, either SO_4^{2-}, or HSO_4^-, or H_2SO_4,[215] and again silver seems to differ from the other metals.

The formation of ordered adlayers of CN^- on Pt(111) was reported by the Weaver and Itaya groups.[221,222] Both groups observed the same overlayer symmetry, which consists of anions adsorbed in atop sites. Sawaguchi et al. studied the structures which form on Au(111) in the presence of $Au(CN)_2^-$ anions and reported two different adlayers depending on the electrode potential.[223]

Strongly adsorbed sulfides were investigated on a Au(111) electrode as early as 1992.[224] The authors imaged different surface structures in dependence on electrode potential. Adsorption on Ni(100) was studied by Suzuki et al.,[183] who observed amorphous multilayers of sulfur formed in the oxidation process.

The STM imaging of self-assembly of inorganic anions was reported by Ge et al.[225] They have reported an STM image of a monolayer α-$SiW_{12}O_{40}^{4-}$ on Ag(111) surface. Such observation are very important because of polyoxometalates function as superacids, corrosion inhibitors, electron transfer reagents, catalysts and photochemical oxidants.

(*iii*) *Imaging of Molecules*

The STM monitoring of molecules in the liquid environment has been always of great importance, but because of known difficulties in the application of STM under *in situ* conditions the high-resolution images were reported later than those known from the UHV environment. The first reports related to this subject were presented in 1991 and 1992.[226,227] Srinivasan et al. presented images of guanine and adenine adsorbed on HOPG (see Fig. 23).[226,227] They studied the process of formation of organic layer

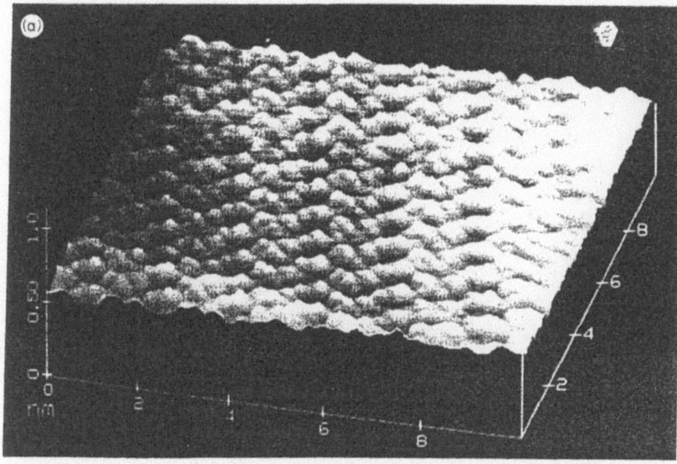

Figure 23. *In situ* STM image of guanine layer on HOPG. Reprinted from Ref. 226, Copyright (1991) with permission from Elsevier.

condensation on electrode surface. Later on, Tao et al. examined the structure of all DNA bases on a metal electrode, Au(111).[228,229]

The monitoring of CO is a subject related to imaging of organics. The first molecular resolution of CO adsorbed on electrode surface was reported by Yau et al.[230] They reported simultaneous STM and IR studies of CO adsorption on Rh(111) substrate, determining two different CO adlattices in dependence on the electrode potential and multiple binding sites for both structures. The first study of CO binding to Pt (111) electrode was by Oda et al.[231] They found four potential-dependent structures of CO adsorbate. Villegas et al. observed similar behavior of this system, which they studied even in greater detail employing the IRS technique, and by which they confirmed non-equivalent binding sites within the same adlayer.[232]

One of the first triumphs of STM imaging in UHV was the resolution of the (3×3) adlayer structure of benzene coadsorbed with CO as a stabilizing agent on the Rh(111) surface.[80] Itaya and coworkers resolved similar structures under *in situ* conditions in the absence of adsorbed CO (cf. Fig. 24).[182] The authors proposed that stabilizing benzene molecules adsorption effect of CO in UHV

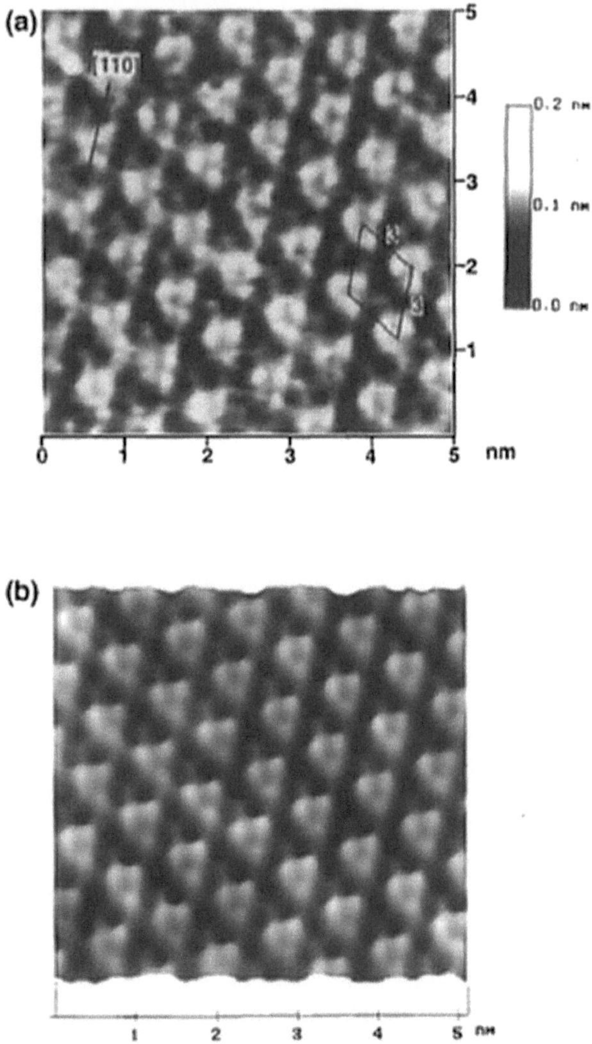

Figure 24. *In situ* high resolution STM image of benzene 3×3 structure on Rh(111). (a) Top view. (b) height-shaded plot. Reprinted with permission from Ref. 182, Copyright (1996) American Chemical Society.

was replaced by water molecule or hydronium ions under *in situ* conditions. Furthermore, the authors observed the dependence of benzene surface coverage on electrode potential, and the replacement of adsorbed molecules by adsorbing hydrogen. Images of benzene adsorbed on Pt(111) are presented in the same paper.[182]

Another STM-UHV imaging success was the resolution of the internal structure of copper-phthalocyanine molecules.[81] It was attained by Ogaki et al. under *in situ* conditions in 1996.[233] They demonstrated an image of a specific modified porphyrin adsorbed on Ag(111)-I surface, revealing the internal structure and molecular orientation in the ordered layer, as shown in Fig. 25.[233]

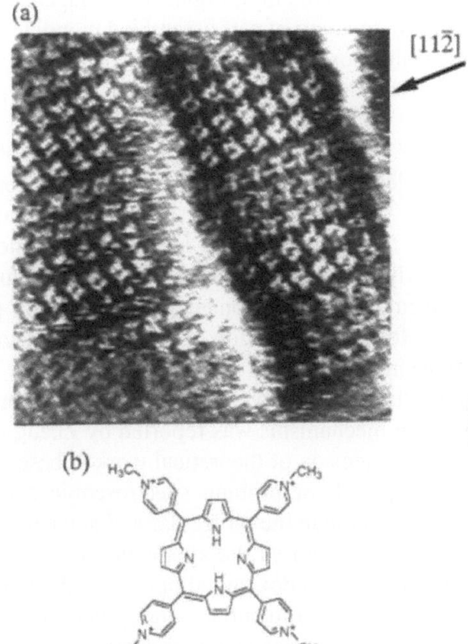

Figure 25. *In situ* high resolution STM image, 24 × 24 nm, of porphyrin on I-Ag(111). (a) Top view. (b) Chemical structure of imaged porphyrin. Reprinted with permission from Ref. 233, Copyright (1996) American Chemical Society.

Itaya's group presented images of benzene, naphthalene and anthracene on Cu(111),[234] and naphthalene and anthracene on Rh(111).[235] Wandlowski and coworkers monitored adsorption of uracil on gold surfaces.[236] They reported imaging chemisorbed molecules as well as physisorbed molecules and determined their adlattice structures. They also made a correlation between the structure and lateral interaction forces of adsorbed molecules. They showed that application of sufficiently positive electrode potentials results in uracil deprotonation, leading to different surface structure and geometric orientation.

A few groups examined larger systems by *in situ* STM. Examples include the deposition of polymers, polypyrrole,[237] polyaniline,[238] and polymethylthiophene.[239] Hagenström et al. reported studies on imaging of self-assembled monolayers (SAM).[240] They showed the possibility of imaging detailed structures, and order-disorder transitions in dependence on electrochemical parameters.

The use of *in situ* STM for the identification of structurally similar molecules on the basis of their redox properties was reported by Tao.[241] By adjusting the substrate potential, he aligned its Fermi level to different energy levels of molecules, which resulted in a nearly ten-fold increase in the tunneling current flowing through molecules due to resonant tunneling. This phenomenon can be used to follow the reaction of individual molecules under *in situ* conditions in the future.

Recently, the application of STM apparatus to probe electron transfer reaction mechanisms was reported by Zhang et al.[242] They provide first an extension of theoretical work[243] based on the small bias voltage required for probing spectroscopic current to large bias voltages, bigger than the molecular and solvent reorganization Gibbs energy. They reported the application of the new model to imaging single-copper redox metalloprotein, showing a constant current over 0.25-V overpotential range, which covers the range where the oxidized and reduced levels are located, extending in this way the spectroscopic applications.

Independently of the results described above, reports on high resolution imaging of organic molecules are still infrequent. It is much more complex compared with the simpler adsorbates of inorganic ions, because organics can exhibit much more complex behavior on the electrode surface. They are more sensitive to metal electrode properties, properties of solution and electrode potential. A lucid example of this complexity was presented by Cunha and Tao. They presented order-disorder transition for 2,2'-bipyridine on Au (111) (see Fig. 26).[244] They observed disordered, ordered, polymerized, standing up and lying 2,2'-bipyridine molecules in the same solution, and these structure changes were induced only by the change of the surface charge.

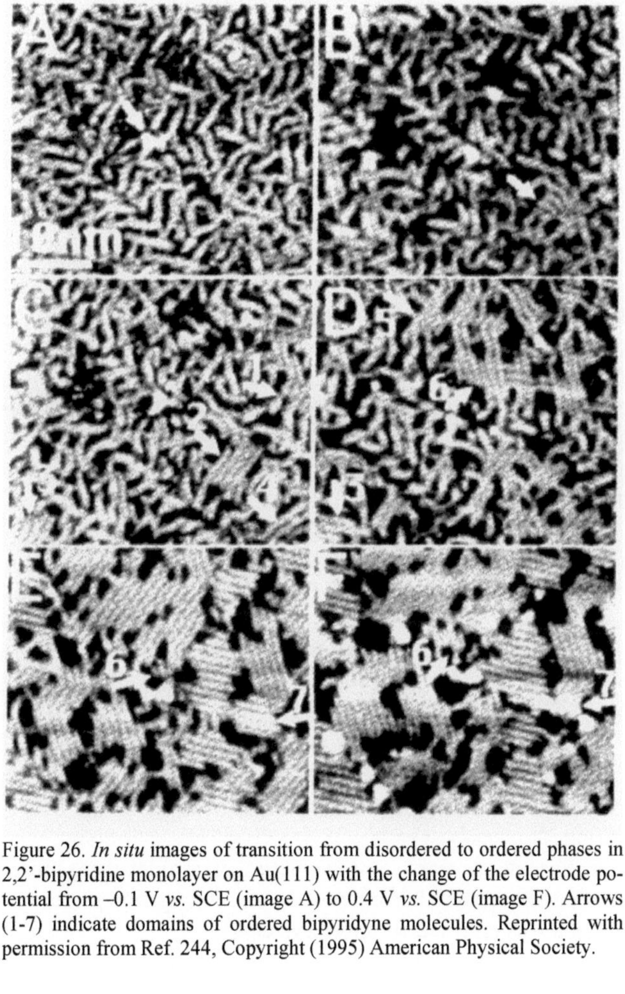

Figure 26. *In situ* images of transition from disordered to ordered phases in 2,2'-bipyridine monolayer on Au(111) with the change of the electrode potential from –0.1 V *vs.* SCE (image A) to 0.4 V *vs.* SCE (image F). Arrows (1-7) indicate domains of ordered bipyridyne molecules. Reprinted with permission from Ref. 244, Copyright (1995) American Physical Society.

(*iv*) *Imaging of Semiconductive Materials*

Much is known about the growth, crystallography and reconstruction of semiconductor (SC) surfaces from STM ultrahigh vacuum experiments. In this environment the first atomic resolution image was obtained for a semiconductor substrate[27,30] before it was done for metal substrate.[36,67] In the case of air and in particular in liquid environment the situation is different. This is because of the presence of the native oxides formed in air on SC surface, and corrosion processes taking place easily in solution.

The early reports on *in situ* imaging of semiconductors surface was published by Itaya and Tomita on studies of n-TiO_2 surface[245] and n-ZnO surface.[246] They performed monitoring of surface morphology created by etching in different inorganic acids. They found that it is necessary to maintain the semiconductor surface polarized at a potential lower than the flat band potential. Further discussion on the conditions required for proper STM imaging of SC in solution was given by Carlsson et al.[247] These authors discussed the influence of doping level and condition of accumulation/depletion on *in situ* STM imaging of n-GaAs and p-GaP surfaces. Szklarczyk et al. used STM to investigate the surface states formed in the electrochemical environment in the p-Si(111) substrate.[248,249] They assigned the onset of tunneling in the band gap region to the presence of the surface states created by solvent or ionic species.

The first application of STM to study the photocorrosion process in the real time was reported by Sakamaki et al. for n-MoS_2,[250] Eriksson et al. for n-GaAs[251] and by *Zhao* et al. for CdS.[252]

Bard and coworkers studied Si(111) in hydrofluoric acid under potential control in 1992.[253] They studied silicon surface etching in dependence on HF concentration and electrode potential. In the cathodic potential range they observed atomic corrugation of silicon surface, cf. Fig. 27, which with time disappeared due to the surface rearrangement because of the hydrogen evolution.[253]

In subsequent years semiconductive electrodes were more extensively studied than before and atomic corrugation under in situ conditions was reported for more systems.

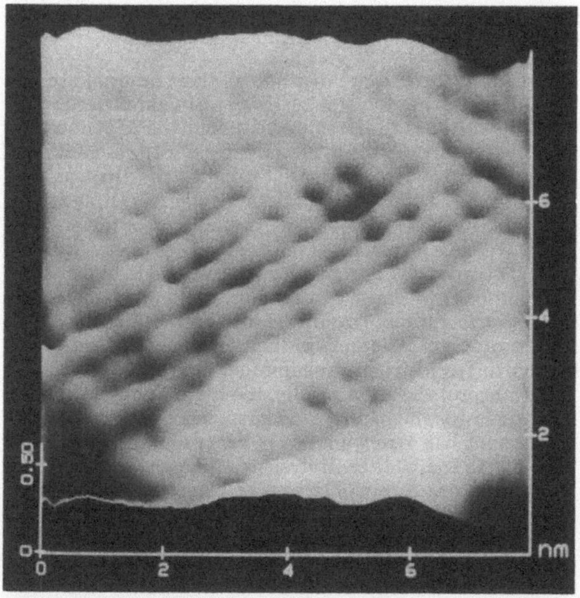

Figure 27. *In situ* image of atomic arrangement of Si(111) surface in HF solution. Reproduced with permission from Ref. 253, Copyright (1992) The Electrochemical Society.

Higgins and Hammers used STM to study both anodic and cathodic dissolution of PbS (001) in aqueous solutions.[254] They observed etching and nucleation at the edges depending on the potential conditions. Atomic resolution images revealed the expected square (001) surface lattice. Furthermore, they reported the atomic corrugation for the Pb redeposited in the form of islands with Pb(111) geometry.

Itaya' group presented atomic resolution images for GaAs(111),[255,256] Si(110),[257] and InP.[258] Their work finally proved that semiconductor surfaces can be prepared for stable *in situ* STM imaging. For GaAs electrode they showed that ideal GaAs(111)A-(1×1) is stable in solution. By comparison of STM results with electrochemical ones they suggested that the hydrogen evolution reaction takes place more favorably on the surface terminated by Ga. For the silicon electrode they reported the dependence of the

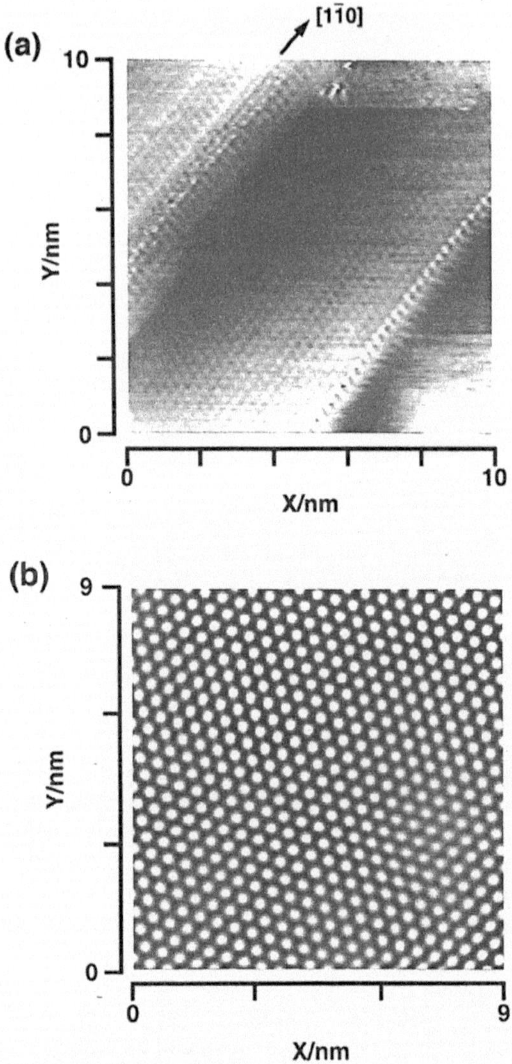

Figure 28. *In situ* atomically resolved STM images of n-InP(111) obtained in H_2SO_4 solution. (a) Unfiltered top view. (b) Image after application of Fourier transform filtering method. Reproduced with permission from Ref. 258, Copyright (1998) The Electrochemical Society.

density of islands and pits formed during electrochemical etching on electrode potential and they also pointed the importance of the electrochemical potential on preparation of well ordered Si surface.[257] For the n-InP substrate they studied the dependence of hydrogen evolution and cathodic decomposition on semiconductor surface plane.[258] They found very slight dependence. In the potential range where decomposition did not occur, they observed clear and well defined atomic lattices (cf. Fig. 28).

(*v*) *Electrochemical Fabrication of Nanostructures: Nanolithography*

Electrochemical STM nanofabrication techniques usually involve either localized etching (dissolution), or plating (deposition) on the surface *via* a sharp STM tip. The first demonstration of nanostructuring was reported by Penner's group.[259,260] They fabricated silver and copper nanoclusters of 20–50 nm wide and 1–7 nm high at predetermined position in the deposition process. In addition to metal clusters the deposition method was used to fabricate nano-sized polyaniline.[238] The authors used a pulse voltage technique and fabricated polyaniline spots 10–60 nm in diameter and 1–20 nm in height.

Crooks' group used STM to locally etch an Au(111) electrode coated with a self assembled thiol monolayer, and they were able to fabricate features from a few tens of nanometers to several micrometers.[261,262] They studied the dependence of nanodeposition on the type of solution and electrode potential.

Kolb's group presented a jump-to-contact technique by which the surface can be decorated with small metal clusters provided there is sufficient interaction between a sample and a tip,[263–265] the effect discussed earlier.[266] By means of this technique, they showed the way to generate different patterns on a different scale. On the nanometer scale they presented the structure made of 12-Cu clusters of 0.8-nm height (see Fig. 29).[264] They reported an interesting observation on the surprising stability against anodic dissolution of the fabricated clusters, and they proposed to explain this effect by quantum confinement of the electrons in the clusters that leads to an energy gap between the electronic states of the clusters.[267]

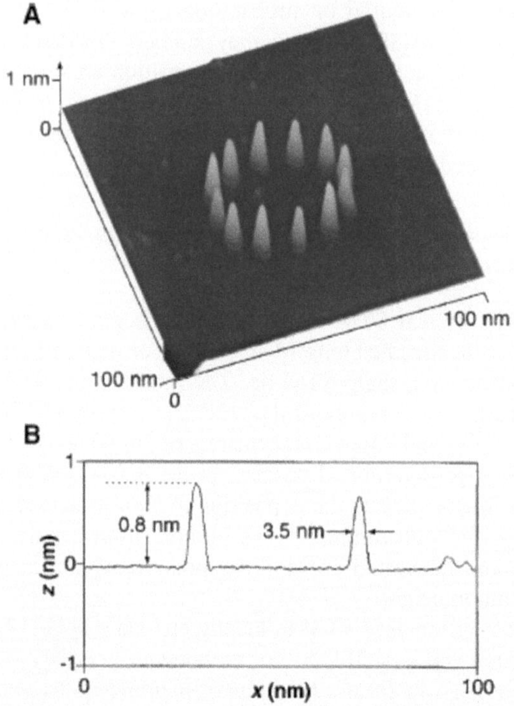

Figure 29. (A) *In situ* STM image of 12 Cu clusters on an Au(111) electrode. (B) Cross section through the center of the circle. Reprinted with permission from Ref. 264, Copyright (1997) AAAS.

The demonstration of nanofabrication on semiconductive electrodes was reported as early as in 1990 by Nagahara et al.[268] These authors fabricated nanostructures on Si(100) and GaAs(100) surfaces via local etching by maintaining a tip-surface bias and tunneling current. They observed nanostructures 20-nm wide and 1.5-nm deep.

The possibility of fabrication of nanowires in the electrochemical environment by means of STM was reported by Tao and coworkers.[269–271] In their method they deposited copper onto a

STM tip that was held at fixed distance (10–150 nm.) from the electrode. With time Cu deposit reached the substrate forming a nanowire between the tip and the substrate. They found that the diameter of the wire was of a few atoms and that such a structure has an electrode potential dependent quantized conductance.

In this Section we presented a short review of some early results on electrochemical fabrication of nanostructures based on STM, more comprehensive reviews related to this topic can be found in Refs. 272 and 273.

III. SUMMARY

During the last twenty years STM has advanced from line scans registered by oscillscopes to sensing of quantum effects. STM studies have proven that this technique is useful in extracting electron transfer properties as well as in probing electronic states of atoms and molecules. Moreover, it identifies structurally similar molecules, senses chemical reactivity, and fabricates new structures on an atomic scale. The real-space, nano- to atomic-scale resolution of STM has changed our way of thinking and working in many areas and created new ones (e.g., nanoelectrochemistry, nanomechanics, nanotribology) with distinct perspectives.

A major extension of the local-probe methods was brought about by the invention of the Atomic Force Microscopy (AFM). The various forces that are mainly used to image are repulsive interatomic, electrostatic, magnetic, van der Waals—all of electronic origin, and lateral forces (friction origin). The other group of techniques developed on the base of STM construction is modern optical techniques, like SNOM (Scanning Near-Field Optical Microscopy). Since 1986, an exponential growth of number of articles is observed. They show the possibility of achieving of high resolution SPM images under a myriad of sample conditions. Presently there are in use more then sixty different variants of SPM techniques (Table 2). Combining them with spectroscopic techniques, like fluorescence spectroscopy, Raman spectroscopy and spin-resonance spectroscopy is of great interest and expands in this way their possibilities.

In electrochemistry, STM and other local probes brought about a new era by advancing the understanding of interfacial

Table 2
List of SPM Techniques

STM related techniques	Scanning Noise Microscopy (SNM) Scanning Noise Potentiometry (SNP) Alternating Current STM (ACSTM) Scanning Chemical Potential Microscopy (SCPM) Scanning Optical Absorption Microscopy (SOAM) Spin Polarized STM (SPSTM) Balistic Electron Emission Microscopy (BEEM) Photon Emission STM (PESTM) Photon Assisted STM (PASTM) Constant Current Mode STM Constant Height Mode STM Barrier Height Imaging Density of States Imaging
AFM related techniques	dc Contact techniques
	Constant Height Mode AFM Constant Force Mode AFM Constant Error Mode AFM Lateral Force Microscopy (LFM) Spreading Resistance Imaging
	ac Contact techniques
	Force Modulation Mode Contact Electric Force Microscopy (EFM) Young's Modulus Microscopy (YMM) Scanning near-field acoustic Force-distance measurements
	Intermittent (tapping) contact techniques (TM- AFM)
	Amplitude modulation Frequency modulation Phase imaging mode Error mode
	Non- contact techniques
	Non- contact mode Frequency modulation
Many-pass techniques	Electric Force Microscopy (EFM) Scanning Capacitance Microscopy (SCaM) Kelvin Probe Microscopy (SKM) DC Magnetic Force Microscopy (DC MFM) AC Magnetic Force Microscopy (AC MFM) Dissipation Force Microscopy Scanning Surface Potential Microscopy (SSPM) Scanning Maxwell Stress Microscpy (SMMM) Magnetic Force Microscopy (MFM) Van der Waals Force Microscopy (VDWFM)

Table 2. Continuation

	Scanning Thermal Microscopy (SThM) Scanning Ion Conductance Microscopy (SICM)
Scanning Near-Field Optical Microscopy (SNOM)	Aperture SNOM (ASNOM) Collection ASNOM (C-ASNOM) Emission ASNOM (E-ASNOM) Evanescent Field SNOM (EF-SNOM) Nonaperture ASNOM (NA-SNOM) Shear Force Microscopy Transmission Mode (TSNOM) Reflection Mode Luminescence Mode
Spectroscopic techniques	Scanning Tunneling Spectroscopy (STS) Frequency-Distance Curves Force-Distance Curves Phase- Distance Curves Full-resonance Spectroscopy Adhesion Force Spectroscopy Amplitude-Distance Curves I(z) Spectroscopy I(v) Spectroscopy
Lithography	AFM Oxidation Lithography STM Lithography AFM Lithography-Scratching AFM Lithography-Dynamic Plowing
Scanning Electrochemical Microscopy (SECM)	

processes on an atomic and molecular level. Working in electrolyte environment, however, extends far beyond classic electrochemistry and its applications. We have witnessed the emergence of atomistic surface electrochemistry. The electrochemists have now tools to monitor on an atomic level structural changes in real time when a process takes place. Even video presentations showing the electrochemical processes in real time are presented.[274,275] The results obtained by the SPM techniques are connected with the UHV microscopic and spectroscopic results on electrochemical systems. The SPM techniques make it possible to study the dependence of substrate reactivity on structure. The understanding of this dependence leads to development of a new way of changing of surface structures and preparation of materials with new properties (e.g., nanocrystals). The liquid-solid interface imaging is also the key to *in situ* studying of biologically active materials.

The prime activity in the STM field, which was focused on super resolution and the understanding of imaging, manipulation and modification processes, is now expanding to the technology domain, where local probes are applied in vacuum, in air and in liquids. They have begun to be applied routinely on one hand as local sensors, and on the other for new product preparation, like for example, in lithography applications. The last decade has witnessed a tremendous success of SPM as a tool for positioning single atoms and molecules with hitherto unprecedented precision. The application in nanomechanics is another important goal for the SPM techniques. This field has several attractive features such as extremely small energies required for production, and small amounts of materials required for manufacturing.

Twenty five years after the birth of SPMs (the local-probe methods) there can be no doubt that this techniques have marked the beginning of a novel field of nanometer scale science and technology, which will offer new opportunities to electrochemistry in the 21st century.

ACKNOWLEDGMENTS

We are grateful to the Ministry of Education and Science for the support of this paper through grant program 3T08A 04128.

REFERENCES

[1]B. Binnig, H. Rohrer, *Rev. Mod. Phys.* **59** (1987) 615.
[2]R. M. Tromp, *J. Phys. Condens. Matter.* **1** (1989) 10211.
[3]M. J. Weaver, X. Gao, Annu. *Rev. Phys. Chem.* **44** (1993) 459.
[4]R. Wiesendanger, *Scanning probe microscopy and spectroscopy: methods and application*, Cambridge University Press, Camridge, UK, 1994.
[5]L. A. Bottomley, J. E. Coury, P.H. First, *Anal. Chem.* **68** (1996) 185R.
[6]G. J. Legget, *Scanning tunneling microscopy and atomic force microscopy*, in *Surface Analysis*, ed. J. C. Vickerman, J. Wiley & Sons, Chichester, 1998.
[7]*Force and tunneling microscopy*, ed. S. Chiang, *Chem. Rev.* **97** (1997) 1017-1230.
[8]K. Itaya, Progr. Surf. Sci. **58** (1998) 121.

[9]*Procedures in scanning probe microscopies,* eds. R. J. Colton, A. Engel, J. E. Frommer, H. E. Gaub, A. A. Gewirth, R. Guckenberger, J. Rabe, W. M. Heckl, B. Parkinson, J. Wiley & Sons, Chichester, 1998.
[10]G. Friedbacher, H. Fuchs, *Pure Appl. Chem.* 71 (1999) 1337.
[11]*Noncontact atomic force microscopy*, eds. S. Morita, R. Wiesendanger, E. Meyer, Springer, Berlin, 2002.
[12]S. Myhra, *Materials characterization by scanned probe analysis*, in *Surface analysis methods in material science*, eds. D. J. O'Connor, B. A. Sexton, R. St, C. Smart, Springer, Berlin, second edition, 2003.
[13]W. A. Hofer, *Progr. Surf. Sci.* **71** (2003) 147.
[14]*Discovering the nanoscale*, eds. D. Baird, A. Nordmann, J. Schummer, IOS Press, Amsterdam, 2004.
[15]*Applied scanning probe methods*, eds. B. Bhushan, H. Fuchs, S. Hosaka, Springer, Berlin, 2004.
[16]J. E. Lilienfeld, *Z. Phys.* **23** (1922) 506.
[17]R. H. Fowler, L. Nordheim, *Proc. Roy. Soc. (London).* **119** (1928) 173.
[18]G. Gamov, *Z. Phys.* **51** (1928) 204.
[19]R. W. Gurney, E. U. Condon, *Nature,* **122** (1928) 439.
[20]L. Esaki, *Phys. Rev.* **109** (1958) 603.
[21]I. Giaever, *Phys. Rev. Lett.* **5** (1960) 147, 464.
[22]J. Bardeen, *Phys. Rev. Lett.* **6** (1961) 57.
[23]B. D. Josephson, *Phys. Lett.* **1** (1962) 251.
[24]P. W. Anderson, J. M. Rowell, *Phys. Rev Lett.* **10** (1963) 230,
[25]R. C. Jaklevic, J. Lambe, *Phys. Rev. Lett.* **17** (1966) 1139.
[26]L. de Broglie, Nature, *Nature,* **112** (1923) 540,
[27]G. Binnig, H. Rohrer, Ch. Gerber, E. Weibel, *Appl. Phys. Lett.,* **40** (1982) 178.
[28]G. Binnig, H. Rohrer, Ch. Gerber, E. Weibel, *Physica B,* **109&110** (1982) 2075.
[29]G. Binnig, H. Rohrer, *Helv. Phys. Acta*, **55** (1982) 726.
[30]G. Binnig, H. Rohrer, Ch. Gerber, E. Weibel, *Phys. Rev. Lett.* **50** (1983) 120.
[31]R. Young, J. Ward, F. Scire, *Phys. Rev. Lett.* **27** (1971) 922.
[32]R. Young, J. Ward, F. Scire, *Rev. Scie. Instr.* **43**(1972) 999.
[33]E. C. Teague, *Ph. D. Thesis (North Texas State University)*, 1978.
[34]E. C. Teague, *Bull. Am. Phys. Soc.* **23** (1978) 290.
[35]U. Poppe, *Verb. Dtsch. Phys. Ges. (VI)* **16** (1981) 476.
[36]G. Binnig, H. Rohrer, Ch. Gerber, E. Weibel, *Phys. Rev. Lett.* **49** (1982) 57.
[37]G. Binning, H. Rohrer, Ch. Gerber, E. Weibel, *Surf. Sci.* **131** (1983) L379.
[38]G. Binnig, H. Rohrer, *Surf. Sci.* **126** (1983) 236.
[39]J. Tersoff, D. R. Hamann, *Phys. Rev. Lett.* **50** (1983) 1998.
[40]A. Baratoff, *Physica B*, **127** (1984) 143.
[41]N. Garcia, C. Ocal, F. Flores, *Phys. Rev. Lett.* **50** (1983) 2002.
[42]E. Stoll, A. Baratoff, A. Selloni, P. Carnevali, *J. Phys. C.* **17** (1984) 3073.
[43]J. Tersoff, D. R. Hamann, *Phys. Rev. B.* 31 (1985) 805.
[44]A. M. Baró, G. Binnig, H. Rohrer, Ch. Gerber, E. Stoll, A. Baratoff, F. Salvan, *Phys. Rev. Lett.* **52** (1984) 1304.
[45]G. Binnig, H. Rohrer, *Proceedings of the 6th General Conference of the European Physical Society, Trends in Physics,* eds. J. Janta, J. Pantoflicek, European Physical Society, Prague, 1984, vol. 1, p. 38.
[46]P. K. Hansma, *Bull. Am. Phys. Soc.* **30** (1985) 251.
[47]G. Binnig, H. Fuchs, Ch. Gerber, H. Rohrer, E. Stoll, E. Tosatti, *Europhys. Lett.* **1** (1986) 31.

[48]R. S. Becker, J. A. Golovchenko, B. S. Swartzentruber, *Phys. Rev. Lett.* **54** (1985) 2678.
[49]R. M. Feenstra, A. P. Fein, *Phys. Rev. B 32* (1985) 1394.
[50]S. A. Elrod, A. Bryant, L. de Lozanne, S. Park, D. Smith, C. F. Quate, *IBM J. Res. Dev.* **30** (1986) 387.
[51]R. J. Behm, W. Hoesler, E. Ritter, G. Binnig, *Phys. Rev. Lett.* **56** (1986) 228.
[52]M. C. Payne, J. C. Inkson, *Surf. Sci.* **159** (1985) 485.
[53]M. D. Pashley, J. B. Pethica, J. Coombs, *Surf. Sci.* **152** (1985) 27.
[54]N. D. Lang, *Phys. Rev. B* **34** (1986) 5947.
[55]N. D. Lang, *Phys. Rev. Lett.* **58** (1987) 45.
[56]A. M. Baró, R. Miranda, J. Alamán, N. Garcia, G. Binnig, H. Rohrer, Ch. Gerber, L. Carrascosa, *Nature* **315** (1985) 253.
[57]R. V. Coleman, B. Drake, P. K. Hansma, G. Slough, *Phys. Rev. Lett.* **55** (1985) 394
[58]R. Sonnenfeld, P. K. Hansma, *Science* **232** (1986) 211.
[59]R. Sonnenfeld, B. C. Schardt, *Appl. Phys. Lett.* **49** (1986) 1172.
[60]M. Ringger, H. R. Hedber, R. Schlogl, P. Oelhafen, H. J. Guntherodt, *Appl. Phys. Lett.* **46** (1985) 832.
[61]R. S. Becker,, J. A. Golovchenko, B. S. Swartzentruber, *Nature* **325** (1987) 419.
[62]G. Binnig, C. F. Quate, Ch. Gerber, *Phys. Rev. Lett.* **56** (1986) 930.
[63]R. P. Feynman, *Eng. and Sci.* **23** (1960) 22.
[64]J. A. Stroscio, R. M. Feenstra, A. P. Fein, *Phys. Rev. Lett.* **58** (1987) 1668.
[65]J. Nogami, S. –Il. Park, C. F. Quate, *J. Vac. Sci. Technol.* A **7** (1989) 1919.
[66]J. J. Boland, *Phys. Rev. Lett.* **65** (1990) 3325.
[67]V. M Hallmark, S. Chiang, J. F. Rabolt, J. D. Swalen, R. J. Wilson, *Phys. Rev. Lett.* **59** (1987) 2879.
[68]R. C. Jaklevic, L. Elie, *Phys. Rev. Lett.* **60** (1988) 120.
[69]J. Wintterlin, J. Wiechers, H. Brune, T. Gritsch, H. Höfer, R. J. Behm, *Phys. Rev. Lett.* **62** (1989) 59.
[70]A. Samsavar, E. S. Hirschorn, T. Miller, F. M. Leibsle, J. A. Eades, T. –C. Chiang, Phys. *Rev. Lett.* **65** (1990) 1607.
[71]H. Niehus, R. Spitzl, K. Besocke, G. Comsa, *Phys. Rev. B* **43** (1991) 12619.
[72]S. Heinze, S. Blügel, R. Pascal, M. Bode, R. Wiesendanger, *Phys. Rev. B* **58** (1998) 16432.
[73]J. C. Dunphy, M. Rose, S. behler, D. F. Ogletree, M. Salmeron, P. Sautet, *Phys. Rev. B* **57** (1998) R12705.
[74]R. M. Tromp, E. J. van Loenen, J. E. Demuth, N. D. Lang, *Phys. Rev. B* **37** (1988) 9042.
[75]Sang-Il Park, J. Nogami, C. F. Quate, Phys. Rev. B 36 (1987) 2863.
[76]T. R. Albrecht, H. A. Mizes, J. Nogami, Sang-il Park, C. F. Quate, Apple. Phys. Lett. 52 (1988) 362
[77]Sang-Il Park, J. Nogami, H. A. Mizes, C. F. Quate, Phys. Rev. B 38 (1988) 4269.
[78]Z Nawaz, T. R. I. Cataldi, J. Kall, R. Somekh, J. B. Pethica, *Surf. Sci.* **265** (1992) 139.
[79]J. K. Gimzewski, E. Stoll, R. R. Schlittler, *Surf. Sci.* **181** (1987) 267.
[80]H. Ohtani, R. J. Wilson, S. Chiang, C, M. Mate, *Phys. Rev. Lett.* **60**(1988)2398.
[81]P. H. Lippel, R. J. Wilson, M. D. Miller, Ch. Wöll, S. Chiang, *Phys. Rev. Lett.* **62** (1989) 171.
[82]P. Weiss, D. Eigler, *Phys. Rev. Lett.* **71** (1983) 3139.
[83]G. C. McGonigal, R.H. Brnhardt, D. J. Thomson,, *Appl. Phys. Lett.* **57** (1990) 28.

[84]T. Sleator, R. Tycko, *Phys. Rev. Lett.* **60** (1988) 1418.
[85]D. P.E. Smith, J. Hörber, Ch. Gerber, G. Binnig, *Science* **245** (1989) 43.
[86]A. Okummura, K. Miyamuro, Y. Ghosh, *J. Vac. Sci. Technol. A 8* (1990) 625.
[87]T. P. Beebe, Jr., T. E. Wilson, D. F. Ogletree, J. E. Katz, R. Balhorn, M. B. Salmeron, W. J. Siekhaus, *Science* **243** (1989) 370.
[88]M. Amrein, R. Dürr, A. Stasiak, H. Gross, G. Travaglini, *Science* **243** (1989) 1708.
[89]S. M. Lindsay, L. A. Nagahara, T. Thundat, P. Oden, *J. Biomol. Struct. Dyn.* **7** (1989) 289.
[90]R. J. Driscoll, M. G. Youngquist. J. D. Baldeschwieler, *Nature*, **346** (1990) 294.
[91]M. A. Voelker, S. R. Hmerroff, J. D. He, E. L. Derenick, R. S. McKuskey, *J. Microsc.* **152** (1988) 548.
[92]M. E. Welland, M. J. Miles, N. Lambert, V. J. M. Morris, J. H. Coombs, J. B. Pethica, *Int. J. Biol. Macromol.* **11** (1989) 29.
[93]D. P. E. Smith, A Bryant, C. F. Quate, J. P. Rabe, Ch. Gerber, J. D. Swalen, *Proc. Natl. Acad. Sci. USA* **84** (1987) 969.
[94]H. Fuchs, S. Akari, K. Dransfeld, *Z. Phys. B* **80** (1990) 389.
[95]Z. Nawaz, J. Kall, T. Richardson, G. G. Roberts, J. B. Pethica, *J. Phys. Condens. Matter.* **3** (1991) 11.
[96]X-D. Wang, H. Lu, T. Hashizume, H. W. Pickering, T. Sakurai, *Surf. Sci.* **67** (1993) 266.
[97]T. Mitsui, M. K. Rose, E. Fomin, D. F. Ogletree, M. Salmeron, *Nature* **422** (2003) 705.
[98]M. F. Crommie, C. P. Lutz, D. M. Eigler, *Science* **262** (1993) 218.
[99]Y. Hasegawa, Ph. Avouris, *Phys. Rev. Lett.* **71** (1993) 1071.
[100]Z. Zhang, M. G. Lagally, *Science* **276** (1997) 377.
[101]D. M. Eigler, E. K. Schweizer, *Nature* **344** (1990) 524.
[102]I. Bartels, C. Meyer, K-H. Rieder, *Phys. Rev. Lett.* **79** (1997) 697.
[103]J. K. Gimzewski, C. Joachim, *Science* **283** (1999) 1683.
[104]B. C. Stipe, M. A. Rezaei, W. Ho, S. Gao, M. Persson, B. I. Lundqvist, *Phys. Rev. Lett.* **78** (1997) 4410.
[105]G. Dujardin, R. E. Walkup, Ph. Avouris, *Science* **255** (1992) 1232.
[106]H. J. Lee, W. Ho, *Science* **286** (1999) 1719.
[107]B. C. Stipe, M. A. Rezaei, W. Ho, *Science* **280** (1998) 1732.
[108]L. C. Venema, J. W. G. Wildöer, J. W. Janssen, S. J. Tans, H. L. J. Temminck Tuinstra, L. P. Kouwenhoven, C. Dekker, *Science* **283** (1999) 52.
[109]O. Pietzsch, A. Kubetzka, M. Bode, R. Wiesendanger, *Science* **292** (2001) 2053.
[110]B. Xu, N. J. Tao, *Science* **301** (2003) 1221.
[111]A. J. Heinrich, J. A. Gupta, C. P. Lutz, D. M. Eigler, *Science* **306** (2004) 466.
[112]M. Lastapis, M. Martin, D. Riedel, L. Hellner, G. Comtet, G. Dujardin, *Science* **308** (2005) 1000.
[113]J. Schneir, R. Sonnenfeld, O. Marti, P. K. Hansma, J. E. Demuth, R. J. Hamers, *J. Appl. Phys.* **63** (1988) 717.
[114]T. R. Albrecht, Ph. D. Thesis, Stanford University, Stanford, California, U.S.A..
[115]U. Staufer, L. Scandella, R. Wiesendanger, *Z. Phys. B* **77** (1989) 281.
[116]D. M. Eigler. C. P. Lutz, W. E. Rudge, *Nature* **352** (1991) 600.
[117]Y. W. Mo, *Science* **261** (1993) 886.
[118]B. C. Stipe, M. A. Rezaei, W. Ho, *Science* **279** (1998) 1907.
[119]K-Y. Kwon, K. L. Wong, G. Pawin, L. Bartels, S. Stolbov, T. S. Rahman, *Phys. Rev. Lett.* **95** (2005) 166101.

[120]J. Repp, G. Meyer, F. E. Olsson, M. Persson, *Science* **305** (2004) 493.
[121]R. P. Feynman, *"There is plenty of Room at the Bottom"*, Am. Phys. Soc. Annual Meeting, 29th, Dec., 1959.
[122]S. Morita, I. Otsuka, T. Okada, H. Yokoyama, T. Iwasaki, N. Mikoshiba, *Jpn. J. Appl. Phys.* **26** (1987) L1853.
[123]O. Lev, F.-R. Fan, A. J. Bard, *J. Electrochem. Soc.* **135** (1988) 783.
[124]P. Lustenberger, H. Rohrer, R. Christoph, H. Siegenthaler, *J. Electroanal. Chem.* **243** (1988) 225.
[125]J. Wiechers, T. Twomey, D. M. Kolb, R. J. Behm, *J. Electroanal. Chem.* **248** (1988) 451.
[126]K. Itaya, E. Tomita, *Surf. Sci.* **201** (1988) L507.
[127]M. P. Green, K. J. Hanson, D. A. Scherson, X. Xing, M. Richter, P. N. Ross, R. Carr, *J. Phys. Chem.* **93** (1989) 2181.
[128]K. Uosaki, H. Kita, *J. Electroanal. Chem.* **259** (1989) 301.
[129]J. Pan, T. W. Jing, S. M. Lindsay, *J. Phys. Chem.* **98** (1993) 4205.
[130]A. Vaught, T. W. Jing, S. M. Lindsay, *Chem. Phys. Lett.* **236** (1995) 306.
[131]W. Schmickler, D. Henderson, *J. Electroanal. Chem.* **290** (1990) 283.
[132]N. J. Tao, C. Z. Li, H. X. He, *J. Electroanal. Chem.* **492** (2000) 81.
[133]Sang-Il Park, C. F. Quate, *Appl. Phys. Lett.* **48** (1986) 112.
[134]S. Morita, K. Itaya, N. Mikoshiba, *Jap. J. Appl. Phys.* **25** (1986) L743.
[135]J. S. Foster, J. E. Frommer, P. C. Arnett, *Nature* **331**(1988) 324.
[136]J. S. Foster, J. E. Frommer, *Nature* **333**(1988) 542.
[137]T. R. Albrecht, M. M. Dovek, C. A. Lang, P. Grütter, C. F. Quate, S. W. J. Quan, C. W. Frank, R. F. W. Peace, *J. Appl. Phys.* **64** (1988) 1178.
[138]C. M. Mate, R. Erlandsson, G. M. McClelland, S. Chiang, *Surf. Sci.* **208** (1989) 473.
[139]H-Y. Liu, F-R. F. Fan, C. W. Lin, A. J. Bard, *J. Am. Chem. Soc.* **108** (1986) 3838.
[140]J. Schneir, R. Sonnenfeld, P. H. Hansma, J. Tersoff, *Phys. Rev. B.* **34** (1986) 4979.
[141]B. Drake, R. Sonnenfeld, J. Schneir, P. H. Hansma, *Surf. Sci.* **181** (1986) 92.
[142]A. A. Gewirth, A. J. Bard, *J. Phys. Chem.* **92** (1988) 5563.
[143]M. H. J. Hottenhuis, M. A. H. Mickers, J. W. Gerritsen, J. P. van der Eerden, *Surf. Sci.* **206** (1988) 259.
[144]K. Shimazu, K. Uosaki, H. Kita, *J. Electroanal. Chem.* **256** (1988) 481.
[145]K. Itaya, S. Sugawara, K. Higaki, *J. Phys. Chem.* **92** (1988) 6714.
[146]F-R. F. Fan, A. J. Bard, *Anal. Chem.* **60** (1988) 751.
[147]L. Vázquez, J. M. Gómez Rodríguez, J. Gómez Herrera, A. M. Baró, N. García, J. C. Canullo, A. J. Arvía, *Surf. Sci.*, **181** (1987) 98.
[148]L. Vázquez, J. Gómez, A. M. Baró, N. García, M. L. Marcos, J. González Velasco, J. M. Vara, A. J. Arvía, J. Presa, A. García, M. Aguilar, *J. Am. Chem. Soc.* **109** (1987) 1730.
[149]J. Gómez, L. Vázquez, A. M. Baró, C. Alonso, E. González, J. González Velasco, A. J. Arvía, *J. Electroanal. Chem.* **240** (1988) 77.
[150]J. Gómez, L. Vázquez, A. M. Baró, C. L. Perdiel, A. J. Arvía, *Electrochimica Acta* **34** (1989) 619.
[151]C. L. Perdiel, E. Custidiano, A. J. Arvía, *J. Electroanal. Chem.* **246** (1988) 165.
[152]J. K. Sass, J. K. Gimzewski, *J. Electroanal. Chem.* **251** (1988) 241.
[153]D. H. Craston, C. W. Lin, A. J. Bard, *J. Electrochem. Soc.* **135** (1988) 785.
[154]B. C. Schardt, S-L. Yau, F. Rinaldi, *Science* **243** (1989) 1050.

[155]R. Christoph, H. Siegenthaler, H. Rohrer, H. Wiese, *Electrochimica Acta* **34** (1989) 1011.
[156]S. Trasatti, *Adv. Electrochem. Electrochem. Engng.* **10** (1977) 213.
[157]M. Szklarczyk, O. Velev, J. O'M. Bockris, *J. Electrochem. Soc.* **136** (1989) 2433.
[158]M. Szklarczyk, J. O'M. Bockris, *J. Electrochem. Soc.* **137** (1990) 452.
[159]P. K. Frolich, G. L. Klark, R. H. Aborn, *Trans. Electrochem. Soc.* **49** (1926) 369.
[160]G. I. Finch, D. L. Layton, *J. Elektrodepositors' Tech. Soc.* **28** (1952) Adv. Copy No. 8.
[161]M. P. Green, M. Richter, X. Xing, D. A. Scherson, K. J. Hanson, P. N. Ross, R. Carr, I. Lindau, *J. Microsc.* **152** (1988) 823.
[162]M. P. Green, K. J. Hanson, P. N. Ross, R. Carr, I. Lindau, *J.Electrochem.Soc.* **137** (1990) 3493.
[163]N. J. Tao, J. Pan, Y. Li, P. I. Oden, J. A. DeRose, S. M. Lindsay, *Surf. Sci.* **271** (1992) L388.
[164]K. Uosaki, H. Kita, *J. Vac. Sci. Technol. A.* **6** (1990) 520.
[165]M. Aguilar P. J. Pascúral, A. Santisteban, *IBM J. Res. Develop.* **30** (1986) 525.
[166]P. H. Schroer, J. Becker, *IBM J. Res. Develop.* **30** (1986) 543.
[167]M. Aguilar, A. García, P. J. Pascúral, J. Presa, A. Santisteban, *Surf. Sci.* **181** (1987) 191.
[168]J. Becker, *Surf. Sci.* **181** (1987) 200.
[169]L. Rosenthaler, H. –R. Hidber, A. Tonin, L. Eng, U. Staufer, R. Wiesendanger, H. –J. Güntherodt, *J. Vac. Sci. Technol. A.* **6** (1988) 393.
[170]J. Clavilier, R. Faure, G. Guinet, R. Durand, *J. Electroanal. Chem.* **107** (1980) 205.
[171]S-L. Yau, M. Carissima, M. Vitus, B. C. Schardt, *J. Am. Chem. Soc.* **112** (1990) 3667.
[172]R.J. Nichols, O. M. Magnussen, J. Hotlos, T. Twomey, R. J. Behm, D. M. Kolb, *J. Electroanal. Chem.* **290** (1990) 21.
[173]O. M. Magnussen, J. Hotlos, R. J. Nichols, D. M. Kolb, R. J. Behm, *Phys. Rev. Lett.* **64** (1990) 2929.
[174]T. Kachiya, H. Hondo, K. Itaya, *J. Electroanal. Chem.* **315** (1991) 275.
[175]K. Sashikata, N. Furuya, K. Itaya, *J. Electroanal. Chem.* **316** (1991) 316.
[176]N. Kimizuka, K Itaya, *Faraday Discuss.* **94** (1992) 117.
[177]T. Hachiya, K. Itaya, *Ultramicroscopy* **42-44** (1992) 445.
[178]S. Sugita, T. Abe, K. Itaya, *J. Phys. Chem.* **97** (1993) 8789.
[179]N. Shinotsuka, K. SashikATA, K. Itaya, *Surf. Sci.* **335** (1995) 75.
[180]J. Inukai, S. Sugita, K. Itaya, *J. Electroanal. Chem.* **403** (1996) 159.
[181]S. Tanaka, S.-L. K. Itaya, *J. Electroanal. Chem.* **396** (1995) 125.
[182]S. –L. Yau, Y. –G. Kim, K. Itaya, *J. Am. Chem. Soc.* **118** (1996) 7795.
[183]T. Suzuki, Y. Hamada, K. Itaya, *J. Phys. Chem.* **100** (1996) 8954.
[184]S. Ando, T. Suzuki, K. Itaya, *J. Electroanal. Chem.* **431** (277) 1997).
[185]J. C. Bondos, A. A. Gewirth, R. G. Nuzzo, *J. Phys. Chem.* **100** (1996) 8617.
[186]K. Itaya, K. Higaki, S. Sugawara, *Chem. Lett.* **(1988)** 421.
[187]M. Szklarczyk, J. O'M. Bockris, *Surf. Sci.* **241** (1991) 54.
[188]R. C. Bhardwaj, A. Gonzalez-Martin, J. O'M. Bockris, *J. Electroanal. Chem.* **307** (1991) 1050.
[189]J. Wang, T. Martinez, D. R. Yaniv, L. D. McCormick, *J. Electroanal. Chem.* **313** (1991) 129.
[190]T. Yamada, N. Batina, K. Itaya, *J. Phys. Chem.* **99** (1995) 8817.

[191]M. Dietterle, T. Will, D. M. Kolb, *Surf. Sci.* **396** (1998) 189.
[192]S. Morin, A. Lachenwitzer, F. A. Möller, O. M. Magnussen, R. J. Behm, *J. Electrochem. Soc.* **146** (1999) 1013.
[193]T. A. Sorenson, T. E. Lister, B. M. Huang, J. L. Stickney, *J. Electrochem. Soc.* **146** (1999) 1019.
[194]L. A. Kibler, A. Cuesta, M. Kleinert, D. M. Kolb, *J. Electroanal. Chem.* **484** (2000) 73.
[195]R. Vogel, I. Kamphausen, H. Baltruschat, *Ber. Bunsen-Ges. Phys. Chem.* **96** (1992)525.
[196]I. Villegas, M. J. Weaver, *J. Electroanal. Chem.* **373** (1994) 245.
[197]W. L. DeSimone, J. J. Breen, *Langmuir* **11** (1995) 4428.
[198]Y. Nagatani, T. Hayashi, T. Yamada, K. Itaya, *Jpn. J. Apel. Phys. Pt. 1* **35** (1996) 720.
[199]J. Inukai, Y. Osawa, M. Wakisaka, K. Sashikata, Y. –G. Kim, K. Itaka, *J. Phys. Chem.* **102** (1998) 3498.
[200]X. Gao, G. J. Edens, M. J. Weaver, *J. Phys. Chem.* **98** (1994) 8074.
[201]D. W. Suggs, A. J. Bard, *J. Phys. Chem.* **99** (1995) 8349.
[202]K. Sashikata, Y. Matsui, K. Itaya, M. P. Soriaga, *J. Phys. Chem.* **100** (1996) 20027.
[203]O. M. Magnussen, J. Hageböck, J. Hotlos, R. J. Behm, *Faraday Disc.* **94** (1992) 329.
[204]O. M. Magnussen, *Faraday Disc.* **94** (1992) 399.
[205]G. J. Edens, X. Gao, M. J. Weaver, *J. Electroanal. Chem.* **375** (1994) 357.
[206]K. Ataka, M. Osawa, *Langmuir* **14** (1998) 951.
[207]H. Ogasawara, Y. Sawatari, J. Inukai, M. Ito, *J. Electroanal. Chem.* **358** (1993) 337.
[208]A. M. Funtikov, U. Stimming, A. Vogel, *Surf. Sci. Lett.* **324** (1995) L343.
[209]A. M. Funtikov, U. Stimming, A. Vogel, *J. Electroanal. Chem.* **428** (1997) 147.
[210]L. –J. Wan, S. –L. Yau, K. Itaya, *J. Phys. Chem.* **99** (1995) 9507.
[211]M. Wilms, P. Broekmann, M. Kruft, Z. Park, C. Stuhlmann, K. Wandelt, *Surf. Sci.* **402** (1998) 83.
[212]W. Li, R. J. Nicholas, *J. Electroanal. Chem.* **467** (1998) 153.
[213]P. Broekmann, M. Wilms, M. Kruft, Z. Park, C. Stuhlmann, K. Wandelt, *J. Electroanal. Chem.* **467** (1999) 307.
[214]L. –J. Wan, M. Hara, J. Inukai, K. Itaya, *J. Phys. Chem.* **103** (1999) 6978.
[215]T. Senna, N. Ikemiya, M. Ito, *J. Electroanal. Chem.* **511** (2001) 115.
[216]L. –J. Wan, T. Suzuki, K. Sashikata, J. Okada J. Inukai, K. Itaya, *J. Electroanal. Chem.* **484** (2000) 189.
[217]N. S. Marinkovic, J. X. Wang, H. Zajonz, R. R. Adzic, *J. Electroanal. Chem.* **500** (2000) 388.
[218]Y. Shingaya, M. Ito, *J. Electroanal. Chem.* **467** (1999) 299.
[219]N. S. Marinkovic, J. S. Marinkovic, R. R. Adzic, *J. Electroanal. Chem.* **467** (1999) 6978.
[220]M. Schweizer, D. M. Kolb, *Surf. Sci.* **544** (2003) 93.
[221]C. Stuhlmann, I. Vilegas, M. J. Weaver, *J. Chem. Phys. Lett.* **219** (1994) 319.
[222]Y. G. Kim, S. –L. Yau, K. Itaya, *J. Am. Chem. Soc.* **118** (1996) 393.
[223]T. Sawaguchi, T. Yamada, Y. Okinaka, K. Itaka, *J. Phys. Chem.* **99** (1995) 14149.
[224]X. Gao, Y. Zhang, M. J. Weaver, *J. Phys. Chem.* **96** (1992) 4156.

[225]M. Ge, B. Zhong, W. G. Klemperer, A. A. Gewirth, *J. Am. Chem. Soc.* **118** (1996) 5812.
[226]R. Srinivasan, J. C. Murphy, R. Fainchtein, N. Pattabiraman, *J. Electroanal. Chem.* **312** (1991) 293.
[227]R. Srinivasan, J. C. Murphy, *Ultramicroscopy* **42** (1992) 453.
[228]S. M. Lindsay, N. J. Tao, J. A. DeRose, A. A. Lyubchenko, Y. L. Harrington, R. E. Shlyakhtenko, *Biophys. J.* **61** (1992) 1570.
[229]N. J. Tao, J. A. DeRose, S. M. Lindsay, *J. Phys. Chem.* **97** (1993) 910.
[230]S. –L. Yau, X. P. Gao, S. C. Chang, B. C. Schardt, M. J. Weaver, *J. Am. Chem. Soc.* **113** (1991) 6049.
[231]I. Oda, J. Inukai, M. Ito, *Chem. Phys. Lett.* **203** (1993) 99.
[232]I. Villegas, M. J. Weaver, *J. Chem. Phys.* **101** (1994) 1648.
[233]K. Ogaki, N. Batina, M. Kunitake, K. Itaya, *J. Phys. Chem.* **100** (1996) 7185.
[234]L. –J. Wan, K. Itaya, *Langmuir* **13** (1997) 7173.
[235]S. –L. Yau, Y. –G. Kim, K. Itaya, *J. Phys. Chem. B* **101** (1997) 3547.
[236]Th. Dretschkow, A. S. Dakkouri, Th. Wandlowski, *Langmuir* **13** (1997) 2843.
[237]K. Sasana, K. Nakamura, K. Kaneto, *Jpn. Appl. Phys., P. 2* **32** (1993) L863.
[238]R. M. Nyffenegger, R. M. Penner, *J. Phys. Chem.* **100** (1996) 17041.
[239]F. Chao, M. Costa, C. Tian, *Analysis,* **22** (1994) M32.
[240]H. Hagenström, M. A. Scheeweis, D. M. Kolb, Langmuir, 15 (1999) 2435.
[241]N. J. Tao, *Phys. Rev. Lett.* **76** (1996) 4066.
[242]J. Zhang, A. K. Kuznetsov, J. Ulstrup, *J. Electroanal. Chem.* **541** (2003) 133.
[243]A. M. Kuznetsov, J. Ulstrup, *J. Phys. Chem.* **104** (2000) 11531.
[244]F. Cunha N. J. Tao, *Phys. Rev. Lett.* **75** (1995) 2376.
[245]K. Itaya, E. Tomita, *Chem. Lett.* (1989) 285.
[246]K. Itaya, E. Tomita, *Surf. Sci.* **219** (1989) L515.
[247]P. Carlsson, B. Holmström, H. Kita, K. Uosaki, *J. Electroanal. Chem.* **283** (1990) 425.
[248]M. Szklarczyk, A. González-Martin, O. Velev, J. O'M. Bockris, *Surf. Sci.* **237** (1990) 305.
[249]M. Szklarczyk, A. González-Martin, J. O'M. Bockris, *Surf. Sci.* **257** (1991) 307.
[250]K. Sakamaki, K. Hinokuma, K. Hashimoto, A. Fujishima, *Surf. Sci.* **237** (1990) L383.
[251]S. Eriksson, P. Carlsson, B. Holmström, K. Uosaki, *J. Electroanal. Chem.* **313** (1991) 121.
[252]X. K. Zhao, L. McCormick, J. H. Fendler, *Chem. Mat.* **3** (1991) 922.
[253]S. –L. Yau, F. –R. Fan, A. J. Bard, *J. Electrochem. Soc.* **139** (1992) 2825.
[254]S. R. Higgins, R. J. Hammers, *Surf. Sci.* **324** (1995) 263.
[255]H. Yao, S. –L. Yau, K. Itaya, *Surf. Sci.* **335** (1995) 166.
[256]H. Yao, S. –L. Yau, K. Itaya, *Appl. Phys. Lett.* **68** (1996) 1473.
[257]J. H. Ye, K. Kaji, K. Itaya, *J. Electrochem. Soc.* **143** (1996) 4012.
[258]H. Yao, K. Itaya, *J. Electrochem. Soc.* **145** (1998) 3090.
[259]W. J. Li. J. A. Nirtanen, R. M. Penner, *Appl. Phys. Lett.* **60** (1992) 1181.
[260]W. J. Li. J. A. Nirtanen, R. M. Penner, *J. Phys. Chem.* **96** (1992) 6529.
[261]C. B. Ross, L. Sun, R. M. Crooks, *Langmuir* **9** (1993) 632.
[262]J. K. Schoer, F. P. Zamborini, R. M. Crooks, *J. Phys. Chem.* **100** (1996) 11086.
[263]R. Ullmann, T. Will, D. M. Kolb, *Ber. Bunsenges. Phys. Chem.* **99** (1995) 1414.
[264]D. M. Kolb, R. Ullmann, T. Will, *Science* **275** (1997) 1097.
[265]D. M. Kolb, R. Ullmann, J. C. Ziegler, *Electrochim. Acta* **43** (1998) 2751.

[266]U. Landman, W. D. Leudtke, N. A. Burnham, R. J. Colton, *Science* **248** (1990) 454.
[267]D. M. Kolb, G. E. Engelmann, J. C. Ziegler, *Angew. Chem. Int. Ed. Engl.* **39** (2000) 1123.
[268]L. A. Nagahara, T. Thundat, S. M. Lindsay, *Appl. Phys. Lett.* **57** (1990) 270.
[268]C. Z. Li, N. J. Tao, *Appl. Phys. Lett.* 72 (1998) 894.
[270]C. Z. Li, A. Bogozi, W. Huang, N. J. Tao, *Nanotechnology* **10** (1999) 221.
[271]C. Shu, C. Z. Li, X. He, A. Bogozi, J. S. Bunch, N. J. Tao, *Phys. Rev. Lett.* **84** (2000) 5196.
[272]R. M. Nyffenegger, R. M. Penner, *Chem. Rev.* **97** (1997) 1195.
[273]W. Wlieth, W. J. Lorenz, (eds.) *Electrochemical Nanotechnology – in situ Local Probe Techniques at Electrochemical Interfaces,* Wiley-VCH, Weinheim, 1998.
[274]O. M. Magnussen, L. Zitzler, B. Gleich, M. R. Vogt, R. J. Behm, *Electrochimica Acta.* **46** (2001) 3725.
[275]www.ieap.uni-kiel.de/solid/ag-magnussen/

6

Modern Applications of Electrochemical Technology

M. Schlesinger

Department of Physics, University of Windsor,
Windsor, Ontario CANADA N9B 3P4

I. INTRODUCTION

Electrochemistry in general and electrochemical technology in particular has over recent decades evolved from an art to an exact science. This development is seen as the main reason for the constantly growing number and widening types of applications of this practical science/engineering. Means and methods of electrochemical technology provide the essential components of modern macro and microelectronics, optoelectronics, optics, sensors of most types and many more. Industries from the automobile to the high-tech industries have adopted methods of electrochemical technology, even where other methods such as evaporation, sputtering, chemical vapor deposition (CVD) are an option, for reasons of economy and/or convenience.

The next four Sections illustrate and discuss in some detail the foregoing technologies.

Section II introduces the LIGA method (a German abbreviation of the process coined by its inventors in Karlsruhe, Germany)

Modern Aspects of Electrochemistry, Number 42, edited by C. Vayenas et al., Springer, New York, 2008.

of which electro-deposition and electroforming are vital components. LIGA, as is well accepted, constitutes an important and interesting alternative/addition to silicon micromechanics and, as such, is common place world wide.

Section III deals with applications of electrochemical methods in the semiconductor and information (non-magnetic, in particular) storage technologies. This is considered one of, if not the most, active areas in modern applications of electrochemistry and electrochemical deposition. More specifically, the introduction of defect-free deposition methods, known as *super conformal deposition*, is a relatively recent, extremely important, addition to the field. As such, it is discussed at some length in this Section.

Section IV deals with the application of electrochemical technologies to magnetic-storage devices. In the field of magnetic information storage, hard discs, floppy discs, and tape drives are widely used. Electrochemical deposition techniques are essential in their production and this is the raison d'etat for its inclusion in this chapter.

The last Section discusses electrochemical technologies in the service of modern medicine. It is felt that the impact of electrochemistry on the modern, 21st-century medicine is not yet properly recognized and appreciated. In all, while these four Sections appear disjointed they are, in actual fact, intimately connected. It is the central role played by electrochemical technologies in each that makes them possible and without which they simply could not be in practical existence.

II. LIGA, AN IMPORTANT PROCESS IN MICRO-SYSTEM TECHNOLOGY

1. Micro Systems

Microelectronics became a reality with the invention of the transistor and the advent of the space age that necessitated drastic scaling down. Electronics had until that time employed techniques and methods some of which, in principle, are still used in some branches of engineering. Specifically, circuit components were combined on a circuit board in accordance with a pre-designed assembly plan. As we are discussing here the period before microelectronics, it should be understood that wide manufacturing tolerances were the norm. Components were arranged in certain classes, and circuits were wired, soldered and tested. What we refer to now days as *trimming elements*, enabled one to approach the originally intended values of the circuit by means of iterative measurement and alignment routines. Naturally, costs were determined almost exclusively by the large amount of manual labor required, and meaningful rationalization of economy was possible only to a very modest extent, if at all.

This was the state of affairs when microelectronics, with its modern manufacturing processes, was introduced. In this approach a circuit is first designed on a computer simulated to an ever improving degree of realism, and optimized. Next, the geometric pattern of the circuit is transferred to a silicon wafer by optical/lithographic means that allows very large volumes of data to be transmitted in parallel, and which is essential for mass production at greatly reduced cost. In addition, the imaging process allows patterns to be very small with details down to the sub-micrometer range. Consequently, a huge volume of details and/or single elements can be packed in the area of a silicon wafer. As indicated above, a product will be cheaper the more components can be made in parallel in one process step. In the four or even five decades since the start of microelectronics, patterns not only have become smaller continuously, but in addition their performance spectrum has improved by several orders of magnitude as a result of shorter leads, lower parasitic capacitances and improved process management. At the same time, miniaturization has reduced the manufacturing costs of each individual component so dramatically that these improved components have in time become cheaper and cheaper.

Fortunately, it is found possible also to apply the methods of microelectronics to, for example, mechanics, optics, fluidics, and, finally, chemistry or biochemistry. Systems may thus be designed with extreme functional density and powered at costs which do decrease by orders of magnitude year by year. The recognition of this prospect was the main reason why methods of manufacturing micro-patterns were devised, and these micro-patterns were combined in micro-systems. Thus, one may state that microelectronics should be considered as the source from which micro-systems engineering sprung. From the start it was clear that, even though a large technical treasury of methodologies in microelectronics is available to draw on, individual development concepts, technologies, materials, and experiences will need to be adapted, modified, or developed.

In the course of this development, avenues for micro-pattern production based on microelectronics opened, perhaps not in detail, but in the basic underlying concepts.

The technique most widely used at the present time is *silicon micro-mechanics*, which to nobody's surprise, bears distinct similarity to methodologies of microelectronics. Indeed, many processes used in microelectronics can be used without change. One major distinction, however, between micromechanics and microelectronics is the three-dimensional character of micromechanics and the largely two-dimensional character of microelectronics. To develop the third dimension, anisotropic etching has been developed in which the wafer, consisting of a single silicon crystal, can be shaped in accordance with certain set of rules. This geometry of shape is provided by the specific crystallographic morphology of the silicon crystal. The anisotropic etching solution produces different etching rates in different directions of the crystal. If this property is exploited in a skillful way, three-dimensional microstructures can be etched out of a crystal whose outer surfaces are limited by the lateral structure of the resist and by crystalographically preferred faces. One important advantage of silicon micromechanics is the dual possibility to accommodate the *micro-pattern* and the *evaluation circuitry*, on the same substrate. In many cases, especially for volume production in, for instance, the automotive industry or in consumer electronics, integration in a monolith will be the solution most advantageous in terms of cost and reliability. In many other cases, however, a hybrid solution will be preferable in which components of different origins must be integrated into a system by special assembly techniques. The literature bears witness to the high-performance standard of silicon micromechanics, especially in the field of sensors and their combination with integrated evaluation circuits.

An important and interesting alternative to silicon micromechanics is the so-called LIGA (German abbreviation of the three major process steps, see bellow) process. It was initially developed at the Institute of Microstructure Technology of the Research Center in Karlsruhe, Germany. In this technique, the material used is not a (silicon) single crystal, but a layer of polymethylmethacrylate (PMMA), several hundred micrometers thick. The method and its applications will be introduced and described in detail in this chapter.

The most important benefits of the technology are:

- The range of materials available for use ranges from plastics through metals and metal alloys to ceramics.
- Patterning is not limited by the morphology of a crystal; consequently, lateral pattern design is completely unrestricted.

Unlike silicon micromechanics, this technique is capable of producing circular and annular patterns, which is of major importance in various actuators.

In addition to these foregoing technologies, variants to the method have been developed by and used in combination with the two major technologies. In this connection, mention must be made of *silicon-surface technology*. Indeed, the basic unit in most practical cases is a silicon wafer, often pre-patterned by anisotropic etching. Instead of penetrating into the crystal from the surface by means of etching, a pattern may also be built up on the surface layer by layer by means of *thin-film electro-deposition* techniques. This technology provides complete freedom to create free lateral patterns as well as the use of an extended range of materials.

A major drawback on the other hand, is the low pattern height of these thin-film patterns, as the name itself implies. In the case of rotating micro-components, precise bearings could not easily be designed in the early days of the process. As a result of short support faces, rotors tended to wobble, thus experiencing excessive wear and exhibiting a correspondingly short service life. For the same reason, no high forces or torques could be transmitted initially, though this state of affairs is rapidly improving now.

2. The LIGA Process

As stated above, the development of the LIGA process began at the Research Center in Karlsruhe (FZK), Germany, in the 70s as a rather inexpensive method of producing very small slotted nozzles of any lateral shape for uranium-isotope separation by the nozzle process. Its usage is now wide spread globally as well as to a much larger number of applications than initially contemplated. Thus, the process is recognized as an important means in the production of microstructures in as diverse areas as the automotive industry and medical technology. This development continues worldwide to this day.

As stated above, LIGA is the German abbreviation of the three major process steps, *lithography* (Lithographie), *electroplating* (Galvanoformung) and *molding* (Abformung). A schematic diagram of the steps involved in the fabrication of basic LIGA structures is shown in Figure 1. Reference 1 is an excellent source of information regarding the LIGA method for readers who may be interested in expanding their familiarity with the process of electro-deposition of microstructures.

In the first of five steps (see Fig. 1) a pattern of an X-ray absorber mask is transferred onto a resist layer hundred or so micrometers in thickness by X-ray shadow projection. In practice often synchrotron radiation is used due to its very high degree of collimation and short wavelength (0.2–1 nm).

The X-ray masks consist of a thin membrane (e.g., 3-μm titanium or 30-μm beryllium) together with absorbers consisting of X-ray opaque gold layers that are more than 10 μm in thickness in order to achieve the required mask contrast. Within the irradiated sections of the resist layer, the polymer chains are destroyed reducing their molecular weight. In most experiments polymethyl methacrylate (PMMA) is used as the X-ray resist.

As a consequence of the second, development step, the exposed resist is selectively dissolved while at the same time the masked and un-irradiated parts remain unchanged.

The third and fourth stages involve *electrodeposition* on a micro-patterned substrate employed to build up a complementary pattern in a metal such as copper, nickel, or gold. This is done by filling the empty spaces of the electrically non-conductive resist starting, as one must if electrodeposition is employed, from the electronically conducting base plate itself. Alternatively, suitably patterned *electroless metal deposition* may be employed. The metal pattern thus produced can be used to create—with the utmost degree of detail and at a relatively reduced cost—any number of plastic copies by a molding processes such as standard-reaction injection molding (RIM) or thermoplastic injection molding (TIM) and hot embossing of thermoplastics. The basic difference between

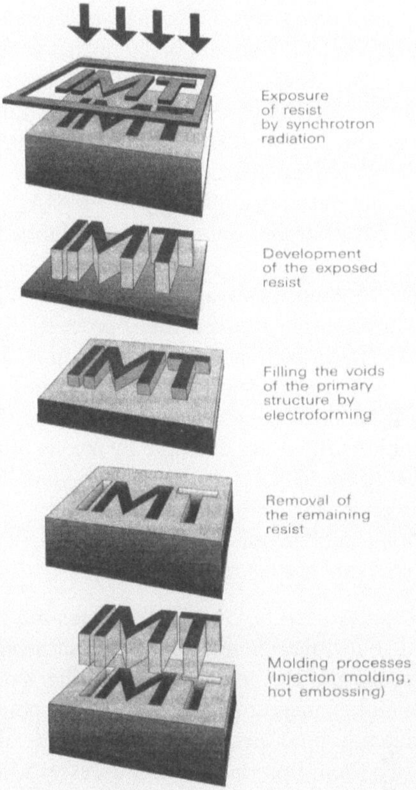

Figure 1. Schematic process steps of the LIGA technology. Reprinted from Ref. 1, Copyright (1996) with permission from Taylor & Francis Books (UK).

RIM and TIM is that in the case of RIM the components are mixed shortly before injection into the mold and are cured by polymerization in the mold. On the other hand, in the TIM method a uniform molding compound is heated above its glass transition temperature to produce a material that can be injected into the mold where it resolidifies as the temperature decreases. We wish to stress that the hot embossing technique for thermoplastics is particularly suitable for molding micro-scale structures on processed silicon wafers, e.g., on microelectronic circuits. Once the wafer carries the protection and metallized layers, it should be laminated with the molding compound. In the next step the wafer and the plastic material are heated and a mold insert is pressed, in vacuum, into the molding compound. Once the compound has cooled the mold insert is removed. A wide variety of plastic materials can be patterned by these molding techniques, including any of PMMA, polycarbonate, polyamide, polyethersulfone, polyvinylidene fluoride, and epoxy resin.

The plastic structures can again be filled with metal in a second electroforming (electrodeposition) process. Thus, metallic microstructures can be fabricated in an economic, low cost and effective way. The plastic microstructures can also serve as so called *lost molds* for the production of ceramic microstructures. In this case, the mold is filled, for example by means of a slip casting process. During firing of the ceramic, the polymer is completely degraded and evaporates.

A recent joint publication from Sandia and Lawrence Berkeley National Labs[2] points to the fact that

> "resist substrates used in the LIGA process must provide high initial bond strength between the substrate and resist, little degradation of the bond strength during X-ray exposure, acceptable undercut rates during development and a surface enabling good electrodeposition of metals. Additionally, they should produce little fluorescence radiation and give small secondary doses in bright regions of the resist at the substrate interface."

To develop a new substrate satisfying all these requirements, workers in those labs have investigated secondary resist doses due to electrons and fluorescence, resist adhesion before exposure, loss of fine features during extended development, and the nucleation and adhesion of electrodeposits for various substrate materials. The result of these studies is a new anodized aluminum substrate and accompanying methods for resist bonding and electrodeposition. The work referred to here appeared in 2005. It serves as "living proof " and indication that LIGA methods are in a constant process of development-improvement as well as in the widest practical use.

3. Microstructures Manufactured by the LIGA Process

(*i*) *The Sacrificial Layer Technique*

A combined fabrication technology made possible through a sacrificial layer technique was developed[3] to obtain partly or totally movable microstructures together with fixed structures on a single substrate. Many practitioners consider this ability to be the hall mark of the LIGA process.

The substrate is first coated, for instance, via electroless or physical vapor deposition with a thin (< 1 μm) metallic layer, which in turn is patterned by photolithography and wet etching. This layer serves two roles, as a plating base and as an electrically conducting layer for the finished structures. In the subsequent step a sacrificial layer, of about 5 μm in thickness, is deposited on the substrate and also patterned by photolithography and wet etching. Titanium is used most often as the sacrificial material because it adheres well to the resist and to the electrodeposited layer and can be etched with hydrofluoric acid that does not attack other materials such as chromium, silver, nickel, copper, and which are usually used in the LIGA process.

The standard LIGA process is then followed: polymerization of the thick X-ray resist directly on to the substrate, exposure to synchrotron radiation through a precisely adjusted mask, development of the resist, and electrodeposition. Some parts of the metallic microstructures are built up on the first metal layer, while other parts lie on top of the sacrificial layer. After stripping the resist, the sacrificial layer is etched selectively against all the other materials.

(ii) Microstructures with Different Shapes in the Third Dimension

The plastic molding process may be combined with deep etch X-ray lithography to fabricate microstructures with different shapes in the third dimension[4] In the first step, only the upper part of a very thick resist layer is patterned with the embossing technique using a mold insert fabricated by the standard LIGA process. The relief structure obtained by this first patterning process is subsequently exposed to synchrotron radiation through a precisely adjusted mask in order to transfer a different pattern into the lower part of the polymer layer. By combining these two processes, a stepped plastic template is produced. A metallic microstructure can then be obtained using a standard electroforming process. Microstructures with inclined sidewalls of arbitrary angle have been successfully fabricated by simply varying the angle at which the synchrotron radiation strikes the resist. If the resist is exposed twice to synchrotron radiation, *conical* structures can be obtained.

III. APPLICATIONS IN SEMICONDUCTOR TECHNOLOGY

1. Cu Interconnections on Chips

One—if not the most—active area of modern application of electrochemical deposition, is the *semiconductor technology*.

At present it is felt that in this context, a fundamentally important recent advance in the silicon-based semiconductor industry is the development of *copper interconnects* on chips. This new technology replaces aluminum or aluminum alloy (e.g., Al-Cu) conductors produced via vacuum-based deposition techniques with copper conductors themselves produced through electroless and/or electrodeposition. Specifically, vacuum-based deposition techniques include *physical vapor deposition* (PVD) and *chemical vapor deposition* (CVD).

Copper has been replacing aluminum since 1999 due to its low bulk electrical resistivity and superior electro migration resistance. The electrical resistivities of pure Al and Cu are 2.9 and 1.7 $\mu\Omega$ cm, respectively, and that for Al-alloys is 3-4 $\mu\Omega$ cm. As well, activation energies for electromigration, using identical structures and experimental conditions, are 0.81 ($\pm$ 0.03) and 1.1($\pm$ 0.1) eV for Al (0.5 wt % Cu) and Cu, respectively.[7-10] These are important properties as lower electrical resistivity results in higher speed of devices . Recall that device speed is, everything else being equal, dependent on the RC (resistivity times capacitance) value of the device. Higher electro-migration resistance (higher activation energy for electromigration) results in higher reliability and thus lower failure rate of interconnect.[11] Again, recall that electromigration is the drift of positive Al atoms due to the positive electron 'wind', both toward the positive direction of an applied electric field. This robs material from one side adding material on the other causing eventual device failure. Finally, it ought to be understood too that, generally speaking, electrochemical deposition is far less cumbersome and more economical than the above-mentioned PVD or CVD method. Specifically, the most frequently used PVD technique is *sputtering*. In the sputtering process of Al a target surface of Al is bombarded by energetic ions of an inert gas, e.g., Ar, Kr.[5,6] The result of this bombardment or collisions between the incident energetic particles and the Al target surface, is the ejection of one or more Al atoms from the target. These ejected Al atoms deposit on the silicon-based semiconductor. Sputtering tools operate under glow discharge conditions.[5,6] CVD deposition process of Al involves surface chemical reaction between the silicon wafer and the reactant-gas species containing an Al compound. This surface reaction results in the formation of an Al thin film.[5]

The reader may justifiably wonder at this point, as to the reasons why Al interconnection were not replaced by Cu ones much earlier. The reason was that copper metal from the interconnection could readily diffuse into the semi-conductor, interfering with (changing) the electronic properties of material. This now can be avoided with the development of new effective *diffusion barriers*.

The number of transistors per chip is increasing continuously (in the case of microprocessors the total number of transistors per chip was 11, 21, and 40 million in the years 1997, 1999, and 2001, respectively) and the physical feature size of individual transistors is decreasing. Concurrently, the dimensions of interconnections (interconnects) on the chip are scaling down.[12] For example, line widths were 0.25, 0.18, 0.15, and 0.10 μm in the years 1997, 1999, 2001, and 2006, respectively. This reduction in size (miniaturization) of the interconnect line width and thickness is accompanied by a positive (increase) change in the ratio of surface to bulk atoms. Thus, an understanding of the physical properties of the evolving new submicron thin films requires deeper understanding of the physics and chemistry of both, bulk and surface.[13]

The required depth of understanding of the physical properties of metal thin films used for interconnects on chips may well be illustrated by the following example. It was found that the performance of conductors on chips, Al or Cu, depends on the morphology of the conductor metal. Vaidya and Sinha[14] reported that the measured median time to failure (MTF) of Al-0.5% Cu thin films is a function of three micro-morphological variables or attributes: its median grain size, the statistical variance (σ^2) of the grain size distribution, and the degree of [111] fiber texture in the film.

2. Deposition of Cu Interconnections on Chips

The change from Al to Cu interconnects, mentioned above, required concurrent significant changes in the fabrication process, from metal-RIE to dielectric-RIE (reactive ion etching). The introduction of copper interconnects creates, in addition, new problems in the fabrication of those interconnects on chips. The most important of those mentioned briefly above, are the possible diffu-

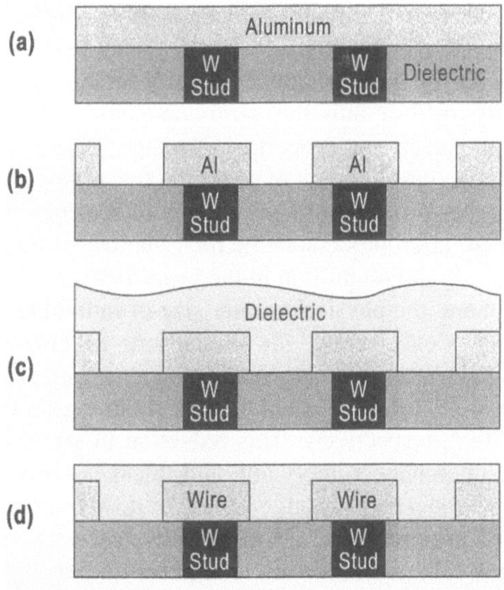

Figure 2. Process steps for forming Al interconnects using the metal-RIE process; a) Al deposition, b) Al RIE through photoresist mask, c) dielectric (for example SiO_2) deposition, d) dielectric planarization by CMP.

sion of Cu into Si, SiO_2 and other dielectrics[8] and the subsequent reaction of Cu with Si forming silicides.[27] Diffusion of Cu through Si results in the poisoning of the devices (transistors), and diffusion through SiO_2 dielectrics leads to the degradation of the dielectrics.

Metal-RIE process was/is used in the fabrication of Al interconnects on chips.[5] This process is depicted in four steps in Fig. 2. The first step in the metal-RIE process is sputter deposition of a blanket thin film of Al (or Al alloys, such as Al-Cu, Al-Si) over a planerized dielectric (e.g., silicon dioxide). In the next step, the unwanted metal is etched away by *reactive ion etching* (RIE) through a photoresist mask. The features produced this way are separated, electrically isolated, metal Al conductor lines. In the RIE process chemicaly active ions such as F or Cl bombard the Al surface and form volatile aluminum fluorides or chlorides, which are then pumped away in the vacuum system.[15,16] After etching, a dielectric is deposited in such a fashion that it fills the gaps between the lines as well as above them. In the last step, the dielectric is planarized using the *chemical mechanical polishing* (CMP) technique.[5]

Multi-level Cu interconnections on chip are fabricated at the present time using a dielectric-RIE process since it is difficult to pattern Cu by RIE. This is so since the vapor pressure of halides is very low at room temperature.[17,18] In this process a blanket Cu deposition is followed by *chemical-mechanical polishing* (CMP) of Cu.[5] This approach is known as a *damascene process* (a process used in Damascus for centuries to form inlaid metal features on jewelry). Figure 3 illustrates the procedure sequence of the single damascene process. This fabrication procedure starts with the deposition of a dielectric layer on a Si wafer, patterning it using photolithography and dielectric-RIE process. After patterning the dielectric, a diffusion barrier metal and a Cu seed layer are deposited using PVD or CVD technique. Finally, Cu is electrodeposited into the recesses, trenches (lines) and holes (vias). The excess of the Cu deposited on the upper surface is removed by CMP.[5] In the dual damascene process, vias and trenches are patterned and filled with Cu at the same time, as depicted in Fig. 4.

The *selective Cu deposition* process was suggested, among others, by Ting and Paunovic.[17] This is an alternative process to

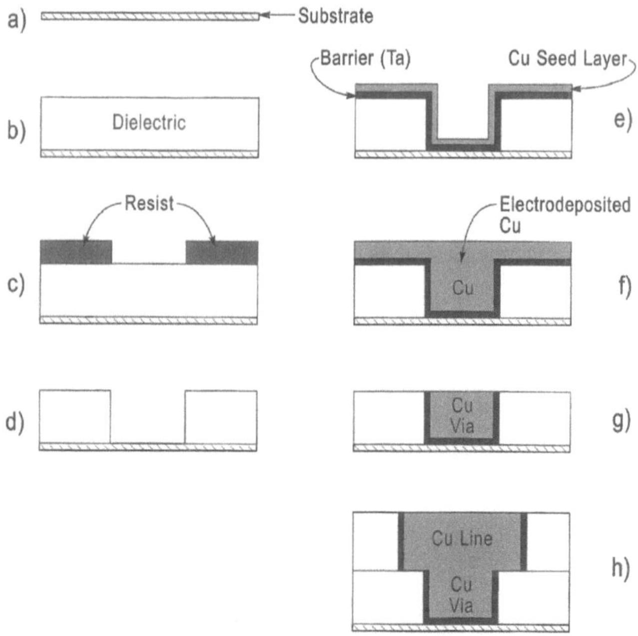

Figure 3. Process steps for forming Cu interconnects using single damascene process (dielectric patterning); a) planarized substrate, b) dielectric deposition, c) dielectric RIE through photoresist mask, d) etched insulator, e) deposition of diffusion barrier (Ta) and Cu seed layer, f) electrodeposition of Cu into via (vertical interconnection), g) CMP of Cu excess, h) patterning and deposition of Cu line (wire).

the fabrication of multi-level Cu interconnections, as shown in Fig. 5. This process is also known as *through-mask deposition* process.[18] The first step in this procedure is the deposition of a Cu seed layer on a Si wafer followed by deposition and pattern of a resist mask to expose the underlyings seed layer in vias and trenches. In the next step, Cu is deposited to fill the pattern. After this the mask is removed, the surrounding seed layer is etched, and the dielectric is deposited. Electroless Cu deposition was suggested for the blanket and selective deposition processes.[19]

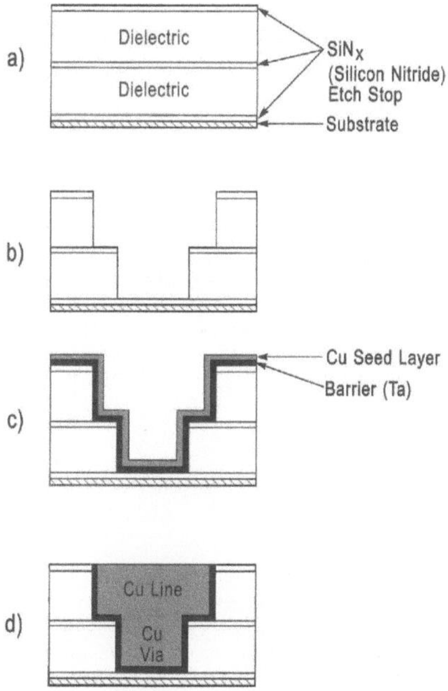

Figure 4. Process steps for the dual damascene process; a) deposition of dielectric, b) dielectric RIE to define via and line, c) deposition of diffusion barrier and Cu seed layer, d) electrodeposition of Cu into via and trenches followed by Cu CMP.

A variation of the selective Cu deposition process, limited to electroless Cu deposition, is the *lift-off process,* known as a *planarized metallization process.*[20] Figure 6 shows a stepwise sequence for this technology.

One important difference between the damascene and the plating through mask procedures is the way the trenches and vias are filled with electrochemically deposited Cu, either through electro or electroless techniques. In multi-level metal structures, the vias provide paths for connecting two conductive regions separated by inter-level dielectric (ILD). In a damascene process the Cu deposit grows from the active bottom and the sidewalls, as shown in Fig. 7a.

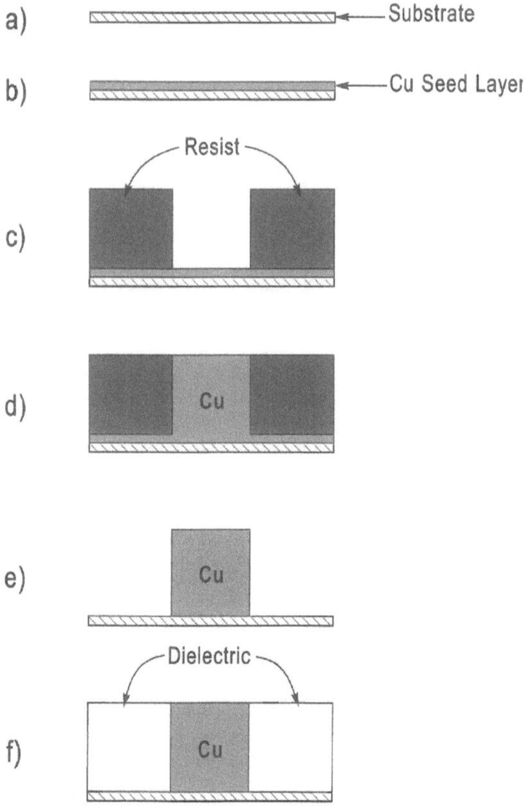

Figure 5. Through-mask deposition process; a) Si substrate, b) Cu seed layer deposition, c) photoresist deposition and patterning, d) through-mask electroless deposition of Cu, e) stripping of photoresist and etching of Cu seed layer outside of line, f) dielectric deposition.

Preferred growth from the bottom may be achieved by the addition of suitable additives.[21,22] In the plating through mask process only the bottom is active while the sidewalls are inactive, resulting in the grow of Cu deposit from the bottom, as depicted in Fig. 7b.[23-25]

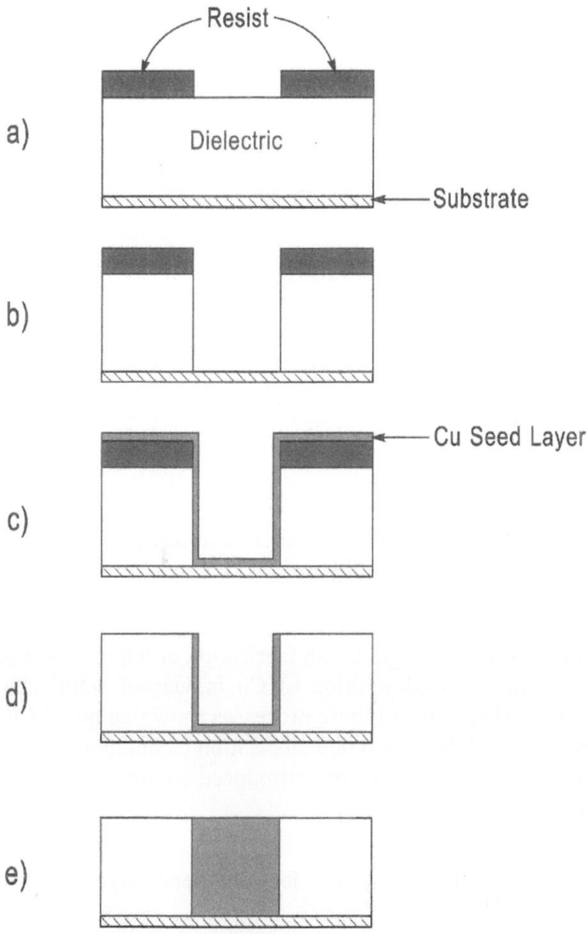

Figure 6. The process sequence for the lift-off process (the planarized metallization process); a) a resist film is patterned on a dielectric film, b) dielectric patterning, c) thin catalytic film layer (PVD or CVD Ti, Al) is deposited, d) a lift-off technique removes the excess material leaving the catalytic layer in the trench only, e) electroless Cu deposition.

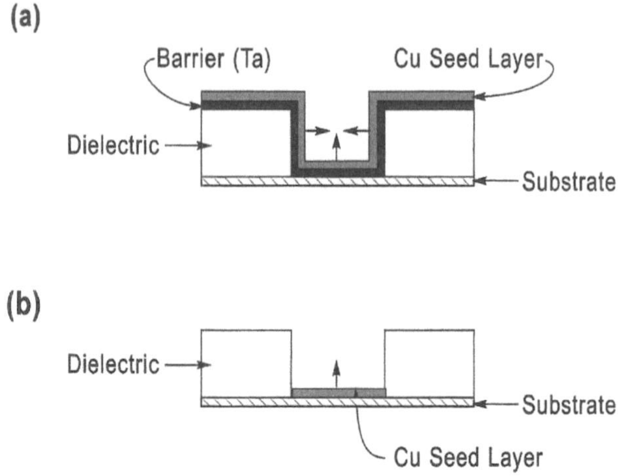

Figure 7. Growth of deposit in vias and trenches during Cu electrodeposition in a) damascene and b) deposition through mask process.

One of the challenges in the fabrication of interconnections on a chip is the electrodeposition of Cu in vias of small diameter (≤ 0.2 μm). Modelings of these processes show that novel Cu electrodeposition solutions and new deposition techniques are required in order to solve the problems introduced by the development of new integrated circuits.[26]

3. Diffusion Barriers and Seed Layer

Diffusion barrier layers are an integral part of the fabrication of copper interconnects (Figs. 3 and 4). Barrier films isolate (encapsulate) Cu interconnects from adjacent dielectric materials. The diffusion barriers most studied to date are Ta,[26,27] Ti, and TiN.[28,29]

Since barrier metals have relatively high/low electrical resistivity/conductivity (Ta 12.4; Ti 80 μΩ cm), it is necessary to cover the barrier layer with a conductive metal layer (seed layer). This conductive metal layer consists of a *Cu seed layer* that is deposited by PVD or CVD techniques (Figs. 3 and 4). When the electrode

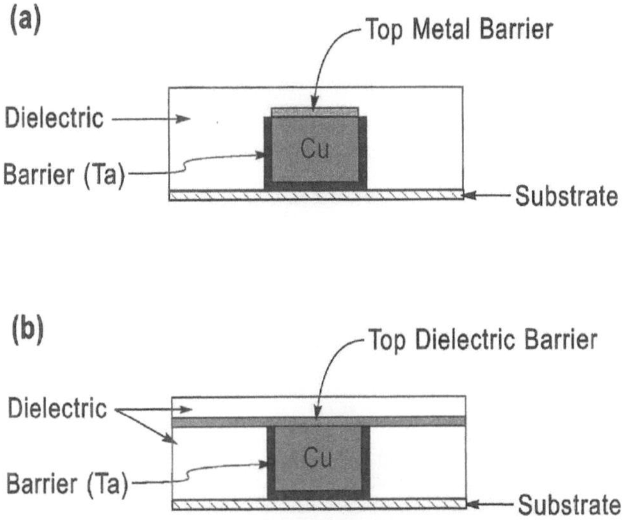

Figure 8. Fully encapsulated Cu line; a) metal (alloy) top barrier, b) dielectric top barrier.

position of the Cu on a barrier/Cu seed layer/bi-layer is completed, vias and trenches are filled with Cu and the excess Cu is removed using CMP (Figs. 3g, 3h and 5d). The exposed Cu in lines needs *capping* with a barrier material to prevent diffusion. Two types of capping material may be used: metals (alloys) and dielectrics. The suggested selectively deposited metal barriers (Fig. 8a) are: electroless Co (P), Co-Ni (P), Co-W (P) and selective CVD W.[32-34] The preferred dielectric barrier material (Fig. 8b) is blanket SiN_x. The result of the capping is a fully encapsulated Cu line (Fig. 8).

It is seen from the discussion above that Cu is electrodeposited in vias and trenches on a bilayer: barrier metal/Cu seed layer. In those cases where the barrier layer itself is composed of two layers (e.g., TiN/Ti), Cu is electrodeposited as trilayer: barrier bilayer/Cu seed layer. This type of underlayer for electrodeposition of Cu raises a series of interesting theoretical and practical questions of considerable significance regarding the reliability of interconnects on chips. In Section II we noted that the reliability of the

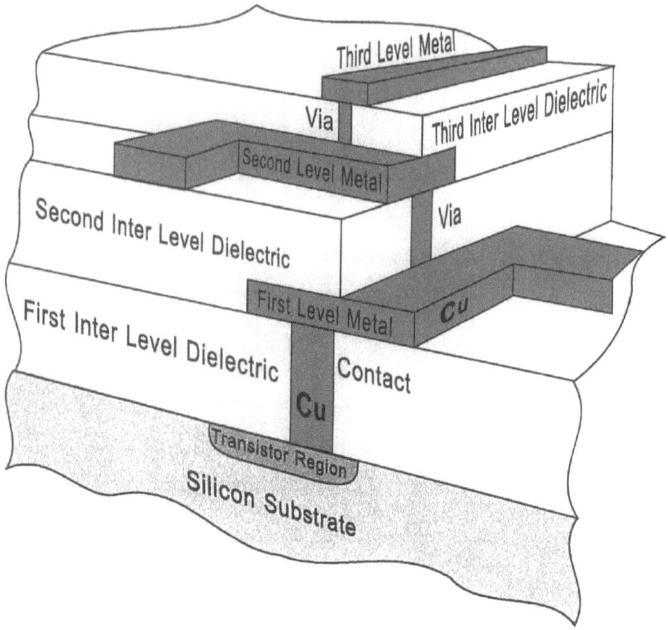

Figure 9. Three-level Cu interconnect IC structure.

interconnects depends on the micro-structural attributes of the electrodeposited Cu (for Cu-based interconnects). These micrstructural attributes—such as grain size, grain size distribution, and texture—determine the mechanical and physical properties of thin films. Thus, an important basic point regarding these questions is the issue of the influence of the underlayer-barrier metal on the microstructure of the Cu-seed layer. Yet another question is the influence of the microstructure of the Cu-seed layer on the structure and properties of electrodeposited Cu.[35] Zielinski et al.[35] demonstrated that the Cu microstructure is sensitive to the texture and microstructure of the barrier metal. For example, textural inheritance was observed on Ta underlayer. In this case there is quasi-grain-to-grain epitaxy. In addition, the presence of a barrier layer can further influence the microstructural evolution upon annealing. Tracy et al.[36] determined that a strongly textured underlayer, such as Ti or Ti/TiN, induces a similar strong texture in the Cu. A strong Ti texture is passed to the TiN, which passes the texture to the Cu (or to Al-Cu). Tracy and Knorr[37] determined that a typical copper film consists of three texture components: [111], [200], and random. Indeed, [200] and [511] texture components are possible under some deposition conditions.

Integration of Cu with a dielectric introduces new problems and/or challenges (1(ii), [34-36] (Fig. 9). For example, when polyimides are used as intermetal dielectrics, reliability concerns include corrosion of the underlying metal and adhesion of the metal films to polyimide underlayers (1(ii)).

The discussion above concerning the influence of the underlayer on the microstructure and reliability of interconnects in IC illustrates that there is a need for an additional in depth understanding of the processes of deposition and physical-mechanical properties of electrodeposited Cu used in IC fabrication.

4. Super-Conformal Electrodeposition of Copper into Nanometer Vias and Trenches

(*i*) *Super-Conformal Electrodeposition*

Defect-free deposits in vias and trenches known as *super-conformal electrodeposition*, may be achieved in the presence of certain additives. In super-conformal deposition, electrodeposition inside vias and trenches occurs preferentially at the bottom.

Sub-conformal and conformal electrodeposition of copper in vias and trenches, on the other hand, occurs in an additive-free acid sulfate solution. These deposits have two types of defects: voids and seams (Fig. 3 in Ref. 22).

(*ii*) *Mechanism of Super-Conformal Electrodeposition*

Two mechanisms for super-conformal deposition in the presence of additives have been proposed: the *differential-inhibition mechanism* and the *inhibition-acceleration mechanism*.

(*a*) *Differential-inhibition mechanism*

It was proposed by Andricacos et al.[22] This mechanism considers one-additive system. It is noted in Ref. 47 (Ch. 10, Sections 10.4 and 10.5) that in general, adsorption of additives (inhibitor) at the cathode affects the kinetics and growth mechanism of electrodeposition. The surface coverage of the additive (inhibitor), θ, is a function of the diffusion controlled rates of the adsorption-desorption processes. In the differential-inhibition mechanism it is assumed that a very wide range of additive fluxes over the microprofile (vias and trenches) exists, that is, extremely low flux in deep interior corners, low flux at the bottom center, moderate flux at the sidewalls, and high flux at shoulders.

This type of concentration variation of the inhibitor (i.e., θ) results in,

- variation of the inhibition of the electrodeposition kinetics in vias and trenches over a very wide range, i.e., several orders of magnitude, and
- preferential electrodeposition on the bottom of the vias and trenches leading to void-free deposits.

The inhibition factor used in the interpretation of leveling cannot describe the shape-change behavior in super-conformal electrodeposition of copper (see Refs. 22, and 47 Ch. 10, Section 10.6).

(*b*) *Inhibition-acceleration mechanism*

Moffat et al.[41] proposed the inhibition-acceleration mechanism in order to explain the experimentally observed corner rounding (inversion of curvature, Ref. 41, Fig 19) and general shape evolution in super-conformal electrodeposition of copper in vias and trenches of nanometer dimensions.[41,42] These authors studied as well a three-additive system composed of two inhibitors and one accelerator. They concluded that super-conformal deposition and corner rounding might be attributed to competitive adsorption of inhibitor and accelerator. This model is based on the assumption of curvature enhanced accelerator coverage in vias and trenches.

(iii) Mathematical Modeling

Two types of modeling have been used to interpret and optimize super-conformal electrodeposition of copper: the *deterministic modeling* and the *stochastic modeling*

(*a*) *Deterministic modeling*

The first model in this category is based on the differential-inhibition mechanism. Andricacos et al.[22] modified Dukovic's computer program that was designed to follow the shape evolution through mask electrodeposition of copper.[43] The essential characteristics of the modified program are:

- the flux of the inhibitor along the profile of the features (vias and trenches) varies over a very wide range, i.e., several orders of magnitude, and
- this variation is determined by a dynamic balance between the arrival of the fresh additive and its consumption (reduction at the cathode).

This mathematical model is able to predict the best combination of current density and inhibitor concentration that will produce super-conformal deposition.

(*b*) *Stochastic modeling*

The second model in this category is based on the inhibition-acceleration mechanism. The basic assumption of this model is that the accelerator is strongly adsorbed on the surface thereby displacing the inhibitor.[42, 44]

Pricer, Kushner, and Alkire[45] used Monte-Carlo type stochastic methods for modeling the filling of vias and trenches by electrodeposition in the presence of a hypothetical blocking additive. Random numbers were used to determine when and if a particle would make a certain move.[46] Each move for a species had a probability value associated with it that depends on the time step. This Monte-Carlo computer program had 14 parameters related to the additive, Cu^{2+} and Cu^{+} ions. An analysis of Monte Carlo *snapshots* during trench filling resulted in a series of conclusions. We are presenting here only two:

- Void size decreases as the concentration of additive increases. This decrease occurs because the additive is adsorbed on the upper part of the trench, blocking Cu^{2+} deposition and allowing the Cu^{2+} to diffuse further into the trench without being consumed.
- At low adsorption rates (10^3 nM/s) large voids can be seen; at higher adsorption rates (10^7 nM/s) the trench fills faster from the bottom and no voids can be seen. Faster copper deposition at the bottom is caused by more of the additive being able to get adsorbed on the upper part of the trench resulting in an increased diffusion of Cu^{2+} ions onto the bottom of the trench.

IV. INFORMATION STORAGE: APPLICATIONS IN THE FIELDS OF MAGNETISM AND MICROELECTRONIC

1. Magnetic Information Storage

Ours is the information age. Consequently, the demand for high performance, low cost, and non-volatile information storage systems is on a constant rise. There are a great variety of information storage systems with varying degree of development and commercialization. Those include magnetic tape drives, hard disc drives, magnetic floppy disc drives, magneto optic (MO) disc drives, phase-change optic disc drives, semiconductor flush memory, magnetic random access memory (RAM) and holographic optical storage. Magnetic information storage technology (i.e., hard disc, floppy disc and tape drives) is the most widely used. Electrochemical deposition techniques are essential in the production of it and this is the raison d'état for the present discussion in this write up. Figure 10 depicts the hierarchy of the information storage in a typical PC in terms of cost vs. performance speed and efficiency.

In general, it is most efficient for the microprocessor to access and store in the semiconductor main memory (DRAM) alone because the DRAM access time is < 100 ns. However, DRAM is volatile and any information in the DRAM is lost once the computer is turned off. Thus, lower level of information storage devices is essential. The latter may include magnetic hard disc, magnetic floppy drives, etc. As a rule, the higher-level storage media are

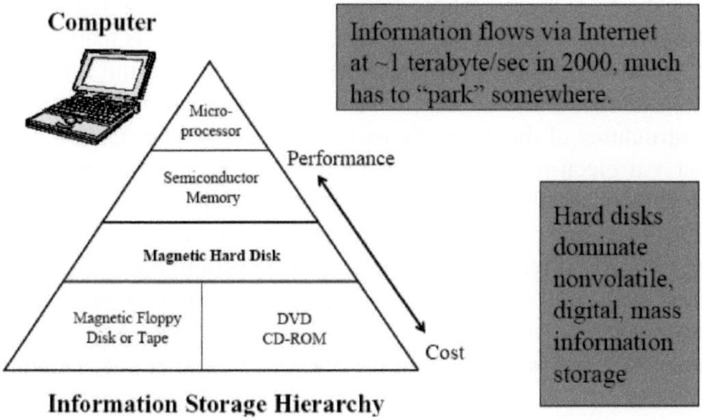

Figure 10. Information storage hierarchy. Reprinted from Ref. 48, Copyright (1999) with permission from Elsevier.

superior in performance but inferior in cost (i.e., cost more). Put it differently, the lower levels of the memory hierarchy are not volatile and are cost wise superior (i.e., cheaper). In all these, as noted before, electrochemical deposition techniques play an important role. To illustrate the point we present Fig. 11, which depicts a schematic cross section of a thin film hard disc.

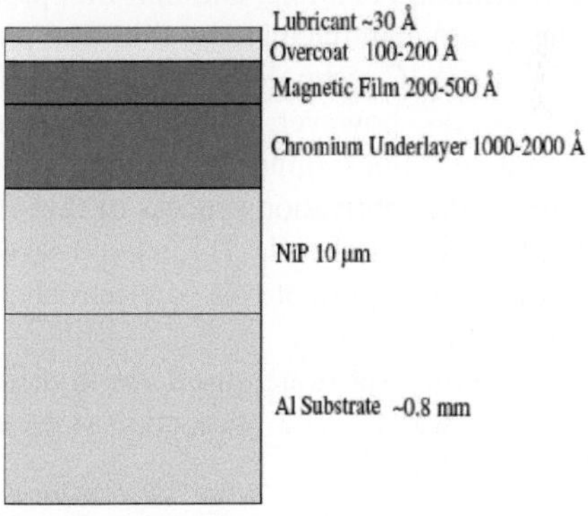

Figure 11. Schematic cross section of a thin film hard disc. Mobile hard disk drives use glass substrates and smaller diameters (2.5" and 1"). Reprinted from Ref. 48, Copyright (1999) with permission from Elsevier.

The Ni-P layer—whose purpose is to render a nearly perfectly smooth, rigid and properly textured under surface—is made usually of an electrochemical-electrolessly deposited thin film. The chromium under layer that controls magnetic properties and microstructures of the magnetic recording layer, was produced in the past via electrochemical deposition. The magnetic layer itself—typically a cobalt based alloy—was also produced via electroless deposition methods. These methods required the addition of additives, mostly proprietary, to the deposition bath in order to produce discs with the right magnetic properties. Also, note that the smoothness provided by the electrolessly deposited Ni-P layer is essential if one is to maintain the read/write head at a fly height of < 40 nm .

More recently, though, the chromium under layer, the magnetic recording layer, and the overcoat are produced mainly via vacuum-deposition methods. The reason for this is economical, that is, to ease the mass production at accurately fixed thicknesses.

2. Read/Write Heads

Combined read/write heads belong to the old technology. Newer technologies follow a trend toward using *separate* heads for reading and writing. [49]

Metal in gap (MIG) heads or ferrite ones are produced with a combination of machining, bonding, and thin film processes. Thin-film inductive heads are manufactured by thin-film processes similar to the semiconductor IC technology (see Section III). Thin-film head production process, however, is rather unique as it involves both, very thin and very thick films. We choose to present here a detailed summary of the fabrication process of thin-film inductive heads with a single-layer spiral coil. This should serve, once again, to illustrate the centrally important role of electrochemical deposition in connection with modern IT.

The most commonly substrates used for thin-film heads are ceramic-alumina TiC wafers. The production steps are described as follows (see Figs. 12 to 14):[3]

1. The substrate made of aluminum-titanium-carbide (ALTIC) is the aerodynamic slider body of the flying head. The first step in the fabrication consists on the sputtering of thick

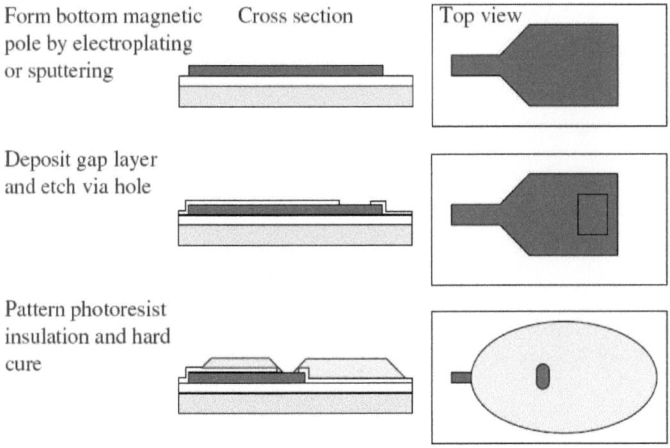

Figure 12. Thin film head fabrication stages (1). Reprinted from Ref. 48, Copyright (1999) with permission from Elsevier.

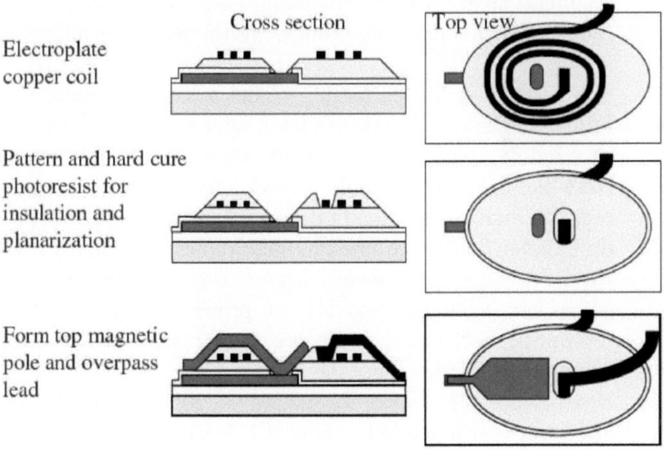

Figure 13. Thin film head fabrication stages (2). Reprinted from Ref. 48, Copyright (1999) with permission from Elsevier.

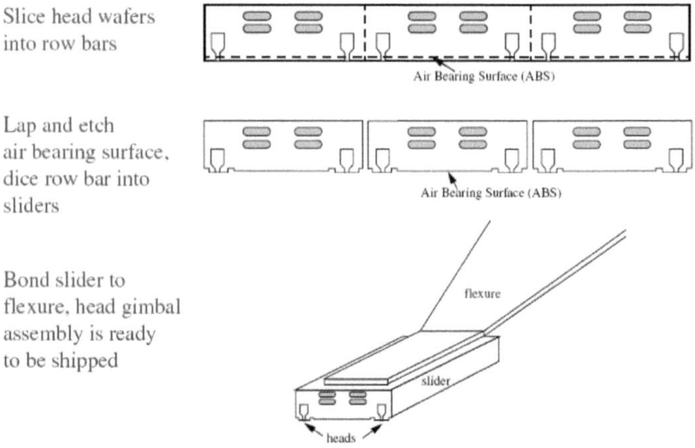

Figure 14. Thin film head fabrication stages (3). Reprinted from Ref. 48, Copyright (1998) with permission from Elsevier.

alumina (~ 15 μm) on the polished substrate. This layer acts as an insulating layer between the substrate and the bottom magnetic layer. A thin conducting undercoat layer of NiFe (50–100 nm) is deposited, often, by sputtering for electroplating.

2. The deposition and patterning of the bottom magnetic pole follows. It is usually electroplated with a through photoresist-window frame mask to a thickness level of 2–4μm. Note that the magnetic pole is made into a *pancake* shape in order to increase the head efficiency, while the narrow pole tip's dimension is the one that determines the narrow track width. As stated, the widely used Co-based alloy magnetic poles are electrodeposited (wet process) and the nanocrystalline FeN based alloys are sputter deposited in a vacuum chamber (dry process).
3. Next comes the sputter deposition of alumina to form the gap layer that is made to about 0.1–0.5 μm thick. The gap-layer thickness greatly affects the linear resolution and side read/write effects of the recording heads.

4. A hard-cured (to 200 ^{0}C) photoresist insulator is patterned between the coil and the magnetic core. The most common material used in this step is a Novolak-type polymer or polyimide that is popular for its high-insulation and photolithographic properties. It is made about 5-μm thick to provide electrical insulation as well as a planar surface to subsequently deposit copper coils.
5. The next layer consists of electrodeposited-spiral copper coils of 3-μm thick, 3–4-μm wide, and 2-μm apart in the region above the magnetic pole. A seed layer is added first through sputtering or any other suitable means. Next, the seed layer must be selectively etched. The coils are made wider outside the magnetic poles to reduce overall coil resistance. It is interesting to note that it was not until ~1997 that the semiconductor IC industry started to electrodeposit copper (Cu) interconnects on chip instead of sputtering aluminum (Al) while already for a long time before that electrodeposited Cu had been used in magnetic recording heads.
6. As in step 4, a hard-cured photoresist insulator of ~5-μm thick is patterned to provide electrical insulation for the copper coils and planarized surface for the top magnetic pole deposition. Leveling of the coated surface is essential to preserve the soft magnetic properties of the top magnetic layer. Leveling is effectively achieved by annealing above the softening temperature of the photoresist. For a multilayered coil steps 5 and 6 are repeated.
7. As in step 2, this step consists of the deposition and patterning of the top magnetic pole (2–4-μm thick) and the overpass lead, which provides electrical connection to the central tap of the coils. The back regions of the magnetic poles are, as a rule, made thicker than the pole tips to achieve high head efficiency and avoid magnetic saturation.
8. The next layers consist of an electrodeposit of 20–40-μm thick copper studs, a deposit of 10–15-μm thick alumina overcoat, and lap open copper studs. The thick alumina layer is required to protect the magnetic and coil structures during machining and lapping processes and to prevent corrosion.

9. In this step soft-gold bonding pads with a seed layer is electrodeposited as described before. The gold pads are required to protect the underlying devices and to facilitate wire bonding. One must note that wafer level testing of thin film heads may be performed at this point. This is useful for diagnosis of possible problems during the production process.
10. Next, the wafer is scribed and sliced into row bars as seen in Fig. 14. They are bonded to tooling bars that hold them during mechanical processing.
11. Then, sample heads are taken randomly for pole-tip geometry measurements after lapping and then good individual sliders are sorted out.
12. In the last step good sliders are bonded to flexures as shown in Fig. 14.

Most thin-film inductive heads employ multilayer spiral coils to achieve large read back signals. This means that the copper plating and the hard cured photoresist steps must be repeated. Electrical connection between the neighboring coil layers must also be implemented. Naturally, these steps greatly add to the complexity of the thin film head production.

3. High Frequency Magnetics

These past two to three decades have witnessed significant increases in switching frequencies. Specifically, switching frequencies have risen from the 75-kHz range to the 1000-kHz (1 MHz) range and beyond. This has facilitated the decrease of physical sizes of electronics objects, such as ICs, board areas, etc. Such trends have been more visible in the power-module arena where very low profile and minimal board area is required.

The physical density of modules is increasing due to a number of factors. Among them is improved semiconductor devices, which allow increased frequency of operation, and consequently, require less energy storage. Thus, a dominating magnetic- and electric-energy storage element may be reduced in size allowing further size reductions of the modules (such as storage discs, etc.). Naturally, with the elevated frequencies of operation come new challenges and developments required of the magnetic components (e.g., inductors, transformers etc.). These challenges and developments are concerned primarily with the increase in losses as well as with the desire to minimize volume, while at the same time managing the mechanical structure to permit efficient heat removal and cost-effective manufacturing techniques.

Generally speaking, magnetic components may be called upon to perform many functions in a power-conversion system (e.g., a radio receiving unit where EM waves are converted to acoustic ones). These functions are broadly divided between *power-handling magnetics* and *signal-handling magnetics*. We focus here on power-handling magnetics since these are of unique concern to power electronic systems. These primarily include *power inductors* and *power transformers*.

To put matters in context with the subject of electrochemical deposition, we observe that *traditional* magnetic devices have usually been made of a set of wire-wound windings on an insulating bobbin/cylinder with insulating tape systems, varnishes, etc., and a magnetic core mounted over the winding set. These construction methods have always led to limitations on size, manufacturability, performance, and repeatability. Recently a strong movement took place toward planar, low-profile magnetics with windings fabricated into standalone or embedded, printed wiring boards (PWB). These offer many advantages such as: very low profiles not possible with wire wound magnetics, reduced leakage inductance, higher degrees of interleaving at no added cost, higher operating frequencies (into the megahertz regime), and further cost reductions by embedding directly into the power train circuit board. However, some issues arise in connection with this movement. Those include maintaining insulation systems, providing sufficient copper metal weights, minimizing line spaces for maximum packing, and tight control on tolerances on heavy copper-metal weights.

Copper weights pose issues altogether new to the PWB industry. Magnetic designs typically hold tight tolerances (~ 10%) on dc resistance. Since heavy weights are achieved by electroplating extra copper over an existing laminate, the tolerances and uniformity that can be obtained are strongly dependent on the specific conditions of the electroplating process and master artwork. Bath chemistry, current density, and deposition rates affect the final weight. A non-uniform thickness is often formed across an entire PWB panel, being invariably heavier toward the panel's outer edges and lighter in the center (practicing electro-platters refer to this as the *dog bone* effect.) Current thieves, chemistry, agitation, and cathode/anode geometry design can be used to improve this discrepancy in deposit thickness. Ultimately, depending on the PWB manufacturer, a certain tolerance on copper thickness and, hence, dc resistance will have to exist from part to part on the same panel and must be taken into account in the design and specification stage of the magnetic device. This effect may lead to the need for specification of PWB copper thickness (in mils or millimeters) rather than as typically done by weight (e.g., in ounces or grams). This will be necessary in order to achieve tighter dc-resistance tolerance, possibly at the risk of higher costs. The etching process to form the conductor patterns also produces minimum conductor-to-conductor spacing that are highly dependent on the copper thickness at hand.

4. Spintronics

Electronic devices which operate using the spin of the electron and not just its electric charge are on the way to become a multi-billion-dollar industry—and may lead to quantum microchips.[51] As the progress in the miniaturization of semiconductor electronic devices leads toward chip features smaller than 100 nm in size, device engineers and physicists are inevitably faced with the fast approaching presence of quantum mechanics—that counterintuitive and to some, mysterious realm of physics wherein wavelike properties control the behavior of electrons. Practitioners in the semiconductor device world are busy coming up with ingenious ways to avoid the quantum world by redesigning the semiconductor chip within the confines of *classical* electronics. Yet it is a fact that we are being offered an unprecedented opportunity to define a radically new type of device that would exploit the quantum world to provide unique advantages over existing information technologies.

One such quantum property of the electron is its *spin*, i.e., its magnetism. Devices that rely on an electron's spin to perform their functions form the foundation of *spintronics* (short for spin-based electronics), also defined as *magneto-electronics*. Information-processing technology has thus far relied on purely charge-based devices, ranging from the now outdated vacuum tube to today's million-transistor microchips. The conventional electronic devices move electric charges around, ignoring the spin of each electron. Nonetheless, magnetism (and hence electron spin) has always been important for information storage. For instance, even the earliest computer hard-disc drives used magnetoresistance—the change in electrical resistance caused by a magnetic field—to read data stored in magnetic domains. It is a fact that the information storage industry has provided the initial successes in spintronics technology. More sophisticated storage technologies based on spintronics are already at an advanced stages where electrochemical deposition serves as a central role.

MRAM (magnetic random-access memory) is a new type of computer memory. MRAMs retain their state of magnetization even when the power is turned off, but unlike present forms of nonvolatile memory, they have switching rates and re-writability that challenge (faster than) those of conventional RAM. In today's read heads as well as those of MRAMs, key features are made of ferromagnetic-metallic alloys (usually, though not always, electrochemically produced). Such metal-based devices make up the first—and most mature—of the different categories of spintronics.

V. APPLICATIONS IN MEDICINE AND MEDICAL DEVICES

1. Background

It is often the case that the impact a given branch of science has upon another requires a rather special vehicle to become common knowledge. Given that, it is hardly surprising that the impact of electrochemistry on medicine is not yet properly recognized and appreciated by all. Thus, this chapter is designed to focus on electrochemistry as it relates to the medical arena. Specifically, the oft overlooked material-science aspects of medical devices and related power sources that make them "tick," are highlighted. This then is the focal point rather than the more often discussed topic of electrochemical sensors in medical devices.

It is worth observing here that—with the advent of relativity and quantum mechanics in the early part of the twentieth century and the development of molecular biology in the second half—it is now widely accepted by the scientific community that mathematics, physics, chemistry, and biology constitute but different parts of the same broader scientific discipline. Such inter-dependency and subsequent confluence of different disciplines may as yet become one of the hallmarks of the twenty-first century. It is also intended that the reader gains an appreciation of the future potential role electrochemistry may play in medical devices, particularly in the creation of *bio-mimetic* (that is, biology mimicking medical) devices.

In general, medical devices are man-made structures or machines that function inside or outside the body and have a role in human functioning either in sensing a physiologic variable or manipulating the same. If they are implanted in a host, they may be a temporary or a permanent device.

Among the rather well-known medical devices are the mechanical heart valve and the pacemaker. Although medical devices are generally perceived as macroscopic and/or permanent ones, many—and in some cases all—of the effects (wanted and unwanted) of the devices are derived from interactions at the surface. Indeed, when the devices are made of metals, many, if not most, of the surface effects are electrochemical in nature. This fact is the main point of this Section.

We intend to briefly outline the manner in which electrochemistry and electrochemical deposition are crucial to both, the mechanical and material stability of medical devices. In some cases, entirely new industries have been created based on the advances in our understanding of the electrochemical processes occurring at surfaces. An additional aim of the present contribution is to touch upon the ways in which electrochemistry can be used to actively modify surfaces to create a more amicable interaction between the medical device and its host. In at least one example (see below), electrochemical deposition itself (called *electroforming*) is used to actually create a medical device.

2. Electrochemical Power Sources

The last four decades have witnessed remarkable progress in the development of battery-driven implantable and external devices for medical detection and treatment. For instance, the medical condition known as *bradycardia* (where the heart beats too slow) has been treated by implanting pacemakers in the patients since the 1950s, when the first pacemaker was developed. Nowadays over half a million pacemakers are implanted annually and with our aging society, that number is expected to grow. Pace makers for the heart in the sixties were relatively large, several inches in size, lasted two to three years, and were capable only of providing pacing signals to the heart. Modern pace makers, on the other hand, are much smaller in size, last six (some up to ten) years, and are capable of detecting the need and set the heart rate automatically to fit the body activity. In addition, some do record medical information and even treat congestive heart failure.

One of the factors determining the life span of implanted medical devices is the life span of its power source. The impressive progress in extending it has been possible by the advancement of both, primary- and secondary-battery technologies. In other words, electrochemical power sources are truly an integral vital part of modern health care.

On the other hand, the medical condition where the heart beats too fast is known as *tachycardia*. If untreated, this condition may lead to ventricular fibrillation, that is, a condition in which the heart stops beating and shakes uncontrollably and is usually fatal. In 1980, a special device was developed and implanted in patients. It could sense the condition and provide a shock that would stop the fibrillation and restore the normal sinus rhythm via an electrode sutured onto the heart. The device was first powered by a lithium/vanadium pentoxide system; later it was replaced by a system using a cathode material of silver vanadium oxide (SVO or $Ag_2V_4O_{11}$). This is the actual system used in modern ICDs (implantable cardioverter/defibrillator). Another material used is the lithium/manganese dioxide system. Also, a new system using a sandwich cathode design with an inner cathode material of carbon monofluoride and an external cathode layer of silver vanadium oxide is in wide use.

The aim of this type of devices is to further extend the operational lifetime of the implanted medical devices to match the life time of the patient. To that end, fuel cell technology will have to be involved.

It may be observed that engineering of fuel cells has greatly improved over the past number of decades, leading for instance to the development of electric automobiles with zero emissions and improved performance. A fuel cell, in general, extracts electrical power directly from a fuel by converting the chemical gradient between fuel and oxygen into a voltage, usually across an electrolytic separator or membrane. Most fuel cells operate using a polymer electrolyte membrane to separate the fuel (pure hydrogen) from the oxidant (usually air) at temperatures around 100°C. In addition, precious metal catalysts such as platinum supported on carbon, are used to speed up (catalyze) both the anode and the cathode reactions [58,59] and to draw off the electric current.

Fuel cells that operate on biological fuels are still in their developmental stage. Methanol can be used as a fuel as an alternative to hydrogen; it has already been used in portable applications such as laptops and mobile phones. However, special anode catalysts are required to prevent poisoning of the electrodes with side-product reactions, and special electrolyte materials are needed to inhibit the leakage of fuel though the membrane. In a paper published by the Journal of the American Chemical Society, it was demonstrated that glucose, a much more complex molecule than either hydrogen or methanol but one that an implanted device may extract from the patient blood, can be used to drive a miniature fuel cell.[60] Surprisingly, this fuel cell was operated without a membrane to separate the oxygen from the fuel, although normally, fuel cells cannot function under these conditions. Miniature fuel cells are not a new concept. They have been used in electrochemical systems to detect alcohol in breathalysers and in oxygen sensors for automobiles. An example is an oxygen sensor that uses a spot of platinum ink of about 1 mm^2 on a zirconia membrane.[61] Conceivably, even finer sensors could be made by making the spot of the platinum smaller. The lower limit to the (minimal) size of such a fuel cell is not known, but it is quite straight forward to make a cell with an area of 0.01 mm^2, some 40 times less than the size reported by Mano et al.[60] So the small dimension of their cell is not particularly striking. Nonetheless, their fuel cell uses two carbon fibres—each 20-mm long—as electrode supports and current collectors, which could easily be made a hundred times shorter if the fuel cell needed to be miniaturized further. The most surprising elements of Mano et al.'s fuel cell are the use of glucose as the fuel and the single-compartment cell geometry. Both of these features required substantial advances in the catalyst materials used to coat the anode and cathode fibers. The two features make fuel cells promising candidates for running implanted medically relevant devices, as discussed above.

3. Electrochemical Deposition in Medical Devices

Not much can be found in electrochemical literature related to ways to improve medical device surfaces using electrochemical (coating) methods aimed to accomplish a medically relevant goal. An exception is the processing of nitinol, an alloy of nickel and titanium that exhibits shape memory, as explained bellow.

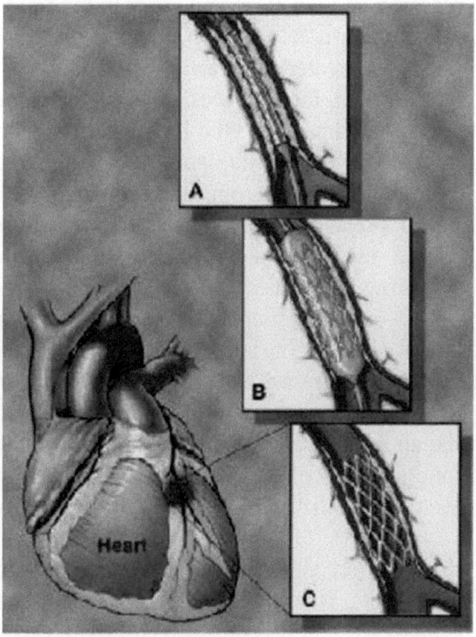

Figure 15. Stents are used to open arteries of the heart blocked by atherosclerotic plaques: (a) a balloon and stent are placed across the plaque; (b) and then the balloon is expanded, leaving the stent to prop open the artery; (c) restenosis is the process wherein scar tissue builds up around the stent, again causing a flow restriction. A balloon is required for stainless steel whereas a nitinol stent will expand on its own due to the shape memory property of nitinol.

Specifically, it is noted that one of the most commonly used medical devices is the *stent* (Fig. 15). Stents are small metallic structures that are expanded in blood vessels, functioning to maintain the patency (freedom from obstruction) of the vessel in which it is placed. Although the first use of stents was in vasculature (blood vessel systems), recent applications include, for example its implantation between two vertebra to increase the rigidity of the spine. A typical vascular stent is placed in its anatomic location and then it is plastically deformed/expanded (stainless steel) or it is allowed to expand to a predetermined size as a consequence of its shape memory (nitinol).

Recent research and development work in the field of medical devices has focused on creating *biomimetic surfaces*. Biomimetic surfaces are surfaces created with an understanding of the patho-physiology surrounding the surfaces, typically created with the intention to somehow manipulate the local environment to make it more accommodating for the implant. As an illustration of the need for a biomimetic surface, one may consider the process of restenosis and its prevention. Restenosis is the process wherein (scar type) tissue in-growth compromises the arterial passage space or lumen, after angioplasty and stenting. This process typically occurs in 30–50% of patients who receive stents in the coronary (heart) blood vessels. These patients often require subsequent surgery and/or further stenting. Several explanations for restenosis have been put forth, including: a foreign body type reaction to the stent materials; an inflammatory type reaction due to the implantation trauma; and a failure of the normal blood vessel lining (endothelial cells) to re-normalize after being perturbed by the stent implantation.

Gold, and more recently iridium, has been considered *ideal* biomaterials for stents. It is highly radio-opaque (it shows clearly in X-ray type analysis) and it is considered inert (corrosion should not be expected of the surface due to its *noble* characteristics.) Indeed, a gold electrodeposited stainless steel stent made it through clinical trials and it was actually approved for clinical use.[62] It was, nonetheless, a failure because its rate of restenosis was higher than stents already on the market. When the production process for the stent was reviewed, it turned out that proper heat treating of the plated surface was not performed. Early animal data[63] showed that changing this one variable reduces the restenosis rate to that of historical data for restenosis. This is yet another example where failure to appreciate surface electrochemistry caused clinical failure of a device.

More recently work has been done to couple organic surfaces to electroplated surfaces. It has been well known for many years that organic molecules containing a thiol group at one end strongly adhere to gold surfaces and form molecular monolayers. Recent work has expanded this knowledge by first electrodepositing a gold layer on the device surface and then applying an organic layer to the electrodeposited surface.[64] In this case, the device was a pacemaker lead. The organic layer was utilized to prevent scarring around the lead and to decrease the resistance that traditionally develops after electrode implantation. As it is well known in the field of electrochemistry, deposition is most comonly accomplished by either electroless and/or electrodeposition. With regard to medical devices—which often are not planar and have submillimeter features—electroless deposition offers many advantages. *Electroless deposition* is a wet electrochemical method in which deposition occurs evenly along the surface of the device including sub-millimeter features and non-planar surfaces. Electroless deposition can be used to evenly coat nanoparticles and to fill submicrometer holes. Electroplating, on the other hand, is very sensitive to the electric field lines between the anode and the cathode. As a result, the coating derived from an electroplating process can be highly irregular.[65] As such, electroless deposition may meet the biocompatability goals. Recent work[66] utilized the inherent porosity of electrolessly deposited nickel to incorporate drugs and provide for their storage and release, thereby opening new doors in device design and surface chemistry.

Another deposition method is the *electroforming* process. During it electrodeposition is performed on a mandrel in a given pattern. When the desired thickness is achieved, the mandrel is etched away from the electroformed stent leaving a free standing structure, i.e., a fully functional stent. In such a process, an entire stent—including structure and surface—can be built with a low energy, low capital electrochemical process. Indeed, Hines [10] developed a process to electroform stents from a gold electroplating solution. Such a process raises the intriguing possibility of having only a monolithic process to form the stent and to modify its surface for optimal biocompatability.

4. Surface Electrochemistry in the Processing of Biomaterials

Electrochemistry has proved to be useful in creating biomaterials and biomimetic enhancements. Understanding of surface electrochemistry at the metal-blood interface undoubtedly proves to be useful, as was the success of nitinol as a medical biomaterial (see also Section V.5). Another example is the study by Gertner and Schesinger demonstrating the biocompatibility and corrosion resistance of coatings that release drugs to its local environment.[66] This study further showed electrochemical methods used to create voids on the surface of the device where pharmaceuticals were loaded and released over a long period.

Electrochemical coating methods can accomplish many of the desirable biomimetic effects described in this Section. This science has the potential to alter surface morphology, release drugs, enhance radioopacity, and prevent corrosion, all with the same electrochemical coating. As discussed earlier in this chapter, electroless deposition is an evenly applied process that can be controlled to create composite coatings of the order of 1 μm in thickness. These type of coatings do not affect significantly the bulk material properties and can rarely been obtained by other methods. An example of this technology is the manufacturing of *nanocomposite coatings*. In this technology a first coating of a hundred nanometers is applied to add radioopacity to the coat; then another one-hundred-nanometer coating that contains and releases the drug is added; finally, a third coating hides the first two from the physiologic environment. To this, it is important to remember that these electrochemical processes are not only highly efficient but also economical with low capital costs. Thus, electrochemistry can be considered a nanotechnology in terms of its efficiency, size scale, and self-assembly properties.

Research of surface electrochemistry processes on metallic implants has focused mainly on corrosion processes and its prevention. These processes are highly sensitive to shape and surface defects. Corrosion studies on *bare* metals are available, although systematic studies of specific devices under conditions of mechanical stress are missing. Scanning electron micrograph (SEM) studies show that polymer coats (nitinol) can help to either enhance or reduce corrosion of the stent. Crevices as large as few μm have been detected, even prior to any expansion or stress applied to the device.

One of the most dramatic examples of a biomimetic coating in clinical use is the stent marketed by the Johnson and Johnson Corporation. The stent had a polymer-based coating containing a drug that inhibited the restenosis process. The drug was released from the polymer at a slow pace over time while the polymer remained on the stent permanently. Results were rather dramatic, with a reduction in the restenosis rate from 40% to 10%.[68] In spite of its success, we should consider that the polymer coat used in this stent might not have been the optimal for the process in consideration.

In the past years extensive research of biomimetic effects of surface morphology on the attachment, spreading, and second messenger systems of cells has been performed. In the field of orthopedic surgery, implants used to replace joints or fix fractures need sintered surfaces, that is, surfaces roughened by fixing 1-mm size metallic beads to it. The activity of osteo-blasts or bone synthesizing cells is dramatically up-regulated with the roughened, porous surface, in such a way that bone cement is not required to maintain fixation of some implants. One orthopedic material recently considered because of its damping properties and shape memory is nitinol.[69] Although adequate bone in-growth is observed, nitinol presents problems that need to be addressed from the point of view of surface electrochemistry, as shown next.

Nitinol achieves its shape memory properties after a series of heat treatment steps. This treatment leaves a thick oxide residue on the surface that is removed by etching, mechanical abrasion, or electropolishing (during electropolishing, the substrate is made the anode and the surface is evened out on a microscopic scale.) The result is a fine layer of titanium oxide (a ceramic) that remains on the surface and that is extremely important for the biocompatibility of nitinol since it prevents corrosion of the nitinol alloy. However, the titanium oxide layer itself is mechanically very brittle. Under mechanical stress such as the shear of blood flow in the aorta or under the bending moments of aortic pulsations, the titanium oxide layer can fracture, exposing the underlying metal to corrosion. Not only is corrosion undesirable as far as biocompatibility (i.e., leaching of nickel and its oxides), but it can lead to deterioration of the mechanical properties at the specific location where corrosion occurs. To summarize, surface defects inherent to the processing of nitinol can cause stress discontinuities in the surface of the titanium oxide film that could lead to film breakdown and localized pitting corrosion; the pitting corrosion in turn could lead to bulk mechanical breakdown at the site.

Optical micrograph of the surface of a commercially available nitinol stent examined prior to implantation showed large surface defects. These defects of the order of 1–10 μm were most likely formed during laser cutting. Once implanted, the nitinol is stressed and bent causing the region around the pits to experience tremendous and disproportionate strain. It is here that the titanium oxide layer can fracture and expose the underlying surface to corrosion.[70]

Sun et al.[71] revealed the heterogeneity of nitinol under various temperature conditions even in a simple lactated Ringer's solution. Lactated Ringer's solution is a mixture of salts and water meant to simulate the tonicity of blood. In this set of experiments, the surfaces were obtained from commercial sources and each sample underwent similar surface processing prior to experimentation. When a nitinol sample was simply placed in the Ringer's solution at constant potential and a given temperature, current transients representing the breakdown and repassivation of the oxide film were detected. Such studies have yet to be done under dynamic conditions (bending) of the nitinol. This dynamic stress would undoubtedly increase the number of passivation and repassivation events. Each event has the potential to release nickel into the surroundings and further the pit formation.

5. Materials Science of Biomaterials

As stated, medical devices comprise a rapidly growing industry where the basic technology changes very quickly. Many of these changes come from the ever-expanding basic knowledge of materials science. The majority of metals and metal alloys used in the device industry today include titanium, nickel-titanium (nitinol), cobalt-chrome, and stainless steel. Significant advances have been made over the years in polymer science (e.g., conducting polymers) and in metallurgy, which enable more than incremental improvements in the devices and sometimes allow for entirely new classes of devices. This is in fact the case for nitinol. However, as will become apparent below, the understanding and consequent exploitation of the chemistry, stability, and durability of nitinol leave still a great deal of room for future improvements before they can be easily manipulated.

Nitinol is the name given to an alloy consisting of approximately equiatomic concentration of nickel and titanium. It was first created in the 1960s and it was quickly discovered to possess a property called *shape memory*, that is, it returns to a preferred (preset) shape upon heating above a critical temperature (T_c). The relative percentages of nickel and titanium determine the actual value of T_c. This property has been exploited in medical devices where T_c is set at the human body temperature of 37°C. When the device made from nitinol is inserted into the body, it re-expands to its preformed shape. The advantages of nitinol's shape memory are several. For example, a pre-shaped device can be compressed into an *introducer* sheath which is many times smaller than the space the device will ultimately fill, e.g., a stent in a blood vessel. After implantation, the device expands to its original shape within minutes. A further advantage is that the device tends to exert a continuous positive force on the blood vessel so that if the blood vessel is compressed externally, the stent will return to its original position. This is extremely important when a stent is placed in a peripheral vessel, for example, a carotid artery.

The lag between the time nitinol was first produced and the time it was used commercially in medical devices was due in part to the fear that nickel would leach from the metal and not be tolerable as a human implant. As it turns out, with the correct understanding of its surface electrochemistry a passivating layer on the surface of the nitiol can be induced by an anodizing process. This layer is comprised of titanium oxide approximately 20-nm thick which acts as a barrier to prevent the electrochemical corrosion of the nitinol itself. Without an appreciation for the electrochemistry at its surface, nitinol would not be an FDA approved biocompatible metal and a whole generation of medical devices would not have evolved. This is really a tribute to the understanding of surface electrochemistry within the context of implanted medical devices.

6. Frontiers: Various Applications in the Field of Medicine

This Chapter is devoted to the subject matter of electrochemistry in the service of, and how it relates to, medicine. It is indeed timely for this kind of topic to be discussed in a medium such as the present. Medicine is the second oldest profession and as such, alone deserves our closer attention. Further, in the ongoing process of *globalization*, we are witnessing not only the tendency of commercial unification of the globe but the rapidly emerging interdependency of different scientific and technical disciplines. Indeed the much bandied-about *n* word—nanotechnology—may well evolve as the underlying thread in this technological melding process.

Finally, an example of what is not included in this contribution is, for instance, the use of scanning electrochemical microscopy (SECM) for imaging pathways of molecular transport across skin tissues. Another topic not included, but one that has been often discussed in the literature, is the topic of electrochemical sensors in medical devices.

It is hoped that this Chapter will nevertheless give the readers a broader view and an appreciation of the tremendous role electrochemistry plays in medicine and medical devices, as well as a glimpse into the future possibilities as both of these now related disciplines develop in time. None of these developments could have become reality if not for the collaboration of device engineers, medical professionals, electrochemists, and battery scientists. Such collaborative efforts require the ability to communicate ideas to professionals outside of their own discipline as well as to understand and appreciate input from other disciplines. It is reassuring to see the emergence of such interdisciplinary and crosscutting programs in universities. This type of trend must be clearly encouraged for the process of *scientific globalization* to further flourish.

VI. CONCLUSION

In conclusion, electrochemistry and electrochemical coatings play a crucial role in biomaterial-host interactions[68] and constitute integral components in medical device design. Electrochemistry should be viewed as a tool that can work for or against the medical device engineer and biochemist. Electrochemical methods, including electro- and electroless plating methods, can be used to create micrometer scale surface morphologies to induce specific types of cellular differentiation or attract and/or repel a specific type of cells. They can be used to increase the radio-opacity of the device; produce nano-composite coatings on a device; enhance or control drug delivery from a device; and accommodate a non-metallic coating. They can even be used to create a complete medical device. The above merits have to be balanced with the fact that electrochemical corrosive processes tend to destabilize surfaces and can undermine the mechanical stability of a device.

There is no doubt that the field of electrochemistry and its continual progress can and will have a substantial impact on the future of medical devices. Devices continue to be scaled down in size, which will necessitate a greater understanding of corrosion processes. As analytical tools for the study of surface chemistry improve and become more widespread, and as nano-architectured control permeates into the medical world, electrochemistry will be viewed as an economical, simple, yet powerful technique to modify and create biomimetic surfaces and medical devices.

REFERENCES

[1]W. Bacher, K. Bade, K. Leyendecker, W. Menz, W. Stark and A. Thommes, *Electrochemical Technology,* Ed. by N. Masuko, T. Osaka, and Y. Ito, Kodansha-Gordon & Breach, 1996, pp. 159–169

[2]S. K. Griffiths et al., *J. Micro Mechanics and Micro Engineering* **15** (2005) 1700–1712.

[3]C. Burbaum, J. Mohr, P. Bley, and W. Menz, *Snsors Mater.* **3** (1991) 751.

[4]M. Harmening et al., *Micro Electro Mechanical Systems* 1992, IEEE Cat No. 0-7803-0497-7/92, 202.

[5](*i*) S. Wolf and R. N. Tauber, *Silicon Processing for the VLSI Era,* Vol. 1: Process Technology, 2nd Ed., Lattice Press, Sunset Beach, California, 2000; (*ii*) S. Wolf, *Silicon Processing for the VLSI Era,* Vol. 2: Process Integration, Lattice Press, Sunset Beach, California, 1990.

[6]R. A. Powell and S. M. Rossnagel, PVD for microelectronics, sputter deposition applied to semiconductor manufacturing, *Thin Films*, Vol. 26, Academic Press, New York, 1999.

[7]J. M. E. Harper, E. G. Colgan, C-K. Hu, J. P. Hummel, L. P. Buchwalter, and C. E. Uzoh, *MRS Bulletin,* **19** (8), 23 (1994).

[8]D. S. Gardner, J. Omuki, K. Kudoo, Y. Misawa, and Q. T. Vu, *Thin Solid Films,* **262**, 104 (1995).

[9]A. J. Learn, *J. Electrochem. Soc.* **123** (1976) 894.

[10]C. -K. Hu, K. P. Rodbell, T. D. Sullivan, K. Y. Lee, and D. P. Bouldin, *IBM Journal. Res. Develop.* **39** (1995) 465.
[11]M. Ohring, *Reliability and Failure of Electronic Materials and Devices,* Academic Press, New York, 1998, Chapter 5.
[12]Y. Taur and T. H. Ning, *Fundamentals of Modern VLSI Devices,* Cambridge University Press, 1998, Chapters 4 and 5.
[13]C. Vickerman, Ed., *Surface Analysis,* J. Wiley & Sons, New York, 1998.
[14]S. Vaidya and A. K. Sinha, *Thin Solid Films* **75** (1981) 253.
[15]W. M. Moreau, *Semiconductor Litography, Principles, Practices, and Materials,* Plenum Press, New York, 1988.
[16]S. P. Murarka, R. J. Gutmann, A. E. Kaloyeros, and W. A. Lanford, *Thin Solid Films* **236** (1993) 257.
[17]C. H. Ting and M. Paunovic, *J. Electrochem. Soc.* **136** (1989) 456.
[18]L. T. Romankiw, *Electrochim. Acta* **42** (1997) 2985.
[19]V. M. Dubin, Y. Shacham-Diamand, B. Zhao, P. K. Kasudev, and C. H. Ting, *J. Electrochem. Soc.* **144** (1997) 898.
[20]C. H. Ting, M. Paunovic, P. L. Pai, and G. Chiu, *J. Electrochem. Soc.* **136** (1989) 462.
[21]Y. Shacham-Diamand and V. M. Dubin, *Microelectronic. Eng*. **33** (1997) 47.
[22]P. C. Andricacos, C. Uzoh, D. O. Dukovic, J. Horkans, and H. Deligianni, *IBM J. Res. Develop* **42** (1998) 567.
[23]E. K. Yung, L. T. Romankiw, and R. C. Alkire, *J. Electrochem. Soc.* **136** (1989) 206.
[24]S. Mehdizadeh, J. O. Dukovic, P. C. Andricacos, L. T. Romankiw, and H. Y. Cheh, *J. Electrochem. Soc.* **139** (1992) 78.
[25]J. O. Dukovic, *IBM J. Res. Develop.* **34** (1990) 693.
[26]S. Mehdizadeh, J. Dukovic, P. C. Andricacos, L. T. Romankiw, and H. Y. Cheh, *J. Electrochem. Soc.* **137** (1990) 110.
[27]R. Stolt, A. Charai, F. M. D'eurle, P. M. Fryer, and J. M. E. Harper, *J. Vac. Sci. Tech.* **A9** (1991) 1501.
[28]L. A. Clevenger, N. A. Bojarczuk, K. Holloway, J. M. E. Harper, C. Cabral, Jr., R. G. Schad, F. Cardone, and L. Stolt, *J. Appl. Phys*. **73** (1993) 300.
[29]M. T. Wang, Y. C. Lin, and M. C. Chen, *J. Electrochem. Soc.* **145** (1998) 2538.
[30]T. Kouno, H. Niwa, and M. Yamada, *J. Electrochem. Soc.* **145** (1998) 2164.
[31]S-Q. Wang, *MRS Bulletin* **19** (1994) 30.
[32]M. Paunovic, P. J. Bailey, R. G. Schad, and D. A. Smith, *J. Electrochem. Soc.* **141** (1994) 1843.
[33]V. Dubin, Y. Shacham-Diamand, B. Zhao, P. K. Vasudev, and C. H. Ting, *Mat. Res. Soc. Symp. Proc.* **427** (1996) 179.
[34]E. G. Colgan, *Thin Solid Films* **262** (1995) 120.
[35]E. M. Zielinski, *J. El. Mat.* **24** (1995) 1485.
[36]D. P. Tracy, D. B. Knorr, and K. P. Rodbell, *J. Appl. Phys.* **76**, (1994) 2671.
[37]D. P. Tracy and D. B. Knorr, *J. El. Mat.* **22** (1993) 611.
[38]P. A. Flinn, *J. Mater. Res.* **16** (1991) 1498.
[39]Y. Morand, *Microelectronic Eng.* **50** (2000) 391.
[40]P. V. Zant, *Microchip Fabrication,* 4th Ed., McGraw-Hill, New York, 2000.
[41]T. P. Moffat et al., *J. Electrochemical Soc.* **147** (2000) 4524.
[42]D. Josell, D. Wheeler, W. H. Huber, J. E. Bonevich, and T. P. Moffat, *J. Electrochem. Soc.* **148** (2001) C767.

[43]J. O. Dukovic, *IBM J. Res. Develop.* **37** (1993) 125.
[44]D. Josel, D. Wheeler, W. H. Huber, and T. P. Moffat, *Phys. Rev. Lett.* **87** (2001) 016102.
[45]T. J. Pricer, M. J. Kushner, and R. C. Alkire, *J. Electrochem. Soc.* **149** (2002) C406.
[46]T. J. Pricer, M. J. Kushner, and R. C. Alkire, *J. Lectrochem. Soc.* **149** (2002) C396.
[47]M. Paunovic and M. Schlesinger, *Fundamentals of Electrochemical Deposition,* 2nd Ed., John Wiley and sons, New York, 2006.
[48]S. X. Wang and A. M. Taratorin, *Magnetic Information Storage Technology*, Academic Press (1999).
[49]C. D. Mee and E. D. Daniel, *Magnetic Recording Technology*, 2nd Ed., McGraw Hill, 1996, p. 6.2.
[50]J. Mallinson, *Magneto resistive Heads, Fundamentals and Applications,* Academic Press, 1997.
[51]S. A. Wolf, D. D. Awschalom, R. A. Buhrman, J. M. Daughton, S. von Molnár, M. L. Roukes, A. Y. Chtchelkanova, D. M. Treger, Spintronics: A spin-based electronics vision for the future, *Science* **294**, Issue 5546 (2001) 1488–1495.
[52]B. Beschoten et al., *Phys. Rev. B* **63** (2001) R121202.
[53]S. A. Crooker et al., *Phys. Rev. Lett.* **75** (1995) 505.
[54]R. Fiederling et al., *Nature* **402** (1999) 787.
[55]B. T. Jonker et al., *Phys. Rev. B* **62** (2000) 8180.
[56]Y. Ohno et al., *Nature* **402** (1999) 790.
[57]M. Schlesinger and M. Paunovic, *Modern Electroplating,* 4th Ed.,
John Wiley and Sons Inc., NY, 2000.
[58]J. Larminie and A. Dicks, *Fuel Cells Explained,* Wiley, New York, 2000.
[59]Proceedings of the Seventh Grove symposium, edited by P.T. Moseley, *J. Power Sources* **106** (2002) 1–411.
[60]N. Mano, F. Mao and A. Heller, *J. Am. Chem. Soc.* **124** 12962–129634.
[61]R. C. Copcutt, A. C. King, and K. Kendall, *Proc. R. Soc. Lond. A* **452** (1996) 2639–2653.
[62]J. Dahl, P. K. Haagar, E. Grube, M. Gross, C. Beythien, E. P. Kromer, N. Cattelaens, C. W.Hamm, R. Hoffmann, T. Reineke, and H. G. Klues, *Amer. J. Cardio.* **89** (2002) 801.
[63]E. R. Edelman, P. Seifert, A. Groothuis, A. Morss, D. Bornstein, and Campbell Rogers, *Circulation* **103** (2001) 429.
[64]M. H. Schoenfisch, M. Ovadia, and J. E. Pemberton, *J. Biomed. Mater. Res.* **51** (1999) 209.
[65]C. Heintz, G. Riepe, L. Birken, E. Kaiser, N.Chakfé, M. Morlock, G. Delling, and H. Imig, J. Endovasc. Ther. **8** (2001) 248.
[66]M. Gertner and M. Schlesinger, *Electrochem. and Solid-State Lett.* **6** (2003) J4.
[67]R. Hines, in *Process for Making Electroformed Stents,* Electroformed Stents, Inc., Stillwell, Kansas, USA, 2000.
[68]M. C. Morice, P. W. Serruys, J. E. Sousa, J. Fajadet, E. B. Hayashi, M. Perin, A. Colombo, G. Schuler, P. Barragan, G. Guagliumi, F. Molnàr, and R. Falotico, *New England. J. Med.* **346** (2002) 177.
[69]A. Kapanen et al., *Biomaterials* **22** (2001) 2475.

[70]M. Schlesinger and M. Paunovic, *Modern Electroplating,* 4th Ed., John Wiley & Sons Inc, New York, 2000.
[71]E. Sun et al., *J. Mater. Sci.: Mater. Med.*, **13** (2002) 959.

Index

MIX
Papier aus verantwortungsvollen Quellen
Paper from responsible sources
FSC® C105338

If you have any concerns about our products,
you can contact us on
ProductSafety@springernature.com

In case Publisher is established outside the EU,
the EU authorized representative is:
Springer Nature Customer Service Center GmbH
Europaplatz 3, 69115 Heidelberg, Germany

Printed by Libri Plureos GmbH
in Hamburg, Germany